INTRODUCTION TO SOIL MICROBIOLOGY

INTRODUCTION TO SOIL MICROBIOLOGY
SECOND EDITION

MARTIN ALEXANDER
Cornell University

JOHN WILEY & SONS

NEW YORK • SANTA BARBARA • LONDON • SYDNEY • TORONTO

Copyright © 1961, 1977, by John Wiley & Sons, Inc.

All rights reserved. Published simultaneously in Canada.

Reproduction or translation of any part of this work beyond that
permitted by Sections 107 or 108 of the 1976 United States Copyright Act without the permission of the copyright owner is unlaw-
ful. Requests for permission or further information should be
addressed to the Permissions Department, John Wiley & Sons, Inc.

Library of Congress Cataloging in Publication Data:

Alexander, Martin, 1930–
  Introduction to soil microbiology.

  Includes bibliographies and index.
  1. Soil micro-organisms. I. Title.
QR111.A49 1977 576′.19′0948   77-1319
ISBN 0-471-02179-2

Printed in the United States of America

10 9 8 7 6 5 4 3 2

To
Renee   Miriam   Stanley

PREFACE

In the last few years, there have been many significant contributions in and from soil microbiology; for example, in the realm of applied science, the studies showing the crucial role of the soil microflora in modifying or destroying environmental pollutants, the finding that the subterranean populations themselves may form toxic products, the large-scale utilization of legume inoculation, and the investigations of the influence of the microflora on the degradation and persistence of agricultural pesticides. There has also been considerable emphasis on the interrelationships between saprophytes and pathogens in the soil as related to plant disease. In the realm of pure science, information on the ecology, function, and biochemistry of the microflora has grown considerably so that a clear picture of soil biology is beginning to emerge.

This second edition is not a definitive monograph but an introduction to soil microbiology. The innumerable developments in recent years make a complete review impossible within the scope of a single volume. Some of the more detailed points have been omitted for the sake of brevity, yet, where conflicts still exist, the contrasting viewpoints are presented. In certain problematic areas, however, I have weighed the evidence and presented the stronger case. Time may change these views, but it is in the very nature of science to be in a constant state of flux and for the errors of one generation to be mended by the next.

Soil microbiology is not a pure discipline. Its origins may be traced through bacteriology, mycology, and soil science; biochemistry and plant pathology have also made their mark, especially in recent years. The last few years have also witnessed the contribution of many facets of environmental science and technology. Any approach to soil microbiology consequently must consider the variety of individuals and disciplines that have created the mold. The approach here is to use microbiology, soil science, and biochemistry as partners. Where possible, each transformation is viewed as a reaction of importance to soil and to crop production, as a biological process brought about by specific microorganisms whose habitat is the earth's crust, and as a sequence of enzymatic steps.

Because these three disciplines are woven together into the fabric of soil microbiology, one should be familiar to some extent with basic principles of soil science, microbiology, and the chemistry of biological systems. In the framework of agriculture and the environmental sciences, the microflora is significant because it has both a beneficial and a detrimental influence on people's ability to feed themselves and on the quality of their environment. For the microbial inhabitant, the soil functions as a unique ecosystem to which the organism must become adapted and from which it must obtain sustenance. But, in the last analysis, the microbiologist can find definitive answers as to how these processes are brought about only through biochemical inquiries.

The book is divided into three general areas. First is a discussion of the major groups of microorganisms, particularly their description, taxonomy, abundance, and their significance and function. Then the major transformations carried out by the microflora are reviewed, including the reactions centered on carbon, nitrogen, phosphorus, sulfur, iron, and other elements. These are presented in terms of agronomic importance, the specific organisms concerned, the possible ecological consequences, and the biochemical pathways involved. Finally, ecological interrelationships affecting the microflora and higher plants are discussed.

References are of great value not only to the research worker but to the advanced student as well. The blind acceptance of secondary sources when the primary material is readily accessible is not the hallmark of the serious student. Where available, reviews are cited at the end of each chapter so that the finer points of each topic may be sought out. Pertinent original citations are similarly included since these permit student and researcher alike to examine the original source, observe the techniques utilized, and draw their own conclusions. Emphasis is given to the more recent papers, but certain of the classical works are included, particularly where the studies have indicated a unique approach. Absence of citations reflects not the lack of quality of an investigation but only the lack of adequate space between the covers of any single book.

As an introduction to soil microbiology, many things are left unsaid. I hope that a groundwork is laid here for a fuller inquiry on the reader's part. If this goal is reached even in a small way, the book will have served its purpose.

I thank the following publishers and journals for the approval to use copyrighted materials: Academic Press (Figs. 10.4 and 11.1); American Society for Microbiology (Figs. 6.3, 16.5, 19.3, and 24.4); American Society of Agronomy (Figs. 14.2, 15.1, 15.3, 16.4, 17.2, 19.2, and 19.5); *Annals of Botany* (Fig.

5.2); Blackwell Scientific Publications (Fig. 25.3); Cambridge University Press (Fig. 17.3); *Canadian Journal of Microbiology* (Fig. 10.3); *Chemistry and Industry* (Fig. 23.1); *Folia Microbiologica* (Fig. 22.4); The Iowa State University Press (Fig. 4.3); *Journal of General and Applied Microbiology* (Fig. 5.1); *Journal of Phycology* (Fig. 6.2); J. B. Lippincott Company (Fig. 3.4); *Macmillan Journals* (Figs. 11.4, 2 !.1, and 23.3); *Ohio Journal of Science* (Fig. 4.2); Oliver and Boyd (Fig. 6.1); Cxford University Press (Fig. 22.3); Pergamon Press (Figs. 3.1 and 23.6); *Plant and Soil* (Figs. 17.4 and 21.2); *Science* (Fig. 12.1); Springer-Verlag (Figs. 2.4, 13.3, 14.3, and 25.1); The University of Chicago Press (Fig. 2.3); Verlag Chemie (Fig. 10.2); and The Williams and Wilkins Company (Figs. 7.2, 13.2, 16.2, and 22.2).

Ithaca, New York, 1976   Martin Alexander

CONTENTS

MICROBIAL ECOLOGY

1
The Soil Environment

In an introduction to the microbiology of soil, it is essential to consider carefully the nature of the environment in which the microorganisms find themselves. The forces that play a role in the dynamics of soil populations and the effects of these populations on their environment are governed to a very great extent by the physical and chemical properties of the soil. It is to these, therefore, that attention must first be drawn.

The term *soil* refers to the outer, loose material of the earth's surface, a layer distinctly different from the underlying bedrock. A number of features characterize this region of the earth's crust. Agriculturally, it is the region supporting plant life and from which plants obtain their mechanical support and many of their nutrients. Chemically, the soil contains a multitude of organic substances not found in the underlying strata. For the microbiologist, the soil environment is unique in several ways: it contains a vast array of bacteria, actinomycetes, fungi, algae, and protozoa; it is one of the most dynamic sites of biological interactions in nature; and it is the region in which occur many of the biochemical reactions concerned in the destruction of organic matter, in the weathering of rocks, and in the nutrition of agricultural crops.

GENERAL DESCRIPTION OF SOIL

The soil is composed of five major components: mineral matter, water, air, organic matter, and living organisms. The quantity of these constituents is not the same in all soils but varies with the locality. Of the inanimate portion, the amount of mineral and organic matter is relatively fixed at a single site; the proportion of air and water, however, fluctuates. Air and water together account for approximately half the soil's volume, the volume so occupied representing the *pore space*. The mineral fraction, contributing generally slightly less than half the volume, originates from the disintegration and decomposition

3

of rocks, but the fraction has become, during the course of time, modified from the rocks from which it was derived. Organic matter usually contributes some 3 to 6 percent of the total. The living portion of the soil body—including various small animals and microorganisms—makes up appreciably less than 1 percent of the total volume, yet it is undoubtedly essential for crop production and soil fertility.

The inorganic portion of soil, because of its influence on nutrient availability, aeration, and water retention, has a marked effect on the microbial inhabitants. In the mineral fraction are found particles of a variety of sizes, ranging from those visible to the unaided eye to clay particles seen only under a microscope. The various structural units are classified on the basis of their dimensions. At the largest extreme are stones and gravel, materials whose diameter exceeds 2.0 mm. Somewhat smaller are the sand particles that have a diameter of 0.05 to 2.0 mm. Structures whose diameter falls between 0.002 and 0.05 mm are classified as silt, and those with a diameter less than 0.002 mm (2 μm) are considered to be clay particles.

The individual particle types differ from one another in other ways in addition to their dimensions. Thus, many more individual structural units are present in one gram of pure clay than in a gram of silt, and more particles are found in a gram of silt than in a like quantity of sand. More important, however, is the far greater surface area exposed per unit of mass of clay than for the larger particles (Table 1.1). Because the chemical properties and activities of the particles are directly related to their surface area, the status of clay as a reactive constituent in the soil body assumes prominence. In turn, the clay fraction is the most influential in terms of microbiological effects. The clay minerals contain silicon, oxygen, and aluminum; also, iron, magnesium, potas-

TABLE 1.1
Size and Surface Area of Soil Particles (Foth and Turk, 1972)

Particle Type	Diameter mm	No. of Particles/g[a]	Surface Area sq cm/g
Very coarse sand	2.00–1.00	90	11
Coarse sand	1.00–0.50	720	23
Medium sand	0.50–0.25	5,700	45
Fine sand	0.25–0.10	46,000	91
Very fine sand	0.10–0.05	722,000	227
Silt	0.05–0.002	5,780,000	454
Clay	<0.002	90,300,000,000	8,000,000

[a] Assumed to have spherical shapes. Calculated on the basis of maximum diameter of the particle type.

sium, calcium, sodium, and other elements may be found to varying degrees. Three of the major clay minerals in soils of the United States are kaolinite, montmorillonite, and illite. Subsequent discussion will reveal, in part, the unique biological role of these and other clay minerals.

By comparison, silts exert a lesser influence on the physical, chemical, and biological properties of soil. The sand particle, a comparatively large unit exposing a small surface area, is of still lesser consequence. Sand, however, does affect the movement of water and air.

For the purposes of description, textural classes have been established. *Texture* is determined on the basis of the soil's content of sand, silt, and clay, and the name of the textural class is ascertained from the triangle shown in Figure 1.1. To obtain the class name, a line originating at the point corresponding to the percentage of silt is drawn inward and parallel to the left side of the triangle. From the point corresponding to the percentage of clay, a second line is drawn parallel to the base of the triangle. The class name is given by the segment in which the two lines intersect.

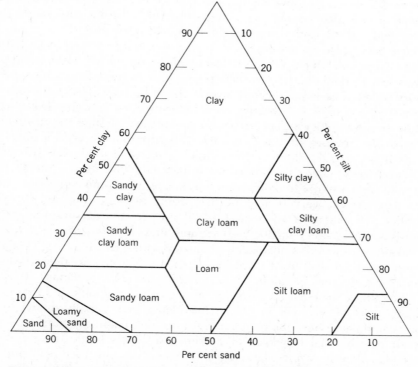

Figure 1.1. The textural triangle from which the names of textural classes are obtained.

The diagram, it will be noted, introduces a new term, *loam*. A loam is a soil not dominated by any of the particle sizes. It may be further noted that considerable emphasis in the textural triangle is placed upon clay content; thus, a soil with less than 40 percent clay may be classified as a sandy clay. This emphasis is a necessary outcome of the great reactivity of clays. Often, the adjectives light and heavy are used in technical or common parlance. Soils dominated by large particles exhibit a coarse texture and are termed light. Heavy soils, by contrast, have a fine texture and are dominated by small particles. It should be borne in mind that textural designations serve a far greater purpose than simple nomenclatural subdivisions because the ease with which a soil is worked, its aeration, and moisture relationships—hence its biological activity—are governed to a great extent by texture.

A vertical section cut into the soil reveals that it possesses a distinct *profile*. In the profile are several horizontal layers known as *horizons*. Even the most casual examination of the layers reveals appreciable differences in structure, texture, and color. These horizons are used in the classification of soils.

As a rule, three major layers make up the profile, the A, B, and C horizons. In addition, an organic horizon may be present, especially in forests. A typical profile may contain (*a*) a shallow or thick surface zone of decaying organic debris, (*b*) an underlying horizon from which certain inorganic constituents have been removed during the long period of soil formation, (*c*) a horizon at greater depth in which is deposited some of the constituents from the upper layers, and (*d*) a bottom layer similar to the original material from which the soil had developed. The organic debris layer is the O horizon. The A horizon, the surface soil, designates the stratum subjected to marked leaching. It is also the layer of greatest biological concern as roots, small animals, and microorganisms are here most dense. In this zone, the concentration of organic matter is highest; hence it is the dominant reservoir of microbial food. The B horizon, the subsoil underlying the A horizon, usually has little organic matter, few plant roots, and a sparse microflora. In it, iron and aluminum compounds often accumulate. At the very bottom of the profile is the C horizon, the layer containing the parent material of the soil proper. In this stratum, organic matter is present in very small quantities, and little life is noted.

No single description adequately characterizes the nature of soil profiles as individual profiles and horizons differ from one another in their thickness, chemical constitution, aeration, color, texture, and water relations. It is not surprising, therefore, that they support microbial communities differing in size and in activity. The attention of the microbiologist is usually drawn to the surface soil because here the population is most dense and the nutrient supply greatest; likewise, the beneficial or detrimental effects of the microflora on higher plants are most pronounced in the A horizon. On the other hand, the subsoil modifies the characteristics of the surface layer as a habitat for both macro- and microorganisms.

DIFFERENCES AMONG SOILS

There are several broad soil belts on the earth's surface. Large areas of the Northern Hemisphere contain *Spodosols,* a group of soils developed in temperate, humid climates in forest areas. These soils usually are poor in organic matter and tend to be acidic. The formation of Spodosols is of considerable interest to the microbiologist since the process is associated with the decomposition of organic matter accumulated at the soil surface and with the downward movement of organic substances formed or released by the subterranean microinhabitants. Soils that are termed *Mollisols* occupy vast areas of land in temperate regions and smaller areas in the tropics. They commonly have a thick A horizon that often is particularly rich in organic matter. In tropical or semitropical zones are found *Oxisols* and *Ultisols*; in these soils, the horizons are usually not distinct. In desert regions are found *Aridisols*, which contain little organic matter, a result of the sparse vegetation in the arid areas. Several representative profiles are shown in Figure 1.2.

The broad soil belts delineated above have been investigated intensively and a classification system established. A review of the classification scheme currently used in soil science is beyond the scope of the present discussion, but the reader can obtain further information from the references cited at the end of the chapter.

The subdivisions within the profile and the common classification schemes are applied to *mineral soils*, whose dominant solid matter is inorganic. *Organic soils* (or *Histosols*)—including the mucks and peats—are widespread and frequently are highly fertile in terms of crop-producing capacity. Organic soils generally have 60 to 95 percent organic matter and, therefore, only a small proportion of mineral constituents. As a result, their chemical and physical properties do not resemble those of the mineral soils. Mucks and peats are formed in bogs and marshes where conditions for the microbiological decomposition of organic matter are poor, and large quantities of carbonaceous substances accumulate. With time, the accumulated residues assume the brown or black coloration typical of organic soils. Because of the way in which mucks and peats are formed, they do not have the usual type of profile. Unfortunately, organic soils have not received their deserved attention so that much of the chemical and microbiological literature is concerned specifically with mineral soils.

Local differences are found among soils. In moving from one area to the next, the depth, color, pH, and chemical composition of the various horizons are found to be dissimilar. The variations may often be traced to the nature of the rock material from which the soil developed, climatic factors, the type of vegetation, and topography. Indeed, it is common to find a single farm situated on several soil types. The differences may be small or they may be appreciable. Physical, chemical, and biological variations need not be measured in kilome-

Figure 1.2. Profiles of three great soil groups. Left: Oxisol; center: Mollisol; right: Spodosol. (Courtesy of W. M. Johnson.)

ters, however, as many differences can be found within a small area. Thus, a poorly drained area a short distance from a well-drained site possesses a somewhat altered community of microorganisms. In later chapters, it will be demonstrated that these differences can be measured even in centimeters; for example, the microorganisms immediately surrounding the root surface are not the same as those just one centimeter away from the plant tissue. For the microbiologist, therefore, the soil type is of great consequence as is the microscopic environment within any one soil.

SOME PHYSICAL CONSIDERATIONS

Solid materials occupy only about half the soil's volume. The remainder is composed of pores filled with air and water, both essential for life. The amount of pore space is dependent on the texture, structure, and organic matter content. In clay soils, the pores are generally small. In sandy areas, on the other hand, the pores are large, but the total quantity of pore space is less than in soils rich in fine particles. The size of individual pores and the total pore space affect the movement and retention of water. In sandy soils, water moves rapidly through the large pores, but little is retained. The numerous micropores of heavier soils, on the other hand, contribute to the greater water retention.

The porosity of heavy soils is affected by the state of aggregation. *Aggregates* are large structural units composed of clay and silt particles. In contrast with sand, silt, and clay, aggregates are temporary structures whose stability varies with land management practices, meteorological conditions, microbial activity, and other factors. They range in size from large bodies that are easily broken apart to small granules of firm consistency. In addition to their effects on water and air movement—which in turn regulate the activities of the microflora—aggregates are of interest microbiologically since the cell material and excretions of bacteria, fungi, and actinomycetes are factors affecting the formation and stability of the granules.

The water relationships of soils and the biological effects of moisture have received much study. In certain regions or during certain parts of the year, the soil is quite wet, and too much water is present for optimum biological action. At other times, the moisture level is low, and microorganisms suffer. Because soil water is derived from atmospheric precipitation, the supply is quite variable, and marked fluctuations in the soil water content are the rule in nature.

Part of the water moves with the pull of gravity, this being called *free* or *gravitational water*. Such water is situated within the larger soil pores that are often filled with air; as a result, gravitational water directly affects aeration. In addition, some water is retained against the gravitational pull; the retention against gravity results from the attraction between the water and other soil constituents. Not all of the water in soil is biologically available, and only part of that portion held against gravitational attraction can be used by living systems. Apparently, the nonbiological constituents of soil compete well with microorganisms for water, an indication of the great binding power of the inanimate materials.

The soil solution contains a number of inorganic salts, but except in arid regions, the solution is quite dilute. The liquid phase is of importance to the subterranean flora because it contains several required nutrients. As needed food materials are found in the soil solution, the downward movement of water removes from the zone of microbial accessibility substances essential for proliferation. Nitrogen, potassium, magnesium, sulfur, and calcium but little phosphorus or organic matter are lost through leaching in this way. The rate and magnitude of such losses are regulated by the quantity of precipitation, the presence and type of vegetation, and the soil texture.

Aeration and moisture are directly related because that portion of the pore space not containing water is filled with gas. Air moves into those pores that are free of water; water in turn displaces the air. The gas found in the profile may be said to constitute the soil atmosphere. This subterranean atmosphere is not identical with that in the air above the earth or that at a point several centimeters from the soil surface. Commonly, the CO_2 concentration exceeds the atmospheric level by a factor of tenfold to one-hundredfold, but O_2 is less plentiful. The difference in the composition of the above-ground and below-ground atmospheres arises from the respiration of microorganisms and plant roots, living organisms consuming O_2 and releasing CO_2. Diffusion of the gases tends to right somewhat the concentration gradient so that the content of O_2 and CO_2 is governed by both the diffusion rate and by the rate of respiration. As a rule, the O_2 content declines and the CO_2 level in the gas phase increases with depth.

Changes in the soil atmosphere alter the size and functions of the microflora as both CO_2 and O_2 are necessary for growth. It is of interest, therefore, to speculate on the possibility of attaining a well-aerated (or more appropriately, oxygenated) soil. A well-aerated soil, from the microbiological viewpoint, is one in which microbial processes requiring O_2 proceed at a rapid rate. However, it is unlikely that soil ever becomes sufficiently aerated to satisfy all of its inhabitants because of the problems of gas movement into the small pores and microenvironments in which the organisms are situated. Hence, a soil that is sufficiently well aerated for the growth of higher plants does not necessarily contain an optimum concentration of O_2 for the microflora.

Improper aeration, the opposite extreme, is associated with poor drainage and waterlogging. Since small pores have a greater tenacity for water than the larger pores, the aeration status of heavy soils, which are dominated by micropores, is often poor; that is, a large part of the volume is occupied by liquid rather than by gas. In conditions where the O_2 supply is inadequate, the rates of many microbial transformations are reduced, and some processes may be eliminated. In O_2-deficient habitats, new microbiological processes may come into play, some of which may be deleterious to plant development; for example,

N_2 or CH_4 is evolved, organic inhibitors appear, and sulfide, ferrous, and manganous ions accumulate during periods of O_2 deficiencies.

SOME CHEMICAL CONSIDERATIONS

Microorganisms obtain many of their nutrients from the inanimate portion of soil so that some consideration needs to be given to the chemical composition of the environment. Certain species get carbon or nitrogen from the atmosphere as CO_2, CH_4, or N_2, but the bulk of these two elements as well as the remaining microbial foods are derived from solid or liquid phases of the soil.

A chemical analysis of two soils is given in Table 1.2. Note from these data that considerable differences exist between the two soils. In general, the chemical composition of soils varies to a great extent, but certain elements are always abundant. Except for the organic soils, whose constitution is entirely different, the dominant substance is silicon dioxide. It often accounts for 70 to 90 percent of the total mass. Aluminum and iron are similarly abundant, and lesser quantities of calcium, magnesium, potassium, titanium, manganese, sodium, nitrogen, phosphorus, and sulfur are found. The organic matter content of mineral soils is variable, the range extending from 0.50 to 10 percent of the total weight. The nitrogen content is usually approximately one-twentieth the organic matter level, that is, 0.025 to 0.50 percent. Except for carbon and potassium, the major elements needed for the synthesis of protoplasm are present to the extent of less than 1 percent of the soil weight.

TABLE 1.2

Chemical Composition of Two Soils (Marbut, 1935)

Hori-zon	Depth (cm)	Percent of Various Constituents								
		Si	Fe	Al	Mn	Ca	Mg	K	Na	P
		Becket Fine Sandy Loam								
A	0–15	24.7	0.75	3.73	0.003	0.64	0.09	1.71	0.30	0.06
A	15–28	38.9	1.18	3.56	0.003	0.39	0.11	2.40	0:34	0.02
B	28–33	32.5	2.78	5.09	0.003	0.46	0.20	2.83	0.34	0.03
B	33–61	33.9	2.49	5.46	0.006	0.44	0.25	2.86	0.50	0.03
C	61–91	36.3	2.19	5.29	0.010	0.39	0.29	3.14	0.41	0.03
		Miami Silt Loam								
A	0–5	33.5	2.03	4.80	0.042	0.58	0.37	1.68	0.79	0.06
A	5–13	36.0	2.14	5.03	0.045	0.45	0.38	1.68	0.76	0.04
A	13–30	36.1	2.24	5.34	0.039	0.38	0.37	1.82	0.86	0.03
B	41–81	32.4	4.12	7.44	0.026	0.50	0.72	1.97	0.72	0.04
B	80–91	30.6	3.90	7.80	0.042	1.12	1.18	2.19	1.03	0.05
C	>91	22.4	2.32	4.52	0.023	9.71	3.63	1.60	0.63	0.04

It is evident from Table 1.2 that depth affects the chemical composition. Generally, the surface soil is richer in silicon dioxide than the subsoil, a result of the downward movement and deposition in the B horizon of other constituents during the long periods required for the formation of a soil. Organic matter and nitrogen are similarly most conspicuous in the surface layer. Conversely, the concentration of calcium and magnesium is frequently less in the A than in the lower horizons.

The statement that a soil has 3.1 percent organic matter or 0.14 percent nitrogen does not mean that these quantities are readily available to microorganisms. Only a small portion of the total organic carbon or nitrogen is utilized by the microflora each year; the rest remains as a slowly available reservoir. Therefore, the level of organic matter or nitrogen reflects more a potential than an actual supply. It is the available nutrient—a common term but one that has frequently eluded chemical characterization—which is of immediate significance to the microscopic inhabitants. By contrast, elements found in lower concentrations than carbon and nitrogen may be present in amounts sufficient to satisfy fully the biological requirement for them. For example, the need for magnesium, sulfur, and potassium is small, and the supply probably exceeds the demand in most circumstances. The same is usually true for those inorganic nutrients that are required in minute amounts. Trace nutrients include zinc, copper, molybdenum, and cobalt. Because of such considerations of food supply and food demand and the meager information on nutrient availability in soil, elemental analyses have only a limited value. Nevertheless, other means are available to determine which substances limit the populations of microorganisms.

A factor that must be considered in a discussion of the nutrient supply is the remarkable ability of soils to retain ions. Cations (positively charged ions) such as NH_4^+, K^+, Ca^{++}, and Mg^{++} are removed from solution by clay minerals, which, because they possess a negative electrical charge, attract positively charged ions. Soil organic matter also retains cations, and its ability to remove charged ions from solution must be considered with that of the clay. In fact, on a weight basis, organic colloids are more active than clays in ion exchange. This retention of positive ions leads to an important soil characteristic, that of *cation exchange*. In cation exchange, one positively charged ion about the clay complex is replaced and released by another type of ion. Ionic exchange, through its effects on nutrient availability and acidity, has a considerable bearing on biological transformations.

An important characteristic of soil is its *cation exchange capacity*, which, as the name suggests, is a measure of the capacity of the clay and organic colloids to remove positive ions from solution. The data are usually expressed in terms of milliequivalents of ions removed per 100 g of soil. The exchange capacity varies with the amount and type of clay and organic matter. Because soils of heavy texture are richer in both clay and organic matter, they tend to have higher cation exchange capacities than light soils; therefore they may remove more of those nutrients that exist in the cationic form.

A number of the inorganic substances assimilated by microorganisms are anionic. These negatively charged ions are represented by bicarbonate, nitrate, phosphate, sulfate, and molybdate. Anion exchange, however, is never appreciable in soil; hence, it is of little importance biologically. Thus, ammonium is readily removed from the soil solution whereas nitrate, its oxidized counterpart, is not strongly retained by the colloidal complex.

THE ORGANIC FRACTION

The *organic fraction* of soil, often termed *humus*, is a product of the synthetic and decomposing activities of the microflora. Since it contains the organic carbon and nitrogen needed for microbial development, it is the dominant food reservoir. Because humus is both a product of microbial metabolism and an important food source, the organic fraction is of special interest to the microbiologist.

When plant or animal remains fall on or are incorporated into the soil, they are subjected to decomposition. From the original residues, a variety of products are formed. As the original material and the initial products undergo further decomposition, they are converted to brown or black organic complexes. At this stage, there no longer remains any trace of the original material. The native organic fraction originates from two sources: the original plant debris entering the soil and the microorganisms within the soil body. The latter work upon the former and synthesize microbial protoplasm and new compounds that become part of the organic fraction.

Humus exists in a dynamic state. It is under continual attack, yet it is being reformed by the subterranean inhabitants from the remains of the land's vegetation. The decomposition leads to a loss of some of the carbonaceous materials; at the same time, new microbial tissue is generated. The rate of carbon loss can be related to the structure and fertility of a soil; it also reflects the level of biological activity.

In undisturbed land, the organic matter content remains relatively constant. Modification of the habitat by cropping or by altering the aeration changes the level of humus as the original equilibrium between the rate of carbon addition and the rate of its volatilization is upset. The large amount of organic matter in peat soils, for example, is associated with a slow decomposition of vegetative remains in poorly drained, poorly oxygenated circumstances. When the area is drained, the O_2 level rises, and the accumulated carbonaceous matter is decomposed and volatilized as CO_2.

Chemists have been attempting to unravel the details of humus composition since the earliest days of soil science, but much is still to be discovered. It has been pointed out above that the organic fraction is derived from (*a*) plant constituents that are modified by the microflora and (*b*) constituents of microbial cells and products of microbial metabolism that are relatively resistant to decay and therefore persist for some time after death of the organism. In terms of specific elements, the organic fraction contains compounds of carbon, hydrogen, oxygen, nitrogen, phosphorus, sulfur, and small amounts of other elements. Only a small portion of the total is soluble in water, but much can be brought into solution by alkali. In terms of types of compounds, humus contains a number of polymerized substances; aromatic molecules, polysaccharides of several kinds, bound amino acids, polymers of uronic acids, and phosphorus-containing compounds can be demonstrated (Table 1.3). It must be emphasized, nevertheless, that no definite composition can be assigned to the organic fraction. Variations in its constitution are observed not only in different localities—a natural consequence of differences in temperature, rainfall, and mineral matter—but also within individual fields. Humus should be considered as a portion of the soil that is composed of a heterogeneous group of substances, most having an unknown parentage and an unknown chemical structure.

TABLE 1.3

Several Constituents of the Organic Molecules Found in Humus[a]

I. Amino acids	VII. Pentose sugars
Glutamic acid	Xylose
Alanine	Arabinose
Valine	Ribose
Proline	VIII. Hexose sugars
Cystine	Glucose
Phenylalanine	Galactose
II. Purines	Mannose
Guanine	IX. Sugar alcohols
Adenine	Inositol
III. Pyrimidines	Mannitol
Cytosine	X. Methyl sugars
Thymine	Rhamnose
Uracil	Fucose
IV. Aromatic molecules	2-0-Methyl-D-xylose
V. Uronic acids	2-0-Methyl-D-arabinose
Glucuronic acid	XI. Aliphatic acids
Galacturonic acid	Acetic acid
VI. Amino sugars	Formic acid
Glucosamine	Lactic acid
N-Acetylglucosamine	Succinic acid

[a] Except for the amino acids and aliphatic acids, which are found only in low concentrations, the constituents rarely exist in free form; rather they are bound in polymers or other poorly defined complexes.

SUMMARY

The physical and chemical characteristics of soil determine the nature of the environment in which microorganisms are found. These environmental characteristics in turn affect the composition of the microscopic community both qualitatively and quantitatively. It is from the soil that the water, air, and the inorganic and organic nutrients are obtained; similarly, the soil serves as a buffer to the drastic changes that occur above the ground. The microbiologist often considers the soil as a vast, dynamic medium for the subterranean inhabitants as well as a site in which substances not available to higher plants are made available through microbial agencies.

In the succeeding chapters, emphasis will be placed on the biological transformations occurring within the soil. It is essential, however, to remember that the microscopic inhabitants do not exist in an isolated state but rather that they are just a part of a highly complex environment regulated by natural forces and, to a lesser extent, by man's activities. An appreciation of soil microbiology can only be gained by viewing the soil system as a dynamic whole, as a natural environment in which microorganisms play an essential and often poorly understood role.

REFERENCES

Baver, L. D., W. H. Gardner, and W. R. Gardner. 1972. *Soil physics*. Wiley, New York.

Bear, F. E., ed. 1964. *Chemistry of the soil*. Reinhold Publishing Corp., New York.

Black, C. A. 1968. *Soil-plant relationships*. Wiley, New York.

Brady, N. C. 1974. *The nature and properties of soils*. Macmillan Publishing Co., New York.

Foth, H. D. and L. M. Turk. 1972. *Fundamentals of soil science*. Wiley, New York.

Marbut, C. F. 1935. Soils of the United States. In O. E. Baker, ed., *Atlas of American agriculture*. Government Printing Office, Washington.

Thompson, L. M. and F. R. Toeh. 1973. *Soils and soil fertility*. McGraw-Hill Book Co., New York.

2
Bacteria

Soils contain five major groups of microorganisms: the bacteria, actinomycetes, fungi, algae, and protozoa. The soil *ecosystem* includes these microbial groups as well as the inorganic and organic constituents of a given site. All of the inhabitants of the particular locality make up the *community*, although it is not uncommon to speak of communities of major categories of organisms. The collection of cells or filaments of the individual species that are represented in the community are considered as distinct *populations*. The bacteria are especially prominent because of the many populations in a given soil and the fact that they are the most abundant group, usually more numerous than the other four combined.

The number of bacterial cells in the soil is always great, but the individuals are small, rarely more than several micrometers in length. Because of the minute size of the bacteria and the large cells or extensive filaments of the other four groups, the bacteria probably account for appreciably less than half of the total microbiological cell mass. In adequately aerated soils, the bacteria and fungi dominate whereas bacteria alone account for almost all the biological and chemical changes in environments containing little or no O_2. Although many transformations similar to those of the bacteria are carried out by the other groups, the bacteria stand out because of their capacity for rapid growth and vigorous decomposition of a variety of natural substrates.

The bacteria isolated from soil can be placed in two broad divisions: the *indigenous* or *autochthonous* species that are true residents and the *invaders* or *allochthonous* organisms. Indigenous populations may have resistant stages and endure for long periods without being active metabolically, but at some time these natives proliferate and participate in the biochemical functions of the community. Allochthonous species, by contrast, do not participate in a significant way in community activities. They enter with precipitation, diseased tissues, animal manure, or sewage sludge, and they may persist for some time in a resting form and sometimes even grow for short periods but never do they contribute in a significant way to the various ecologically significant transformations or interactions.

Among the indigenous populations are bacterial species that flourish dramatically when readily available organic nutrients are added. These actively metabolizing bacteria need nutrients provided from without for their rapid growth, but the supply is readily exhausted; thus, they respond promptly to soil amendment, become and remain numerous as long as the nutrients are available, then decline once their food source is depleted. Other autochthonous populations characteristically grow using the soil organic fraction, resistant constituents of plant residues, or components of other microbial cells as nutrients. Because these nutrients are less readily available and are present for extended periods, such organisms grow slowly, and their abundance is not subject to marked fluctuations (1).[1]

Bacteria are also divided on a systematic or taxonomic basis by the system proposed in *Bergey's manual of determinative bacteriology*. Other schemes provide for physiological differentiation using a variety of nutritional and metabolic characteristics including the identity of the energy source, the carbohydrates used for growth, the capacity to utilize N_2 as nitrogen source, etc. The ability to grow in the absence of O_2 is an important biochemical trait that has led to three separate and distinct categories: *aerobes,* which must have access to O_2; *anaerobes,* which grow only in the absence of O_2; and *facultative anaerobes*, developing either in the absence or presence of the gas.

Cell structure also serves as a means of bacterial characterization. Among the major morphological types are the *bacilli* or rod-shaped bacteria, which are the most numerous (Figure 2.1), the *cocci* or spherical-shaped cells, and the *spirilla* or spirals. The latter are not common in soil. Some of the bacilli persist in unfavorable conditions by the formation of endospores that function as part of the normal life cycle of the bacterium. These endospores often endure in adverse environments because of their great resistance to both prolonged desiccation and to high temperatures. Spore-forming genera are present among the aerobic and anaerobic bacteria. The endospore can persist in a dormant state long after the lack of food or water has led to the death of vegetative cells. When conditions conducive to vegetative growth return, the spore germinates and a new organism emerges. In addition to these three morphological types, bacteria exhibiting distinctly different shapes have been found, but many of these organisms have yet to receive adequate study.

DISTRIBUTION AND ABUNDANCE

A determination of the numbers of viable bacteria in pure culture is a relatively simple procedure by plate counting or other means. The situation is far more

---

[1] The numbers in parentheses refer to the literature cited at the end of the chapter.

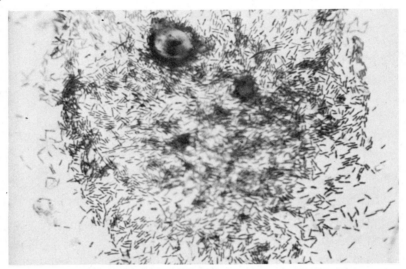

Figure 2.1. Rod-shaped bacteria developing in soil. (Courtesy of S. Ishizawa.)

difficult in a highly heterogeneous ecosystem such as the soil, where conventional microbiological techniques only estimate a portion of the total number of bacteria. No one medium is adequate nutritionally for all the species present since the growth requirements for many strains are unknown, and the observed count represents only a fraction of the total. A second limitation arises from the fact that bacteria frequently occur in the soil as colonies, and these may not disintegrate when the soil dilutions are shaken so that estimates tend to be low.

The standard methods of examining soils for viable counts often give variable numbers, and the errors in sampling and in sample preparation are frequently far greater than the variations inherent in the counting procedure itself. This limitation can be minimized by the use of many composites prepared from numerous borings made in the field. It is far better to use many subsamples than numerous replicate plates per dilution since the variation among duplicate soil samples is far greater than the variation between replicate plates or replicate dilutions (10). Not uncommon is the observation of greater variation between estimates made at the same time in different areas of a field than between counts made at a single location at various times of the year. The uninitiated must always bear in mind the vast microbial diversity that exists in this highly heterogeneous environment. The presence of a single rootlet or particle of plant debris may cause a microecological effect of magnitude sufficient to change the counts ten- or a hundredfold. The microflora should be viewed, therefore, in microecological rather than in gross terms, for apparently

minor deviations in moisture, organic matter, or pH may have drastic influences at points one centimeter apart, no less one meter or one kilometer apart.

Various procedures have been proposed for the direct microscopic examination of bacteria in soil. One such technique entails the incorporation of weighed amounts of soil in melted agar and the addition of drops of the agar infusion to a calibrated hemocytometer. The suspension is stained and then examined microscopically (12). If the amount of soil, the volume of agar, and the area over which the agar is spread are known, then the bacterial numbers can be determined quantitatively. A procedure developed independently by Rossi and Cholodny and commonly known as the Rossi-Cholodny buried slide or contact slide method has been used extensively for qualitative studies. A microscope slide is buried in the soil, and after appropriate periods the slide is removed, the larger debris carefully dislodged, and the microbial film adhering to and developing upon the glass surface stained with phenolic rose bengal. The Rossi-Cholodny slide allows microorganisms to develop in the physical posture typifying their normal position and associative relationships with their neighbors. In addition to conventional staining techniques and light microscopy, a number of other procedures have been devised for direct examination of bacteria and other inhabitants of soil; these include the use of microcapillaries incubated in contact with soil (3), the examination of soil suspensions by electron microscopy (22), the application of fluorescence microscopy with an appropriate agent for the fluorescence technique (13), and the viewing of cells by transmission electron microscopy (4).

The abundance of many bacterial, algal, and protozoan types cannot be estimated by the methods described above; for example, some organisms never produce recognizable colonies on agar media. For these, the dilution or most probable number technique is used, a method of estimating microbial density without direct enumeration. Following the inoculation of known volumes of a tenfold soil dilution series into flasks of nutrient media suitable for the specific organism under study, growth will commence providing the inoculum contains one or more cells. Thus, if growth occurs in the culture prepared from the 10^5 but not the 10^6 dilution, the population of that microbial type is taken as between 10^5 and 10^6. Quantification is achieved by inoculating five or more replicate flasks of medium with each dilution, recording the number of flasks showing turbidity or growth at each dilution, and making use of appropriate most probable number tables (2).

Estimates of bacterial numbers vary according to the means of determination. Plate counts usually give values ranging from several hundred thousand up to 200 million bacteria per gram of dry soil, the abundance being a reflection of the many environmental forces acting on these minute inhabitants. Plating, however, probably underestimates the true bacterial population density as many soil bacteria fail to develop upon conventional media. Estimates by direct microscopy provide values of the order of 10^8 to 10^{10} bacteria per gram of dry soil. The viable counting techniques thus give no more than 10 percent and occasionally even less than 1 percent of these values; a range of 1 to 10 percent approximates the percentage of the total count that is observed by cultural means.

Many estimates have been made of the total mass of bacteria, as well as of other microorganisms, in diverse localities. As an example of how the *biomass* is calculated, each cell may be assumed to have a volume of one cubic micrometer. Then, in a fertile soil containing approximately 10^8 bacteria per cubic centimeter of space, the bacteria would occupy

$$\frac{10^8 \text{ bacteria} \times 1.0 \ \mu m^3}{1 \text{ cm}^3} = \frac{10^8 \ \mu m^3}{10^{12} \ \mu m^3}$$

0.01 percent of the total soil volume. Using a microscopic count of 10^9 per cubic centimeter, 0.1 percent of the total volume would be bacterial protoplasm. If the viable and microscopic counts are taken as 10^8 and 10^9 per gram, respectively, and the average bacterial cell is considered to weigh 1.5×10^{-12}g (wet weight), some 300 to 3000 kg live weight of bacteria is calculated to be present in each hectare of surface soil; that is, 0.015 to 0.05 percent of the total mass. Estimates with various soils and often by dissimilar methods yield values ranging from about 100 to 4000 kg/ha for the bacteria on a live weight basis. These values are equivalent to about 0.010 to 0.40 percent of the total soil mass. The bacterial or microbial biomass has also been estimated by determining the quantity of certain cell constituents in soil, and these assessments give values comparable to those based on cell numbers.

Bacteria are rarely free in the liquid phase of the soil because most cells adhere to clay particles and humus. A large part of the microflora is probably segregated into definite colonies developing in favorable microecological sites or into distinct masses associated with slimy bacterial excretions. Bacteria and the inanimate colloidal particles or even pure clays are attracted to each other; this effect is in part an electrostatic attraction of the soil for the bacteria. The adsorption results in a diminution in the number of bacteria passing through soil in moving water and causes a greater retention by the soil. Where adsorption is prominent, biochemical activities are affected.

The numbers and types of bacteria are governed to a large extent by soil type and cultivation practices. The numbers in most grasslands, for example, are greater than in comparable arable land, a result of the greater root density and the larger supply of utilizable organic matter coming from root decomposition and from plant debris. The abundance also increases in going from cooler to warmer climatic areas. Bacterial density is influenced to a large extent by the organic matter content of the habitat. Bacterial numbers are similarly higher in cultivated than in virgin land, but exceptions to this rule are not difficult to find; cultivation usually makes conditions more favorable for bacterial proliferation by means not as yet fully understood.

It is interesting to observe ecological extremes, noting the similarities and differences in biological composition and activity of various environments. In temperate zones, bacteria overwinter in frozen soil and, although there may be some dying off and partial selection for strains able to withstand low temperature, the effect is minor because the community following the thaw is similar in size and makeup to that of the previous autumn. The circumstances are somewhat different, however, in areas of the arctic that are frozen 9 to 10 months of the year. Here, in localities that never reach a temperature of more than 10°C, counts in excess of a million per gram are observed even when the soil temperature remains below the freezing point for a period of several months. Such bacteria are undoubtedly in a state of dormancy awaiting the spring thaw for activity to recur.

Desert soils present another extreme. Even in those soils that exist almost in an oven-dry state, bacteria are present. Spore-forming bacilli are often prominent, a likely result of the unfavorable conditions for vegetative development.

ENVIRONMENTAL INFLUENCES

Environmental conditions affect the density and composition of the bacterial flora, and nonbiological factors can frequently alter greatly the community and its biochemical potential. The primary environmental variables influencing soil bacteria include moisture, aeration, temperature, organic matter, acidity, and inorganic nutrient supply. Many lesser variables such as cultivation, season, and depth have been described and are of undoubted significance, but their influence arises from combinations of the primary determinants.

Moisture governs microbial activity in two ways. Since water is the major component of protoplasm, an adequate supply must be available for vegetative development. But, where moisture becomes excessive, microbial proliferation is suppressed not by the overabundance of water, which is not deleterious per se, but rather because the oversupply limits gaseous exchange and lowers the available O_2 supply, creating thereby an anaerobic environment. The maximum bacterial density is found in regions of fairly high moisture content, and the optimum level for the activities of aerobic bacteria often is at 50 to 75 percent of the soil's moisture-holding capacity. Furthermore, the bacterial population of various soils is closely correlated with their moisture contents. Even the periodic variations in community size of a single soil occurring with time are directly associated with fluctuations in moisture (11), emphasizing thereby the key biological role of water supply.

Waterlogging brings about a decrease in the abundance of bacteria developing in air—sometimes following an initial but brief rise in the number of aerobes—and a parallel stimulation of the strict anaerobes. This change from an aerobic to a largely anaerobic flora is effected by the disappearance of free O_2 as a result of its utilization by O_2-requiring microorganisms so that only bacteria tolerant of low O_2 levels or complete anaerobiosis are capable of proliferation.

Temperature governs all biological processes, and it is thus a prime factor of concern to the bacteria. An association between community size and temperature has been shown (11), but such quantitative effects are in addition to distinctive qualitative changes. Each microorganism has an optimum temperature for growth and a range outside of which development ceases. The temperature range and the optimum for proliferation serve as a means of delineating three microbial groups. Most microorganisms are *mesophiles* with optima in the vicinity of 25 to 35°C and a capacity to grow from about 15 to 45°C. Mesophilic types constitute the bulk of the soil bacteria. Certain species develop best at temperatures below 20°C, and these are termed *psychrophiles*. True psychrophilic bacteria are not common in soil; the bacteria even in winter are cold-tolerant mesophiles rather than psychrophiles. However, bacteria growing at temperatures from just above freezing to 5°C are numerous, although most proliferate more rapidly at higher temperatures, and these represent individuals of many genera (8). *Thermophiles* are ubiquitous. These are organisms that grow readily at temperatures of 45 to 65°C, and some, the obligate thermophiles, are incapable of multiplying below 40°C. Beyond its microbiological effects, temperature governs the rates of biochemical processes carried out by the bacterial flora, an increase stimulating the rate of reaction up to the point of optimum temperature for the transformation. For the temperatures found in most climatic regions, a warming trend favors the biochemical changes brought about by the microbial inhabitants.

Community size in mineral soils is directly related to organic matter content so that humus-rich localities have the largest bacterial numbers. The addition of carbonaceous materials also has a profound influence on bacterial numbers and activities, and the plowing down of green manures or crop residues initiates a ready microbiological response. This stimulation is most pronounced during the first several months of decomposition and largely disappears after the first year.

Highly acid or alkaline conditions tend to inhibit many common bacteria as the optimum for most species is near neutrality. The greater the hydrogen ion concentration, the smaller generally is the size of the bacterial community. It follows, therefore, that the liming of acid environments would greatly increase bacterial abundance. Nevertheless, soils of pH 3.0 contain many bacteria.

Although organic carbon is the major constituent of the food supply, inorganic nutrients are required, and it is not surprising that the flora is sometimes affected by the application of inorganic fertilizers (9). These substances serve a dual function since they supply both the plant and the microorganism with the needed inorganic nutrients. Often fertilizers exert no beneficial effects, and in such instances the explanation is probably the same for macro- and microorganisms; that is, the supply of the nutrients in the soil exceeds the biological demand. A not uncommon observation, on the other hand, is the suppression of bacteria by ammonium-containing fertilizers. This is not because of the added nitrogen but rather is the result of the acidity generated through the microbial oxidation of ammonium to nitric acid.

Cultivation practices also exert numerous direct and indirect biological effects. Plowing and tillage operations are drastic environmental treatments that usually cause marked bacteriological alterations. Such changes vary with the kind of operation, the soil depth, and especially the type of crop residues that may have been turned under. The effects noted seem to arise from improving the soil's structure and porosity, favoring the movement of air, altering the moisture status, and exposing inaccessible organic nutrients to bacterial action.

Season of year is a secondary ecological variable, secondary in the sense that it is composed of several well-defined primary variables. The net influence created by season, moreover, is not simple, compounded as it is by temperature, rainfall, crop remains, and the direct and indirect effects of plant roots. In temperate regions, a burst of activity occurs in the spring as the soil becomes warm and the organic materials from the previous fall and winter become accessible for decay. Usually, cell numbers are greatest during the spring and the autumn, and a decline occurs during the hot, dry summer months. In autumn, there is an increase in numbers after the low point in summer because of the more favorable moisture status and the availability of residues of root or above-ground tissues. The numbers commonly diminish in winter, and the cells remain in a state of biochemical inactivity; the bacteria are not eliminated during prolonged periods of freezing so that the microflora is ready for reactivation in the spring (14). Meteorological conditions during any single year may alter the usual seasonal sequence, for example, a hot, rainy summer or a cool, dry autumn. The seasonal changes in numbers of bacteria are closely related to fluctuations in moisture and temperature (11). Alterations in moisture and temperature during the year may influence the bacteria directly; alternatively, the climatic factors may in part operate indirectly through the surface vegetation which is the source of the carbonaceous nutrients reaching the microflora as root excretions, sloughed-off subterranean tissues, or as crop debris.

Depth is another secondary ecological variable that affects the bacteria. In temperate zones, these organisms are almost all in the top meter, largely in the upper few centimeters. At the very surface of cropped land, the community is sparse as a result of the inadequate moisture and the possible bactericidal action of sunlight. Examination of a typical profile from the surface to the C horizon reveals an increase as one goes from the very surface down a few centimeters, but the numbers decline with greater depth. Some typical values for mineral soils are presented in Table 2.1. In contrast with field soils, in which the greatest number of bacteria is typically found several centimeters below the upper crust, the highest number in shaded land of forest, orchard, or meadow is frequently in the top 1 to 2 cm. In organic soils, the abundance of bacteria often fails to

TABLE 2.1

Distribution of Microorganisms in Various Horizons of the Soil Profile (29)

Depth cm	Organisms/g of Soil $\times 10^3$				
	Aerobic Bacteria	Anaerobic Bacteria	Actinomy-cetes	Fungi	Algae
3–8	7800	1950	2080	119	25
20–25	1800	379	245	50	5
35–40	472	98	49	14	0.5
65–75	10	1	5	6	0.1
135–145	1	0.4	—	3	—

decline appreciably with depth, sometimes being greater at 160 cm than at the surface (6). Most of the changes associated with position in the profile are explained in terms of microbiological alterations produced by variation in the quantities of available organic carbon and O_2. Moisture, pH, and inorganic nutrients are probably of lesser consequence, but CO_2 concentration may be of some significance.

MORPHOLOGICAL AND GENERIC GROUPS

Bacteria can be subdivided into a number of morphological groupings, such divisions providing convenient means for describing the organisms indigenous to a given locality or responding to external influences. Because of the vast number and variety of bacteria in soil, it has not been possible to describe all types or to determine the generic placement of all strains. Difficulty is even encountered in the characterization of soil bacteria on the basis of their morphology since many of the dominant strains exhibit several shapes in culture depending on the age of cells and the medium used; for example, a common bacterium is a gram negative rod when young but becomes coccoidal and gram positive as the culture ages.

One approach to grouping the bacteria into morphological categories involves examining stained preparations made of the organisms growing as colonies on agar plates inoculated with soil dilutions. By such methods, non-spore-forming rods of various sizes, spore-forming bacilli, cocci, and short rods changing into cocci are found to be prominent. Many soil bacteria in culture are pleomorphic; that is, they change their shape during growth. The short rods that change to cocci with age are frequently representatives of *Arthrobacter*, and the spore-forming rods growing on agar incubated in air are typically species of *Bacillus*. The morphological assessments can also be combined with the gram stain, a typical investigation of such a study being presented in Table 2.2. The

TABLE 2.2

Morphological Categories of Microorganisms in two New Zealand Soils (18)

	Percent of Total Isolates	
Category of Microorganisms	Hastings Soil	Napier Soil
Rods, non-spore-forming		
Gram negative	19–24	17–29
Gram positive	2–6	3–7
Rods, spore-forming	12–18	10–15
Pleomorphic bacteria	36–46	31–41
Actinomycetes	11–25	12–32

ranges of values depicted in the table are derived from the use of different media and tests made on different dates.

The shapes and sizes of many bacteria in soil itself, however, appear to be different from those observed when the organisms are grown in culture media. Furthermore, a high percentage of the bacteria in soil are surprisingly small, and many have diameters of less than 0.3 μm and some smaller than 0.1 μm. Thus, a significant portion of the cells in soil would not be seen by light microscopy. In addition, direct examination of soil suspensions by electron microscopy reveals the presence of organisms, many of which are presumably bacteria, that exhibit morphologies vastly different from those that appear rapidly and abundantly on the usual agar media; some of these microorganisms bear stalks, others have surfaces with bristlelike appendages, and still others are shaped like stars or show surfaces with minute bulges (5, 21).

The bacteria are also frequently differentiated on the basis of the changes they bring about. Biochemically active strains may be investigated or isolated by the *elective culture* method in which a small quantity of soil is inoculated into a culture solution designed to favor one physiological group over another, for example, a medium with cellulose as sole carbon source or with protein as nitrogen source. To obtain pure cultures, several serial transfers are made in the elective medium followed by plating on agar. Such techniques delineate biochemical classes which, although not necessarily abundant, are important for fertility and crop production. Included are the nitrifying, ammonifying, urea-hydrolyzing, cellulose-utilizing, and protein-decomposing bacteria. Alternatively, the abundance of such physiological groups can be estimated by diluting soil suspensions, plating the suspensions on agar containing suitable selective media, and counting the number of colonies that appear.

Considerable attention has been given to the taxonomic groups into which bacteria fall. The genera that are either especially common or have attracted particular interest include *Acinetobacter, Agrobacterium, Alcaligenes, Arthrobacter, Bacillus, Brevibacterium, Caulobacter, Cellulomonas, Clostridium, Corynebacterium, Flavobacterium, Hyphomicrobium, Metallogenium, Micrococcus, Mycobacterium, Pedomicrobium, Pseudomonas, Sarcina, Staphylococcus, Streptococcus,* and *Xanthomonas.* Some of these appear abundantly on plates containing almost any one of many media inoculated with highly dilute suspensions of soil, but others require specialized cultural conditions. A few are rarely detected by plating techniques but are common and sufficiently distinctive in morphology to be recognized when soil preparations or suspensions are examined microscopically.

Numerous studies have been designed to identify the bacteria that produce colonies on standard laboratory media, although it is clear that many of the indigenous organisms may be entirely overlooked by such methods because of specialized conditions needed to show their abundance. Generalizing from the results of such surveys is difficult because of the vast differences in bacterial communities in dissimilar soils, but the data do show the more frequently encountered genera. Thus, of the cells from different soil types yielding colonies on agar, 5 to 60 percent are *Arthrobacter*, 7 to 67 percent are *Bacillus*, 3 to 15 percent are *Pseudomonas*, up to 20 percent are *Agrobacterium*, 2 to 12 percent are *Alcaligenes*, and 2 to 10 percent are *Flavobacterium*. Usually less than 5 percent of the colonies are derived from cells of *Corynebacterium, Micrococcus, Staphylococcus, Xanthomonas, Mycobacterium,* and *Sarcina*. On the basis of such ranges, one can easily list dominant genera but not assign percentages applicable to soils in general.

The prevalence of pseudomonads has been demonstrated in both cropped and uncropped land (Figure 2.2), and values in excess of a million per gram are not uncommon. The physiology and metabolism of this microbial group has been the subject of considerable inquiry. On the other hand, soils contain many arthrobacters, which because of their very abundance must play a dominant and key function, but the role of this genus in chemical transformations in nature is as yet unrecognized. The boundaries of the genus *Arthrobacter* are difficult to delineate, and there is considerable overlap with related genera. The most important of the neighboring genera is *Corynebacterium*, and saprophytic corynebacteria are quite common. The term coryneform has been proposed to include strains of *Corynebacterium* and *Arthrobacter*; the coryneforms are considered to be non-spore-forming, gram positive, nonmotile bacteria with a tendency toward irregular shapes.

Related to the coryneforms are the true acid-fast bacteria of the genus *Mycobacterium*. These microorganisms are comparatively less common, and their significance is likewise not known. However, they sometimes give counts of nearly a million per gram of soil, and they may constitute a significant portion of the bacterial community in regions of the tundra (32).

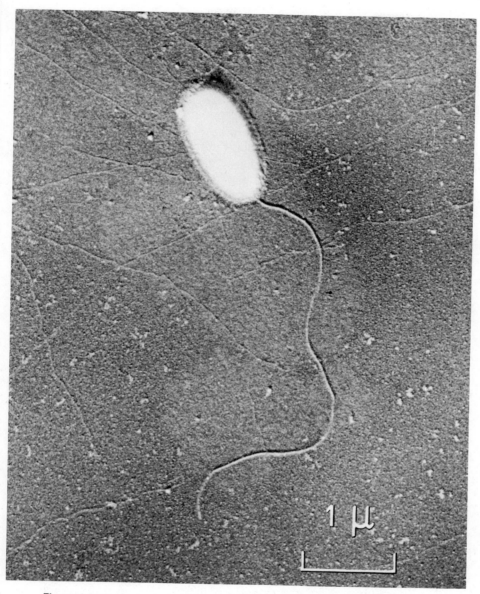

Figure 2.2. Electron photomicrograph of *Pseudomonas* sp. (Courtesy of H. Veldkamp.)

The isolation of *Bacillus* spp. is an easy task since pasteurization of soil suspensions at 80°C for 10 to 20 minutes destroys vegetative cells but not endospores, and subsequent aerobic incubation eliminates the only other common spore formers, the clostridia. The well-characterized nature of the genus, aerobic to facultatively anaerobic, spore-forming rods, has encouraged intensive investigation. The numbers of *Bacillus* are commonly quite high, often varying from about 10^6 to 10^7 or more per gram, but the size of the population is misleading since viable counts do not indicate whether the colony developed from a spore or a vegetative unit. In areas not recently amended with organic matter, *Bacillus* is frequently chiefly in the spore state, persisting in this dormant condition for many years. Only when specific nutrient conditions are provided do the spores become active, and it is as a result of the rare population burst that the soil becomes inhabited for many years by the dormant endospores. Thus, although the vegetative individuals may sometimes be abundant, it is not unknown to find that 60 to 100 percent of the *Bacillus* cells in soil exist as spores (20, 27).

Strictly anaerobic bacteria classified in the genus *Clostridium* occur in the most fertile areas in spite of the apparent availability of O_2. Complete aerobiosis, however, is not an actuality in natural conditions since, where microbial activity is high, the aerobes and facultative anaerobes consume O_2 and replace it with CO_2, lowering thereby the partial pressure of O_2 and permitting the proliferation of obligate anaerobes. This is probably especially true within soil aggregates and in areas with poor drainage. Plate counts show that from about 10^3 to 10^7 *Clostridium* cells occur per gram of different soils. To separate and obtain in pure culture the spore-forming *Clostridium*, advantage is taken of two physiological characteristics, the formation of heat resistant endospores and the capacity for anaerobic growth. The soil suspension is heated at 80°C for 10 minutes to kill the vegetative cells, and the remaining bacteria are plated directly on agar or further enriched by anaerobic incubation.

Evidence has recently appeared that soils are rich in bacterial cells bearing a semirigid appendage with a diameter smaller than that of the mature cell. This appendage is found in the stalked bacteria of the genus *Caulobacter* and also in the bud-forming bacterium *Hyphomicrobium*. The abundance of *Caulobacter* is often quite low, but sometimes as many as 250,000 cells are found per gram (7). Another group of as yet poorly described soil bacteria have small cells, up to 1.5 μm long, that have rows of small rounded protuberances on the surfaces that make the rod-shaped organism look like a corn cob (23).

Common to both soil and animal droppings is a group of microorganisms known as myxobacteria. The vegetative forms of these organisms are flexible rods that move by gliding. Most possess a resting stage during the life cycle in which the resting cells typically are borne in specialized fruiting bodies. The life cycle is completed when the rods emerge from the fruiting bodies and resume their active metabolism (Figure 2.3). Most frequently found are *Myxococcus,*

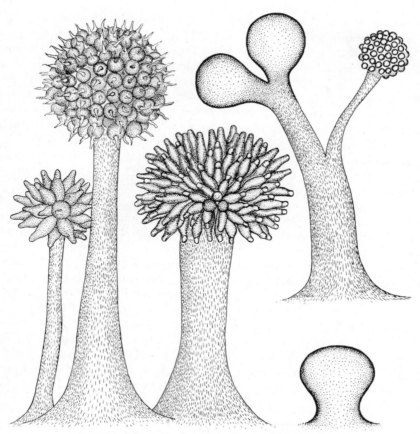

Figure 2.3. Drawing of a common myxobacterium (31).

Chondrococcus, Archangium, and *Polyangium*, but other genera can be obtained with little difficulty (19).

Myxobacteria can be isolated by adding a small quantity of soil to the center of an agar plate previously seeded with a bacterial suspension. Following incubation, the fruiting bodies become apparent to the naked eye. This technique relies upon the ability of myxobacteria to lyse the cells of true bacteria and to use them as a food source. The mechanism of the lysis involves the excretion of extracellular enzymes which dissolve the bacteria. The killing, lysing, and feeding on the simpler bacteria by the myxobacteria is possibly a significant factor in governing the composition of the flora. Myxobacteria have been found in all arable soils and most grassland samples examined with counts ranging from about 2000 to 76,000 per gram (28). A greater population is encountered in moist environments as these organisms do not seem to be tolerant of arid conditions.

Members of the genus *Bdellovibrio* are present in soil but not in abundance, yet they have been the subject of interest because of their unique relationship to other bacteria. The bdellovibrios are small curved rods, hence the suffix *vibrio* in the genus name, which in nature exist as obligate parasites, attaching themselves to and then living at the expense of the cells of larger bacteria. In culture, they fail to have a significant impact when the cell density of the host population is small, but *Bdellovibrio* feeds avidly as the host frequency rises during the growth cycle of the latter, ultimately causing a massive decline in the host population (Figure 2.4). The significance of such parasitism in nature is largely unknown, however.

Various pathogens of humans, livestock, cultivated plants, and other species of animals and plants are found in soil. Sometimes their presence is shown merely by the appearance of symptoms on a suitable host that comes into contact with the soil or on a plant that is growing in it, but often highly selective media and techniques are required to demonstrate their existence at a specific location or to enumerate them. By such means, species of *Agrobacterium*, *Erwinia*, and *Pseudomonas* causing diseases of plants have been found in diverse regions, some of the species being indigenous to soil, others only existing for brief periods as a result of recent contamination with tissues or oozes from diseased plants. Those that persist for some time may reinfect hosts grown on the same land the following season. The incidence of *Clostridium botulinum*, *Clostridium tetani*, and *Bacillus anthracis*—among the bacteria causing maladies of humans and animals—has been the subject of some study because their spores are long-

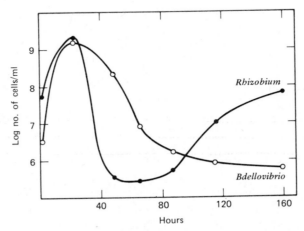

Figure 2.4. The effect of *Bdellovibrio* on the population of a strain of *Rhizobium* growing in solution (15).

lived and hence their existence below ground may lead to occasional episodes of botulism, tetanus, or anthrax. Some evidence exists, moreover, that *B. anthracis* may develop vegetatively in soil under certain circumstances (33). Other human and animal pathogens that have been found in soil include *Listeria monocytogenes*, *Erysipelothrix rhuziopathiae*, and additional species of *Clostridium*.

Because of the frequent contamination of soil with animal droppings, sewage containing disease agents, and tissues of diseased plants, the longevity of many of the introduced bacteria has been investigated frequently. For example, *Coxiella burnetii* (the rickettsia causing Q fever) can be recovered from soil newly contaminated by infected animals, *Salmonella* and pathogenic *Streptococcus* may be found after recent manure additions, and plant-invading species of *Agrobacterium*, *Corynebacterium*, *Erwinia*, *Pseudomonas*, and *Xanthomonas* are repeatedly introduced with parasitized plant tissues. Many of the aliens are soon eliminated and cause few problems, but some endure for considerable periods of time and thus pose a threat to potential hosts. The development of methods for the bacterial control of insects has also prompted assessments of the persistence of the microorganism that is used for the control measure; spores of *Bacillus thuringiensis*, for example, are known to remain viable for several weeks after their application in the field (26).

NUTRITION OF THE DOMINANT FLORA

It has been pointed out that bacteria may be subdivided into taxonomic, morphological, and physiological categories. Since physiological differentiation provides a functional approach, it has the greatest pedological significance, but the individual physiological groups are not mutually exclusive, for a single culture may be able to utilize N_2, decompose pectin, and hydrolyze cellulose. A scheme based on mutually exclusive types has been proposed using nutritional complexity as the determinative characteristic (17, 30). The nutritional approach has the further advantage of demonstrating for various treatments and environmental changes the specific growth requirements of the dominant bacteria.

To classify microorganisms by their nutritional habits, many colonies are picked in a nonselective manner from dilution plates containing a suitable agar medium. The growth of each of the isolates is then tested in media varying in complexity. On the basis of their development in these several media, the bacteria are divided into groups that require the following for maximum growth: no preformed growth factors, one or several amino acids, B vitamins, both amino acids and B vitamins, and a complex mixture of growth factors. Typical studies show that only about one-tenth of the bacteria in certain sites are able to grow readily in minimal media, the remaining nine-tenths requiring some growth substance for maximum development; about 10 percent need amino acids, a like number require B vitamins, and about 30 percent need a complex mixture of growth factors. A surprisingly high percentage of bacteria

TABLE 2.3

Percentage Incidence of Bacteria in Soil Requiring and Excreting Vitamins (25)

Vitamin	Percent of bacteria	
	Excreting Vitamin	Requiring Vitamin
Thiamine	28.0	44.9
Biotin	14.0	18.7
Pantothenic acid	32.7	3.7
Folic acid	26.2	1.8
Nicotinic acid	30.8	5.6
Riboflavin	27.1	1.8
Pyridoxin	18.7	1.8
Vitamin B_{12}	14.0	19.6
One or more vitamins	37.4	54.2

needs vitamins for replication (Table 2.3). These and similar results demonstrate that the nutrition of the soil bacteria varies considerably from the simple to the highly complex.

The ubiquity of amino acid and vitamin-requiring strains poses a problem as to the source of these substances; they must undoubtedly be produced continually in order to support such organisms. The answer seems to be in the ability of a large percentage of the bacteria developing in the absence of growth factors to synthesize such compounds and to excrete them into the surroundings (16, 24). Table 2.3 shows the frequency of species excreting B vitamins. Because of this degree of biological interdependence, the utilization by one bacterium of substances synthesized by its more versatile neighbor, it is not unlikely that amino acids, B vitamins, and possibly more complex substances are important in governing the microbiological equilibrium, serving to explain in part why there is an indigenous microflora of soil distinctive from that found in other habitats.

CHEMOAUTOTROPHIC BACTERIA

Microorganisms are divided into two broad classes with respect to their energy and carbon sources: *heterotrophic* (or *chemoorganotrophic*) forms, which require preformed organic nutrients to serve as sources of energy and carbon, and *autotrophic* (or *lithotrophic*) microorganisms, which obtain their energy from sunlight or by the oxidation of inorganic compounds and their carbon by the assimilation of CO_2. Fungi, protozoa, all animals, and most bacteria are heterotrophs. It should be mentioned that many and possibly all heterotrophs assimilate small quantities of CO_2, but the autotrophs alone use CO_2 as the sole carbon source. Autotrophs are of two general types: *photoautotrophs* (or *photolitho-trophs*) whose energy is derived from sunlight, and *chemoautotrophs* (or *chemolitho-trophs*), which obtain the energy needed for growth and biosynthetic reactions from the oxidation of inorganic materials. Algae, the higher plants, and a few bacterial genera have a photoautotrophic nutrition. Chemoautotrophy, on the other hand, is limited to relatively few bacterial species, yet it is of vast agronomic and economic importance.

The unique character of chemoautotrophs rests on two attributes, their ability to utilize the energy obtained by the transformation of inorganic materials and their capacity to make use of CO_2 to satisfy the entire carbon needs. Some species are limited exclusively to inorganic oxidations and are considered to be obligate or strict chemoautotrophs; others, the facultative autotrophs, may obtain energy from the oxidation of either inorganic materials or organic carbon.

The obligate chemoautotrophs are specific for their energy sources and utilize only one or a small group of related compounds; for example, nitrite for *Nitrobacter*, ammonium for *Nitrosomonas*, and certain inorganic sulfur compounds for species of *Thiobacillus*. Heterotrophs, on the other hand, have a more diversified nutrition.

The term autotrophy itself signifies the self-feeding habit of these bacteria. Nutritionally, they are the most primitive microorganisms since they have all the attributes of life yet they fulfill their needs from a completely inorganic environment. Physiologically, chemoautotrophic bacteria are very complex because they have within the confines of the cell wall all the enzymes, vitamins, coenzymes, carbohydrates, and other protoplasmic constituents typical of the heterotroph, and these they create from the most primitive conditions. Consequently, their synthetic powers must be truly great.

The chemoautotrophic bacteria may be subdivided on the basis of the element whose oxidation provides the energy for growth and cell synthesis.

    I. Nitrogen compounds oxidized.
        A. Ammonium oxidized to nitrite. *Nitrosomonas*
        B. Nitrite oxidized to nitrate. *Nitrobacter*
    II. Inorganic sulfur compounds converted to sulfate. *Thiobacillus*
    III. Ferrous iron converted to the ferric state. *Thiobacillus ferrooxidans*
    IV. H_2 oxidized. Several genera

Members of genera in addition to those listed also grow as chemoautotrophs, oxidizing compounds containing the same elements. Many other bacteria also have been considered as chemoautotrophs, but the work is either too limited, lacks confirmation, or is open to question. Strict chemoautotrophy has been demonstrated unequivocally in *Nitrosomonas*, *Nitrobacter*, and certain *Thiobacillus* species. The hydrogen bacteria are characterized by a facultatively autotrophic nutrition as they use the oxidation of either organic molecules or H_2 for energy. Most chemoautotrophs are strict aerobes, while those capable of proliferating in the absence of O_2 require the presence of an oxygen-rich substance, nitrate for *Thiobacillus denitrificans* and CO_2 for the hydrogen-oxidizing methane producers. The oxidized substances are converted to reduced products in the process.

The energy-yielding reactions in the metabolism of these organisms include the following:

$$NH_4^+ + 1\tfrac{1}{2} O_2 \rightarrow NO_2^- + 2\,H^+ + H_2O \quad \textit{Nitrosomonas} \qquad \text{(I)}$$
$$NO_2^- + \tfrac{1}{2} O_2 \rightarrow NO_3^- \qquad\qquad \textit{Nitrobacter} \qquad \text{(II)}$$
$$S + 1\tfrac{1}{2} O_2 + H_2O \rightarrow H_2SO_4 \qquad \textit{Thiobacillus} \qquad \text{(III)}$$
$$2H_2 + O_2 \rightarrow 2H_2O \qquad\qquad \text{Many organisms} \qquad \text{(IV)}$$

Chemoautotrophic populations are important in nature because of the energy-yielding reactions they catalyze, and several of the processes for which they are responsible are significant in crop production. The formation of nitrate provides the plant with an important inorganic nutrient in an easily assimilable form. Elemental sulfur has been used to control potato scab and to bring alkali soils into production, both actions a consequence of equation III. Other autotrophic transformations are known that have broad geochemical or biological implications.

REFERENCES

Reviews

Alexander, M. 1971. *Microbial ecology.* Wiley, New York.
Clark, F. E. 1967. Bacteria in soil. In A. Burges and F. Raw, eds. *Soil biology.* Academic Press, New York, pp. 15–49.
Gray, T. R. G. and D. Parkinson, eds. 1967. *The ecology of soil bacteria.* Liverpool Univ. Press, Liverpool.
Skinner, F. A. 1975. Anaerobic bacteria and their activities in soil. In N. Walker, ed., *Soil microbiology: a critical review.* Halsted Press (Wiley), New York, pp. 1–19.

Literature Cited

1. Alexander, M. 1964. *Annu. Rev. Microbiol.,* 18: 217–252.
2. Alexander, M. 1965. In C. A. Black, ed., *Methods of soil analysis.* American Society of Agronomy, Madison, Wis., pp. 1467–1472.
3. Aristovskaia, T. V. 1963. *Mikrobiologiya,* 32: 663–667.
4. Bae, H. C. and L. E. Casida, Jr. 1973. *J. Bacteriol.,* 113: 1462–1473.
5. Bae, H. C., E. H. Cota-Robles, and L. E. Casida, Jr. 1972. *Appl. Microbiol.* 23: 637–648.
6. Beck, T. and H. Poschenrieder. 1958. *Zent. Bakteriol.,* II, 111:672–683.
7. Belyaev, S. S. 1968. *Mikrobiologiya,* 37:925–929.
8. Druce, R. G. and S. B. Thomas. 1970. *J. Appl. Bacteriol.,* 33:420–435.
9. Eno, C. F., P. J. Westgate, and W. G. Blue. 1956. *Proc. Soil Crop Sci. Soc. Fla.,* 16:165–175.
10. James, N. and M. L. Sutherland. 1939. *Can. J. Res.,* C, 17:97–108.

11. Jensen, H. L. 1934. *Proc. Linnean Soc. N. S. W.,* 59:101–117.
12. Jones, P. C. T., J. E. Mollison, and M. H. Quenouille. 1948. *J. Gen. Microbiol.,* 2:54–69.
13. Kaczmarek, W., H. Kaszubiak, and H. Guzek. 1973. *Polish J. Soil Sci.,* 6:133–139.
14. Katznelson, R. S. and V. V. Ershov. 1957. *Mikrobiologiya,* 26:468–476.
15. Keya, S. O. and M. Alexander. 1975. *Arch. Microbiol.* 103:37–43.
16. Lochhead, A. G. 1957. *Soil Sci.,* 84:395–403.
17. Lochhead, A. G. and F. E. Chase. 1943. *Soil Sci.,* 55:185–195.
18. Loutit, M. and J. S. Loutit. 1966. *New Zeal. J. Agr. Res.* 9:84–92.
19. McCurdy, H. D. 1969. *Can. J. Microbiol.,* 15:1453–1461.
20. Mishustin, E. N. and V. A. Mirsoeva. 1967. In T. R. G. Gray and D. Parkinson, eds., *The ecology of soil bacteria.* Liverpool Univ. Press, Liverpool, pp. 458–473.
21. Nikitin, D. I. 1964. *Soviet Soil Sci.* No. 6, pp. 86–91.
22. Nikitin, D. I. 1973. *Bull. Ecol. Res. Comm. (Stockholm),* 17:85–92.
23. Orenski, S. W., V. Bystricky, and K. Maramorosch. 1966. *Can. J. Microbiol.* 12:1291–1292.
24. Payne, T. M. B., J. W. Rouatt, and A. G. Lochhead. 1957. *Can. J. Microbiol.,* 3:73–80.
25. Rouatt, J. W. 1967. In T. R. G. Gray and D. Parkinson, eds., *The ecology of soil bacteria.* Liverpool Univ. Press, Liverpool, pp. 360–370.
26. Saleh, S. M., R. F. Harris, and O. N. Allen. 1970. *J. Invert. Pathol.,* 15:55–59.
27. Siala, A., I. R. Hill, and T. R. G. Gray. 1974. *J. Gen. Microbiol.,* 81:183–190.
28. Singh, B. N. 1947. *Proc. 4th Intl. Cong. Microbiol.,* Copenhagen, pp. 465–466.
29. Starc, A. 1942. *Arch. Mikrobiol.,* 12:329–352.
30. Stevenson, I. L. and J. W. Rouatt. 1953. *Can. J. Bot.,* 31:438–447.
31. Thaxter, R. 1897. *Bot. Gaz.,* 23:395–411.
32. Umarov, M. M. 1971. *Vestn. Mosk. Univ. Biol., Pochvoved.,* 26(3):108–110.
33. Van Ness, G. B. 1971. *Science,* 172:1303–1307.

3
Actinomycetes

The true bacteria are distinctly different from the filamentous fungi, and many morphological characteristics separate the two broad types. There is, however, a transitional group between the simple bacteria and the fungi, a group with boundaries overlapping its more primitive and its more developed neighbors. These are the actinomycetes.

The term actinomycete has no taxonomic validity since these organisms are classified as bacteria in a strict sense, all being members of the order Actinomycetales, but not all genera of the Actinomycetales are considered to be actinomycetes in common parlance. The actinomycetes are microorganisms that produce slender, branched filaments that develop into a *mycelium* in all soil genera except for the genus *Actinomyces*. The actinomycete filament may be quite long, although it is short in a few groups, and it may fragment into much smaller units. The individual *hyphae* or filaments appear morphologically similar to the fungal filaments but are much less broad, usually 0.5 to 1.0 μm in diameter, a dimension analogous to that of the bacterial cell (Figure 3.1), but the hypha in certain genera may be up to 2.0 μm in diameter. Many of the soil actinomycetes produce single, pairs, or chains of asexual spores known as *conidia* on their hyphae, and a few of the soil inhabitants bear their spores in a specialized structure known as a *sporangium*.

Despite their placement together with the bacteria, the relation of the actinomycetes to the fungi is apparent in three properties: (*a*) the mycelium of the higher actinomycetes has the extensive branching characteristic of the fungi; (*b*) like the fungi, many actinomycetes form an aerial mycelium as well as conidia; and (*c*) the growth of actinomycetes in liquid culture rarely results in the turbidity associated with unicellular bacteria, instead it occurs as distinct clumps or pellets. On the other hand, the morphology and size of hyphae, conidia, and of the individual fragments of species whose mycelium undergoes

36

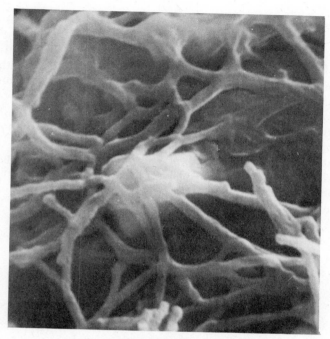

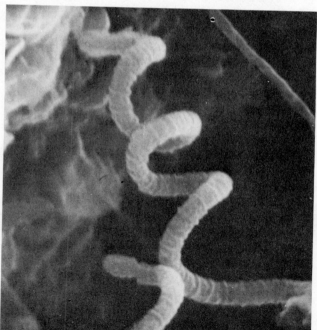

Figure 3.1. Top: mycelium of typical actinomycetes in soil ($\times$ 11,000). Bottom: spiral chain of *Streptomyces* conidia developing on a root ($\times$ 11,000) (10).

segmentation are similar to structures found among the bacteria. In addition, some actinomycete genera produce no aerial mycelium, and they closely resemble *Mycobacterium* and the coryneform bacteria in general morphology, staining reactions, and physiology. Other unique points for the taxonomic placement with the bacteria are the presence in some genera of flagella that resemble those of the true bacteria, similarities in cell wall composition, and sensitivities to antibacterial but not antifungal inhibitors.

Recognition of colonies of the most abundant actinomycetes on agar media is relatively simple providing the incubation period is sufficiently long. Whereas the usual bacterial colony consists of a large population of individuals derived from a single cell by binary fission, that of the actinomycete (Figure 3.2) prior to sporulation consists of but one organism, a mycelium derived from a single propagative unit. The colonies of some genera developing on the agar surface

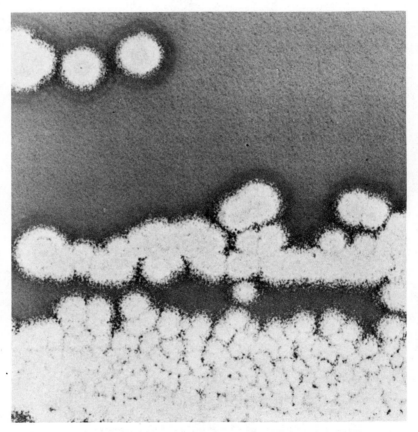

Figure 3.2. *Streptomyces* colonies on agar. (Courtesy of H. Veldkamp.)

may have a firm consistency and adhere tenaciously to the solidified substratum; in certain of these genera, the surface appears powdery and often becomes pigmented when the aerial spores are produced. In the organisms having a simple mycelium, the colony has a more mealy consistency and often crumbles when touched.

DISTRIBUTION AND ABUNDANCE

Actinomycetes are numerous and widely distributed not only in soil but in a variety of other habitats including composts, river muds, and lake bottoms. They are present in surface soil and also in the lower horizons to considerable depths. In abundance, they are second only to the bacteria, and the viable counts of the two are sometimes almost equal. Particularly in environments of high pH, a large proportion of the total community consists of the actinomycetes. Saprophytic existence is the rule among the free-living forms, but a few species can cause diseases of plants, domestic animals, and even humans.

A variety of microscopic or plating methods have been used in ecological investigations, but only the latter techniques are truly quantitative for these microorganisms. Despite the frequent use of plating, it is remarkable that the counts do not seem to be greatly affected by the composition of the medium, an indication that the organisms can utilize a variety of organic nutrients. Counts may be made upon the same plates that are used for bacterial enumeration, but special media are often preferable. In either instance, the period of incubation must be somewhat longer than for the bacteria because of the slow growth characteristics of the actinomycetes. Particularly useful selective media for enumeration are those containing chitin (7) because this polysaccharide is used by a high percentage of the more common actinomycetes and by a relatively lower percentage of bacteria and fungi. Media supplemented with antibacterial compounds are also sometimes employed (16).

In their normal habitats, the actinomycetes may occur as conidia or as the vegetative hyphae, and both forms can give rise to colonies on agar media. Thus, a determination of colony numbers on agar media will not differentiate between propagative units derived from a single conidium, an unbroken cluster of conidia, or a hyphal fragment, and the onset of sporulation in soil will result in high counts despite the lack of appreciable change in total protoplasmic mass. Consequently, data suggesting large communities may be merely indicative of species producing numerous conidia, and the results of plating must then be considered as far from unequivocal in giving a true representation of the active mass of biological material, reflecting only the number of fragments that multiply when placed in proper circumstances.

The size of the community depends on the soil type, particularly on certain of the physical characteristics, organic matter content, and pH of the environment. Plating estimates give values ranging from 10^5 to 10^8 per gram in temperate zones, but lower figures have been found in regions of Antarctica, acid peats, the tundra, and in waterlogged soils, and counts in excess of 100 million have been encountered occasionally. By and large, actinomycetes make up from 10 to 50 percent of the total community determined by plating in both virgin and cultivated land. In alkaline areas, especially when dry, the relative abundance is spectacularly high. Johnstone (6), for example, reports that actinomycetes accounted for 95 percent of the organisms in certain localities on the Bikini Atoll in the Pacific Ocean, a result probably due at least in part to the alkalinity.

Certain generalizations can be made with regard to the role of specific environmental characteristics in determining actinomycete abundance. By comparison with the true bacteria, actinomycetes are less common in wet than in dry areas. The population is likewise greater in grassland and pasture soils than in cultivated land, and the abundance in cultivated fields often exceeds that in adjacent virgin sites. Unfavorable, however, are peats, waterlogged areas, and environments whose pH is less than about 5.0. Soils in warm climatic regions are more conducive to an extensive actinomycete flora than those in cooler areas, and the size of the community in temperate latitudes tends to increase as one comes closer to the tropics. Consequently, the total number and percentage incidence of actinomycetes in the northern hemisphere increases moving from north to south (Table 3.1). The cell density will, of course, vary markedly in similar soils of any one locality, and it will be further influenced by season of year and by cultural practices.

TABLE 3.1
Abundance of Actinomycetes in Various Soils (12)

Zone	Condition	No./g $\times 10^3$		% Actino-mycetes
		Total	Actino-mycetes	
Tundra,	Virgin	2140	30	1.4
taiga	Cultivated	4850	84	1.6
Forest-	Virgin	1090	90	8.1
meadow	Cultivated	2620	790	28.
Meadow-	Virgin	3630	1300	35.
steppe,	Cultivated	4530	1570	35.
steppe				
Dry steppe	Virgin	3480	1200	35.
	Cultivated	6660	2100	32.
Desert	Virgin	4490	1550	36.
steppe,	Cultivated	7380	2380	34.
steppe				

Measurement of mycelial density on microscope slides buried in the soil demonstrates that the hyphae appear infrequently, suggesting thereby that the high values observed in the enumeration of viable units are the result of colonies derived from conidia. Furthermore, as their filaments are always slender, actinomycetes contribute little protoplasmic mass despite the high viable counts. Nevertheless, hyphae can be found by use of the buried slide technique, and their abundance is often affected by the season of year. The filaments, moreover, are sometimes seen to be bound to the larger soil aggregates and to penetrate these aggregates (22).

Current evidence suggests that the dominant conidia-forming groups exist to a large extent in the spore form and that the conidia serve a survival function for the organism. Most of the spores of *Streptomyces*, the numerically dominant genus, fail to germinate when added to soil, a failing that results from effects associated with representatives of the established populations because germination takes place in sterilized soil. Under natural conditions, the spores probably germinate when new organic substrates become available, the hyphae developing on the fragments of organic matter. The spores then are formed and become abundant at those microsites where the available nutrients existed, and the mycelium disappears as it is digested by its own enzymes or enzymes produced by nearby populations (9, 10).

The dominance of conidia points to the resistance of these structures to deleterious environmental conditions. Many strains possess spores resistant to desiccation, these conidia persisting for many years in air-dry soil. The resistance of conidia to destruction allows the organisms to persist when the habitat becomes unfavorable for vegetative activity. The conidia of species of the few genera tested also exhibit a greater tolerance to heat than the hyphae, resisting temperatures that totally destroy the reproductive capacity of the mycelium. This thermotolerance of the actinomycete conidium is not as marked as the resistance to heat of bacterial endospores, and only a few degrees higher than the lethal temperature for hyphae results in inactivation of the conidia.

ENVIRONMENTAL INFLUENCES

In qualitative and quantitative terms, the actinomycete flora is governed by the surrounding habitat. The stage of the life cycle that predominates, the size of the community, its biochemical transformations, and the genera and species found are determined by the forces acting within the ecosystem. Any one biological system is, in the last analysis, a reflection of the other biological systems functioning in association or in opposition and of the physical and chemical characteristics of the ecosystem. For the actinomycetes, the primary ecological influences include the organic matter status, pH, moisture, and the temperature. Season of year and depth in the profile are also of no little consequence, but the role of these two variables seems to be largely an outcome of interactions among the primary determinants.

Actinomycetes are affected directly by the presence of available carbon, and their number is especially great in land rich in organic matter. This is true whether examination is by plate counting or by one of several microscopic techniques. In general, sites high in carbonaceous materials and humus have larger numbers than habitats poor in organic matter. Amendment with organic nutrients such as protein derivatives, crop residues, and animal manure increases the abundance of actinomycetes. The population may sometimes reach 10^8 per gram with crop residue turn-under, especially in environments reaching high temperatures. It is not uncommon for manured soil to have more actinomycetes than adjacent unmanured sites, and the relative abundance of these microorganisms is sometimes increased as well as the actual numbers. Upon organic matter additions, the bacterial and fungus flora usually proliferate initially, particularly if nitrogen is plentiful, and the response of the actinomycetes frequently does not become pronounced until the later stages of decay. This suggests that the greater growth rates and biochemical versatility of certain bacteria and fungi make them the initial agents of destruction whereas the actinomycetes only appear when the more readily available compounds have been metabolized and competitive stress has diminished.

As a group, these microorganisms are not tolerant of low pH, and the community size is inversely related to the hydrogen ion concentration. Most strains of *Streptomyces* and related forms fail to proliferate or have negligible activity below pH 5.0, and the actinomycetes in highly acid environments frequently make up less than 1 percent of the total viable count. Acid-tolerant strains can be demonstrated with ease (20), but their scarcity suggests a minor biochemical significance. It is not uncommon to find actinomycetes in soil having pH values below that which allows for their growth in culture, and it may be that these populations multiply only in microenvironments where the pH rises because of alkaline products excreted during the development of immediately adjacent organisms (21). The fact that the limiting pH for most strains is in the vicinity of pH 5.0 has practical application in the control of certain plant diseases produced by *Streptomyces*; that is, acidification of the soil is used to suppress the pathogen (Figure 3.3). Even continuous applications of ammonium fertilizers without lime suppress the actinomycetes since the ammonium is oxidized to nitric acid by microbial action, and the resultant fall in pH leads to unfavorable growth conditions. Liming generally has a beneficial effect because vegetative development is favored by neutral or alkaline conditions, the population being most abundant in soils of about pH 6.5 to 8.0.

Moisture content is another critical environmental determinant. Under conditions of waterlogging or where moisture is above the microbiological optimum, for example at 85 to 100 percent of the water-holding capacity, these microorganisms appear only rarely. This is a consequence of the aerobic metabolism of all common soil actinomycetes and the consequent inability to develop and spread when free O_2 is lacking. On the other hand, actinomycetes

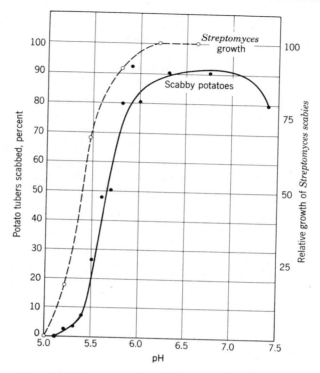

Figure 3.3. The incidence of potato scab as influenced
by pH of the soil (3).

are not as greatly influenced by semidry conditions as are the bacteria, and the filamentous group tends to be favored by low moisture levels both in vegetative development and in conidia formation. Consequently, the numbers of actinomycetes remain high as soils dry out while the relative incidence of bacteria diminishes because of their lack of tolerance to arid conditions. Meiklejohn (11) has recorded the case of a severe drought in Kenya during which time the actinomycetes, initially representing less than 30 percent of the colonies on dilution plates, made up more than 90 percent of the viable organisms. Even in certain true deserts, these microorganisms often dominate the microscopic life, but moisture effects of this type are associated with the persistence of conidia because the vegetative hyphae undoubtedly require appreciable moisture for biochemical activity. Such data suggest that actinomycete spores possess a high degree of resistance to desiccation and persist for longer periods than other taxonomic groups.

Examination of hyphal density on glass slides incorporated in the soil shows that there is little growth of mesophilic actinomycetes at 5°C and essentially none at 39°C. Increasing the temperature from 5 to 27°C leads to greater development, and the optimum range is generally from 28 to 37°C.

Moisture, temperature, and the availability of organic matter from roots and plant residues to a large extent determine the microbiological influence of season. Counts are frequently highest in spring and in autumn, the increase in the latter time of year usually being attributed to the return of plant residues to the soil. The scarcity in winter in the temperate zone presumably results from frost killing. During hot, dry periods of the year, a decline similarly occurs but, because of their great tolerance to desiccation, the relative proportion of actinomycetes often is highest in the dry months.

These filamentous microorganisms are present in the A horizon as well as at considerable depths below the surface, but the cell density estimated by plating techniques progressively declines with depth in the profile. In many but not all soils, the percentage of actinomycetes in the total microflora becomes greater with depth so that they make up a larger segment of the subsurface community. No explanation for this anomalous behavior is yet available although it may be associated with the downward movement of conidia with water or a differential effect of O_2 or CO_2 on bacteria and actinomycetes. Even in the C horizon, counts may range from 10^3 to 10^5 per gram.

TAXONOMY

Many previously unrecognized groups of actinomycetes have been characterized in the past few years, and it is now clear that soils contain a surprisingly large array of distinctly different genera (Figure 3.4). These can be divided into only a few families, however.

  I. *Streptomycetaceae*. Hyphae usually do not fragment. Extensive aerial mycelium and chains of spores with 5 to 50 or more conidia per chain. *Streptomyces, Microellobosporia, Sporichthya.*

 II. *Nocardiaceae*. Hyphae typically fragment to yield small rounded or elongate structures. *Nocardia, Pseudonocardia.*

III. *Micromonosporaceae*. Hyphae do not fragment. Conidia exist singly, in pairs, or in short chains. *Micromonospora, Microbispora, Micropolyspora, Thermomonospora, Thermoactinomyces, Actinobifida.*

 IV. *Actinoplanaceae*. Spores are formed in sporangia. Diameter of hyphae may be 0.2 to sometimes more than 2.0 μm. *Streptosporangium, Actinoplanes, Planobispora, Dactylosporangium.*

  V. *Dermatophilaceae*. Hyphal fragments divide to form large numbers of round, motile structures. *Geodermatophilus.*

 VI. *Frankiaceae*. Inhabit root nodules in certain nonleguminous plants. Not grown apart from plant host. *Frankia.*

VII. *Actinomycetaceae*. No true mycelium is produced. Usually strictly to facultatively anaerobic. *Actinomyces.*

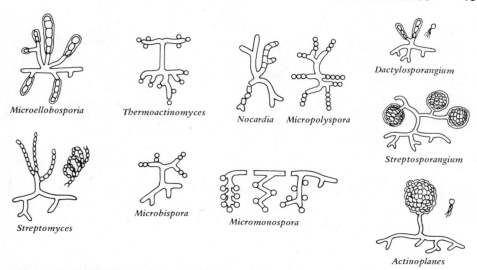

Figure 3.4. Major groups of actinomycetes (from H. A. Lechevalier and D. Pramer, 1971,
***The microbes,* Lippincott, Philadelphia.)**

Although a reasonable number of genera thus inhabit soil, species of only a few genera dominate the actinomycete colonies appearing on agar media inoculated with dilute suspensions of soil. Almost invariably, *Streptomyces* is numerically dominant, often accounting for up to 70 to 90 percent of the actinomycete colonies on most agar media. Rarely is their relative frequency low, but sometimes as little as 5 percent of the actinomycetes in some soils may represent species of this genus. *Nocardia* is usually the second most abundant, and commonly 10 to 30 percent of the actinomycete colonies are nocardias. Species of *Micromonospora* are usually the third most frequently encountered, and from less than 1 to 15 percent of the actinomycetes growing on solid media are members of this morphologically distinctive group. Some typical data are given in Table 3.2. Species of *Actinomyces* are considered to be uncommon, but an aerobic species of *Actinomyces* may be a frequent inhabitant of some soils (4).

All the other genera exist in sparse numbers. In one survey of more than 5000 actinomycete isolates, strains of *Thermomonospora, Actinoplanes, Microbispora, Thermoactinomyces, Streptosporangium, Micropolyspora, Pseudonocardia,* and *Microello-bosporia* accounted for 0.2 percent or less of the total (8). Studies of several Japanese soils similarly showed that *Microbispora* and *Thermomonospora* counts were less than 10^3 per gram (14, 15).

Streptomyces differs from *Nocardia* in that the former possesses a mycelium that does not divide into segments but that gives rise to conidia. The streptomycetes form a well-developed mycelium and aerial hyphae bearing numerous conidia in distinct chainlike arrangements. These spores are formed

TABLE 3.2

Abundance of Actinomycete Genera in Some Soils of U.S.S.R. (17)

Genus	Mollisol		Aridisol	
	No. of Strains	% of Total	No. of Strains	% of Total
Streptomyces	132	76	49	98
Nocardia	41	24	1	2
Micromonospora	1	0.6	0	0
Total no. isolates	174		50	

by division of the hyphae, the divisions progressing from the tip to the proximal end of the filament. When fully formed, the conidia have an oval to rod shape and resemble the cells of true bacteria in size and morphology. In the hyphae, growth is chiefly confined to the apical portion while the rest of the filament remains largely dormant. Turbidity is not formed in stationary liquid culture; rather the cells appear at the surface in a distinctive, flaky manner. In aerated liquid media, streptomycete growth is also not homogeneous and diffuse as with most bacteria, but typical mycelial pellets or clumps develop. Colonies on agar media tend to be tough and have a leathery consistency, and they resist destruction by mechanical force. The colony is frequently pigmented, but not uncommon is the production of water-soluble pigments that diffuse into the medium.

An unforgettable attribute of the streptomycetes is the musty odor they elaborate, an odor reminiscent of freshly turned soil, and it is not unlikely that the rich, earthy smell in newly ploughed land is the consequence of the presence of these microorganisms. The compound or compounds responsible for the earthy odor have been characterized, and the odorous streptomycete metabolite that has attracted most interest has been termed geosmin; however, other volatile products elaborated by the streptomycetes may also be responsible for the characteristic smell (2).

The colonies of *Nocardia* and true bacteria bear a marked resemblance to one another in general features and in consistency. Because of this similarity, most population estimates of bacteria inadvertently include the nocardias so that the so-called "actinomycete" numbers represent in reality almost only the streptomycetes. Species of *Nocardia* have early in their life cycles a rudimentary mycelium that soon fragments into short, rodlike cells.

Mesophilic actinomycetes that have a single conidium at the tip of specialized hyphal branches are placed in the genus *Micromonospora*. The morphology of the micromonosporas is unique, each hypha being 0.3 to 0.8 μm in diameter, while the spores are oval to round, 1.0 to 1.2 by 1.2 to 1.5 μm, and are produced singly at the terminus of the specialized conidiophores. The thermophilic counterpart of this genus, *Thermoactinomyces*, is common in heating compost heaps.

ACTIVITY AND FUNCTION

The actinomycetes develop far more leisurely than most fungi and bacteria, a characteristic suggestive of their inability to be effective competitors and of their lack of prominence when the nutrient level is high and the pressure of competition great. The feeble competitive powers may explain their relative scarcity during the initial stages of plant residue decomposition. When nutrients become limiting and the pressure of the more effective competitors diminishes, the actinomycetes become more prominent.

Actinomycetes are heterotrophic feeders, and their presence is therefore conditioned by the availability of organic substrates. Utilizable carbon sources include simple and highly complex molecules from the organic acids and sugars to the polysaccharides, lipids, proteins, and aliphatic hydrocarbons. Cellulose is decomposed by many species in pure culture, but the rate of decomposition is invariably slow. Many strains can also degrade proteins, lipids, starch, inulin, and chitin. Chitin hydrolysis is especially characteristic of frequently encountered species of *Streptomyces*, and the addition of this polysaccharide to soil serves as a means for the marked stimulation of streptomycete proliferation (13). The widespread capacity of streptomycetes, as well as certain other actinomycetes, to utilize chitin as a carbon source has led to the use of media containing the polymer for the isolation of members of the group. The metabolism of paraffins, phenols, steroids, and pyrimidines is well documented for species of *Nocardia* whereas *Micromonospora* strains decompose chitin, cellulose, glucosides, and hemicelluloses. Certain isolates develop well in media deficient in carbon, these being termed *oligocarbophilic* microorganisms.

The order Actinomycetales has received special attention because many strains have the capacity to synthesize toxic metabolites. As many as three-fourths of the streptomycete isolates may produce the antimicrobial agents known as *antibiotics*. The antibiotic substances produced in culture by actinomycetes are able, in culture at least, to inhibit the growth or cause the elimination of populations of bacteria, yeasts, and fungi of many taxonomic categories. The percentage of actinomycetes producing antibiotics varies with the soil and the season of year, and some test organisms are sensitive to compounds produced by many and some are inhibited by metabolites excreted by only a few actinomycetes (Table 3.3). Despite the great industrial and therapeutic value of these chemicals, there is still no clear picture of the significance of such compounds in natural processes. In addition to the production of antimicrobial

TABLE 3.3

Relative Frequency of Actinomycetes Producing Antibiotics Inhibitory to Various Test Organisms (5)

Test Microorganism	Percent of Actinomycete Isolates Producing Antibiotics	
	Soils Sampled in April	Soils Sampled in October
Rhizoctonia solani	23	19
Fusarium oxysporum	13	7
Candida albicans	10	10
Bacillus subtilis	24	17
Arthrobacter simplex	10	10
Escherichia coli	3.1	1.9

metabolites, many species of *Streptomyces* liberate extracellular enzymes which lyse bacteria (18). The possession of enzymes of this type may be important in the microbiological equilibrium in soil.

Most actinomycetes are mesophiles with an optimum temperature in the range of about 25 to 30°C; thermophilic cultures, however, are not uncommon. These are generally facultative thermophiles, growing at 55 to 65°C as well as at 30°C, but the former range is often the more favorable. By contrast, obligate thermophiles fail to proliferate at the lower temperature. Thermophilic actinomycetes are common in soil, manure, heating hay, and compost heaps, and their presence has been established even in soils that never become warm. Isolation requires the inoculation of a suitable source material into nutrient solutions maintained at elevated temperatures followed by plating of the enrichments on agar media incubated at these temperature extremes. In heating manure piles, the population of thermophilic actinomycetes becomes enormous, for example, up to 10^{10} of these microorganisms are present per gram at 50 to 65°C. The more numerous strains in thermophilic processes are classified in the genus *Thermoactinomyces*, but species of *Streptomyces* may often dominate.

The activities of the actinomycetes in soil transformations still are not clearly defined. Because microscopic examination reveals few actinomycetes in the mycelial stage and since present evidence indicates that the high plate counts are largely the result of conidial persistence, it seems that the actinomycetes have a lesser biochemical importance than the bacteria and fungi. Nevertheless, there is evidence for these microorganisms participating in the following processes:

   a. Decomposition of certain of the resistant components of plant and animal tissue. Actinomycetes do not respond immediately to the addition of natural carbonaceous materials but rather several weeks thereafter, suggesting that they fare poorly in competition with bacteria and fungi during the period when simple carbohydrates are present. They are usually effective competitors only when resistant compounds remain.

   b. Formation of humus through the conversion of plant remains and leaf litter into the types of compounds native to the soil organic fraction. Many strains can produce in culture media the kinds of complex molecules that are assumed to be important in the humus fraction of mineral soils.

   c. Transformations at high temperature particularly in the rotting and heating of green manures, hay, compost piles, and animal manures (Figure 3.5).

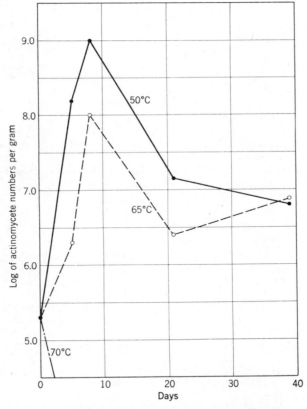

Figure 3.5. Effect of temperature on the development of actinomycetes in a manure compost (19).

In these conditions, the thermophilic actinomycetes may be the dominant group, sometimes to the extent that the surface of the compost pile takes on the white or gray color typical of this group. Here, *Thermoactinomyces,* certain streptomycetes, and species of spore-forming bacteria have the competitive advantage.

d. Cause of certain soil-borne diseases of plants; for example, potato scab and sweet potato pox, for which the causal agents are *Streptomyces scabies* and *S. ipomoeae*, respectively.

e. Cause of infections of humans and animals; for example, *Nocardia asteroides* and *N. otitidis-caviarum.*

f. Possible importance in microbial antagonism and in regulating the composition of the soil community. This role in the ecosystem may be a consequence of the ability of many actinomycetes to excrete antibiotics or their capacity to produce enzymes that are responsible for lysis of fungi and bacteria. Noteworthy in this connection are observations that amendment of soil with a substance, such as chitin, that favors hyphal development of actinomycetes sometimes leads to a marked suppression of fungi causing diseases of higher plants (1, 13).

REFERENCES

Reviews

Porter, J. N. 1971. Prevalence and distribution of antibiotic-producing actinomycetes. *Advan. Appl. Microbiol.*, 14:73–92.
Sykes, G. and F. A. Skinner, eds. 1973. *Actinomycetales*. Academic Press, New York.
Waksman, S. A. 1967. *The actinomycetes,* The Ronald Press, New York.

Literature Cited

1. Buxton, E. W., O. Khalifa, and V. Ward. 1965. *Ann. Appl. Biol.,* 55:83–88.
2. Collins, R. P., L. E. Knaak, and J. W. Soboslai. 1970. *Lloydia*, 33:199–200.
3. Dippenaar, B. J. 1933. *Sci. Bull. 136*, Dept. Agric., Union of South Africa.
4. Gledhill, W. E. and L. E. Casida, Jr. 1969. *Appl. Microbiol.,* 18:114–121.
5. Ishizawa, S., M. Araragi, and T. Suzuki. 1969. *Soil Sci. Plant Nutr.,* 15:214–221.
6. Johnstone, D. B. 1947. *Soil Sci.,* 64:453–458.
7. Kuznetsov, V. D. and I. V. Yangulova. 1970. *Mikrobiologiya,* 39:902–906.
8. Lechevalier, H. A. and M. A. Lechevalier. 1967. *Annu. Rev. Microbiol.,* 21:71–100.
9. Lloyd, A. B. 1969. *J. Gen. Microbiol.,* 56:165–170.
10. Mayfield, C. I., S. T. Williams, S. M. Ruddick, and H. L. Hatfield. 1972. *Soil Biol. Biochem.,* 4:79–91.
11. Meiklejohn, J. 1957. *J. Soil Sci.,* 8:240–247.
12. Mishustin, E. N. 1956. *Soils Fert.,* 19:385–392.
13. Mitchell, R. and M. Alexander. 1962. *Soil Sci. Soc. Amer. Proc.,* 26:556–558.
14. Nonomura, H. and Y. Ohara. 1971. *J. Ferment. Technol.,* 49:887–894.
15. Nonomura, H. and Y. Ohara. 1971. *J. Ferment. Technol.,* 49:895–903.

16. Ottow, J. C. G. 1972. *Mycologia,* 64:304–315.
17. Teplyakova, Z. F. and T. G. Maksimova. 1957. *Mikrobiologiya,* 26:323–329.
18. Villanueva, J. R., S. Gascon, and I. Garcia Acha. 1963. *Nature,* 198:911–912.
19. Waksman, S. A., T. C. Cordon, and N. Hulpoi. 1939. *Soil Sci.,* 47:83–113.
20. Williams, S. T., F. L. Davies, C. I. Mayfield, and M. R. Khan. 1971. *Soil Biol. Biochem.,* 3:187–195.
21. Williams, S. T. and C. I. Mayfield. 1971. *Soil Biol. Biochem.,* 3:197–208.
22. Zvyagintsev, D. G. 1964. *Soviet Soil Sci.,* pp. 307–310.

4
Fungi

In most well-aerated, cultivated soils, fungi account for a large part of the total microbial protoplasm. Although enumeration procedures used with other microbial groups tend to suggest that fungi are not major soil inhabitants, they do in fact make up a significant part of the biomass because of the large diameter and the extensive network of their filaments. Especially in the organic layers of woodland and forest soils do the fungi dominate the microbial protoplasm contained within the decomposing litter, but acid environments in general have the fungi as major agents of decay.

Characteristically, the fungi possess a filamentous mycelium network of individual hyphal strands. The mycelium may be subdivided into individual cells by cross walls or *septa*, but many fungal species are nonseptate. The hyphae of the nonseptate fungi are continuous and multinucleate, the filaments bearing no cross walls. The hypha itself is rather broad and has a diameter appreciably greater than that found in the common actinomycete filaments. Individual hyphae may be vegetative or fertile, the fertile filaments producing either sexual or asexual spores. In nature, the conidia or asexual spores are abundant and widespread, the sexual spores relatively uncommon. In culture medium, the mycelium is usually colorless while the asexual spores frequently are strikingly colored. Size, shape, structure, and cultural characteristics are important in taxonomy since, in contrast with the bacteria, the fungi can be effectively differentiated into genera and species on the basis of morphology.

DISTRIBUTION AND ABUNDANCE
Several techniques have been developed for the study of the fungal flora, each with its own advantages. No single procedure, however, adequately describes the entire generic composition of the flora nor does any one method depict accurately the mass or biochemical capacities of the hyphae. The approach most

frequently used for enumeration is the plate count, in which dilutions of a soil specimen in sterile water are plated on a suitable agar medium. Because bacteria and actinomycetes are usually more numerous than fungi, conventional laboratory media cannot be used as the development of fungal colonies on the petri dishes will be suppressed. Early microbiologists overcame the problem of suppression on solid media by acidifying the agar to pH 4.0, a reaction at which few bacteria and actinomycetes but most fungi develop. On the other hand, acidification is not necessary provided that appropriate bacteriostatic agents are included in the counting medium. Penicillin, novobiocin, rose bengal, and streptomycin have thus been used to inhibit bacteria and actinomycetes.

Population estimates of fungi based on plate counting are open to serious criticism. Since colonies appearing on the agar may be derived from a spore or a fragment of vegetative mycelium, the active or dormant nature of the viable unit in the original sample is unknown. Furthermore, the readily sporulating genera appear in large numbers on the agar plates because each individual spore may give rise to a colony. It is not surprising that fungi sporulating profusely, for example, *Penicillium* and *Aspergillus* spp., are isolated frequently. The mere act of shaking often introduces an error into the population estimate, for the agitation tends to rupture the mycelium and sporulating body into an indeterminate number of fragments each of which may produce a single colony. For these and other reasons, the results of plate counts must be interpreted with considerable care, bearing in mind the numerous shortcomings of the technique.

Fungi may be investigated in a number of ways not involving soil dilutions. Each additional technique serves to help in the characterization of the composition of the flora. Procedures for direct microscopic observation of the upper soil crust *in situ* have been used occasionally, but the special apparatus required and the sparse data obtained have limited the widespread adoption of these methods. The Rossi-Cholodny buried slide procedure, on the other hand, is a convenient means for the observation of microorganisms in situations not too different from the natural state; by this method, visualization of the hyphae and the spatial arrangement of microorganisms in the habitat is possible (Figure 4.1). Fluorescent microscopy is also sometimes employed. Other methods involve observing fungi developing in agar which has been poured over and mixed with soil contained in a petri dish or isolating organisms growing in an agar medium contained in perforated glass or plastic cylinders that have been buried in the field. Some of the procedures do not unduly emphasize the sporulating genera as is the case with dilution counts. As a consequence, it has been possible to demonstrate the abundance of many genera having no extensive capacity for sporulation but that are probably of considerable significance as agents of biochemical change. Because of the diversity of methods, it is not surprising that the results obtained by the various approaches frequently do not agree.

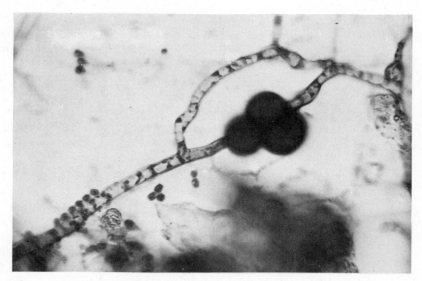

Figure 4.1. Fungus filaments and chlamydospores developing in soil treated with starch. (Courtesy of S. Ishizawa.)

For the purposes of enumeration, conventional plate counts have been most widely used since, although the results are far from unequivocal, this procedure permits a degree of quantification. Such estimates of microbial density reveal the presence in soil of populations typically ranging from as few as 20,000 to as many as 1,000,000 fungal *propagules* per gram, the propagule being considered as any spore, hypha, or hyphal fragment that is capable of giving rise to a colony. At best, plate count values of fungi rarely amount to more than a few percent of the bacterial count; however, as pointed out above, these estimates are misleading.

A variety of methods have been devis d to assess the length of hyphae or the total fungal biomass in horizons of diverse soils. These estimates vary with the procedure used and also with the soil, depth, and season of year. The length of fungal mycelium has been reported to range from 10 to 100 m per gram of surface soil, but values of up to 500 and sometimes in excess of 1000 m have also been obtained. Assuming the filament has an average diameter of 5 μm and a specific gravity of 1.2 and taking the range of 10 to 100 m per gram, it would appear that the weight of fungi ranges from approximately 500 to 5000 kg per hectare of surface soil. Thus, the filaments make up a significant part of the soil mass, although a major portion of the hyphae is quite likely nonviable.

The mycelium is closely associated with particles, and many filaments are physically attached to and frequently penetrate soil aggregates. Some species appear to be commonly growing in or on particles of organic debris, but many seem to be frequently linked to mineral particles (13).

Individual genera and species have been recorded in diverse and highly dissimilar habitats. Listings are thus available for the inhabitants of peats, flooded soils planted to rice, regions with low and extremely high salt contents, locations in many deserts, sites in Antarctica, as well as the tundra. The residents are frequently quite different, but cosmopolitan genera are also in evidence in widely different ecosystems.

ENVIRONMENTAL INFLUENCES

The abundance and physiological activity of the fungus flora of different habitats vary considerably, and the community and its biochemical activities undergo appreciable fluctuation with time at any single site. Both the generic composition and the size of the flora vary with the type of soil and with its physical and chemical characteristics. Whether a given microorganism will be able to survive, adapt itself, and become established in a specific habitat will be determined by the surrounding environment. The major external influences imposed on the fungus community include the organic matter status, hydrogen ion concentration, organic and inorganic fertilizers, the moisture regime, aeration, temperature, position in the profile, season of year, and composition of the vegetation.

Fungi are heterotrophic in nutrition, and neither sunlight nor the oxidation of inorganic substances provides these microorganisms with the energy needed for growth; fungal distribution is consequently determined by the availability of oxidizable carbonaceous substrates. In a general sense, the numbers of filamentous fungi in soil vary directly with the content of utilizable organic matter, but this microbial group is still present and of importance in areas low in organic matter. Improving the nutrient status by the incorporation of crop residues, green manures, or other carbonaceous materials into soil has the anticipated effect of increasing the size of the community. At the same time, application of organic substrates alters the composition of the flora, and the relative dominance of genera such as *Penicillium, Trichoderma, Aspergillus, Fusarium,* and *Mucor* is markedly affected. The stimulation of fungal action by supplemental organic matter is greatest during the initial period of decomposition, and the carbonaceous debris seems to be permeated with a hyphal network. Certain species become abundant immediately on addition of the carbon sources, but their numbers rapidly decline following the initial increase. Other species, however, maintain high population levels for relatively long periods after the incorporation of plant residues (20). The response varies with the chemical composition of the substrate and with certain environmental characteristics, the fungi usually dominating the microflora following carbonaceous amendments of acid environments supplied with adequate nitrogen.

The hydrogen ion concentration is another of the major variables governing the activity and composition of the flora. Many species can develop over a wide pH range, from the highly acid to the alkaline extremes. In culture, the capacity to grow readily at pH values as low as 2.0 to 3.0 is not rare, and numerous strains still are active at pH 9.0 or above. Because the bacteria and actinomycetes are uncommon in acid habitats, the microbial community in areas of low pH is dominated by the fungi. This is not the result of the fungi finding their optimum in acidic conditions but instead it is a consequence of the lack of microbiological competition for the food reserve. Thus, because of the insensitivity of many of the fungi to high hydrogen ion concentrations and the narrow pH range of most bacteria and actinomycetes, the fungi make up a larger percentage of the community and are responsible for a considerable portion of the biochemical transformations in acid habitats.

Individual species, however, may show a lesser tolerance to high hydrogen ion concentrations than the general fungus community. A pH sensitivity can be of profound importance to soil-borne plant pathogens that, as a consequence of the pH range for vigor and growth, are more destructive at acid, neutral, or alkaline reaction. Pathogens such as *Plasmodiophora brassicae* fare best in acid habitats, and the disease produced by it is uncommon or mild in land of pH greater than 7.5. Other plant pathogens grow optimally in soils near neutrality whereas certain species are prolific in alkaline localities, and acidification may therefore become a practical control measure. Liming or acidification practices are to no avail for those disease-producing organisms that are not markedly influenced by pH.

The application of inorganic fertilizers may modify the abundance of filamentous fungi, but such alterations are frequently more the result of acidification than of nutrient addition. Treatment with fertilizers containing ammonium salts increases numbers because microbial oxidation of the nitrogen leads to the formation of nitric acid, and the repeated annual addition of ammonium fertilizers favors the fungal and diminishes the bacterial and actinomycete counts.

All living things demand adequate moisture, and it is not surprising, therefore, that soil water has a direct effect upon the abundance and functions of fungi. Their capacity for catalyzing chemical changes is poor or lacking entirely when the water supply is low. Improvement in the moisture status of the environment favors fungal numbers so that, at suboptimal water levels, the count is often positively correlated with moisture (Table 4.1). Nevertheless, these organisms may persist in relatively semiarid conditions, and they may be metabolically active in localities with low water contents (15). At the opposite extreme, when moisture is excessive, diffusion of the O_2 necessary for aerobic metabolism is inadequate to meet the microbiological demand, and the fungi are among the first to suffer. Many genera are affected detrimentally as the water level increases, but certain of the Mucorales become more numerous (26).

The filamentous fungi as a group are strict aerobes although exceptions

TABLE 4.1

Changes in Population of Fungi in Grassland at Various Moisture Levels (12)

Avg % Soil Moisture	Fungi/g × 10³			Spores as % of Total
	Total	Hyphal Units	Spores	
8.9	99	60	39	39
11.2	89	57	32	36
18.5	142	113	29	20
24.2	149	133	16	10
27.1	173	153	20	12

are known, and some species that are commonly considered to be obligate aerobes grow to a modest extent when O_2 is no longer present (17). Even among the obligate aerobes, however, part of the mycelium may penetrate into locales where O_2 is absent, but much of the hyphal mass must have ready access to air. This dependence on O_2 probably explains to a great extent the concentration of fungi in the surface few inches. The requirement for O_2 also may be a major cause of the virtual absence of fungi from the lower levels of undrained peats and from swamps and bogs. In waterlogged mineral soils, numbers are similarly reduced to such an extent that the biochemical activities of these filamentous microorganisms in excessively wet environments are negligible. Some organisms will still persist for long periods in such unfavorable circumstances, a fact possibly associated with the production of resistant spores. Once the flooded soils are drained, the fungi return rapidly to a position of importance (30).

Most species are mesophilic in their temperature relationships, and thermophilic growth is uncommon. A few thermophilic strains can be demonstrated in normal soil, but such variants become particularly abundant only during the heating of rotting composts. The thermophiles will multiply at 50° and sometimes at 55° but not at 65°C, and they are absent from composts that reach high temperatures. Organisms actively growing at about 37°C seem to be localized in the surface horizons, where heating is appreciable during the summer months of the year. Heating by the sun is marked in tropical and subtropical soils, but the temperature of the surface layers of the soil may become sufficiently high by solar heating that thermophiles may develop even in temperate latitudes. Among the terrestrial thermophiles are species of *Aspergillus, Chaetomium, Humicola, Mucor,* and many other genera (4). Agar plates incubated at 6°C, on the other hand, contain mainly *Cylindrocarpon, Mucor,*

Penicillium, and *Cladosporium* species, and the abundance of these genera relative to total fungi becomes greater with depth in the horizon (6). These observations show a selection within the profile according to the optimal temperature range of individual genera of the microflora.

In cultivated soil, fungi are most numerous in the surface layers, but high counts are often observed in the B horizon of grass sod (Table 4.2). The population frequently remains large in the subsoil, and it may be appreciable to a depth of more than one meter. A greater number of individual species occurs near the surface than deeper in the profile. At the same time, the dominant species change. The organism concentration in the upper layers of the profile is to a large extent the result of the greater abundance there of readily available organic matter. The explanation of the qualitative changes within the fungus community of different horizons, on the other hand, may be associated with a strain adaptation to development at the low partial pressures of O_2 or the high concentrations of CO_2 in the deeper sites. An important selective effect of CO_2 becomes evident when the flora is divided into categories on the basis of vertical distribution: (*a*) fungi common throughout the profile; (*b*) those most numerous in upper layers or in surface litter but uncommon below 5 cm; and (*c*) strains rare in the upper regions but relatively common in lower depths. The surface fungi, group *b*, are inhibited by CO_2 while the microorganisms characteristic of

TABLE 4.2

Distribution of Fungi in the Profiles of Two Canadian Soils (25)

Depth	Horizon	Fungi/g $\times$ 10^3				
		May 30	June 30	July 20	Aug. 30	Sept. 30
		Cultivated Field				
0–7	A	35	6	10	15	22
7–14	A	30	6	6	4	5
14–28	A	3	2	3	3	6
33–52	A	2	2	1	5	5
52–68	B	1	6	0	3	5
68–84	B	0	0	1	2	5
		Grass Sod				
0–7	A	19	15	38	44	7
7–14	A	12	7	13	10	4
14–28	A	13	4	5	5	4
33–52	A	6	19	7	19	21
52–68	B	4	18	17	12	25
68–84	B	9	18	37	21	14

the subsurface sites have a greater tolerance to the gas. Types a and c are thus apparently dominant below the surface because of their relative insensitivity to CO_2 rather than their capacity to proliferate at low partial pressures of O_2 (8). With many fungi, mycelial growth and the formation or germination of various spore types are inhibited by CO_2, but the same gas is stimulatory to other species; thus, CO_2 may serve as an agent for the selection of the inhabitants of a particular site. The influence of depth upon the abundance and species composition of the fungal flora may therefore be associated both with the concentration of organic matter and the composition of the soil atmosphere.

Season of year exerts its influence in many ways. The warmth of spring usually is beneficial, but periods of summer drought or winter cold take their toll. The availability of organic matter is markedly influenced by season, and carbonaceous nutrients are abundant in the fall in the form of dying roots and plowed-under crop residues. As a result, counts tend to be high in the autumn and spring and often decline during dry periods in the summer (Table 4.2). Some localities may occasionally have active communities in the summer, but fungal action is always at a low ebb in regions having cold winters. With the return of the warm weather in spring, the fungus life of the soil becomes rejuvenated once more.

The dominance of one or another group is frequently related to the type of vegetative cover. Certain of the microorganisms are associated with definite plant communities while others seem to be unaffected by the kind of vegetation. In quantitative terms, for example, fields cropped continuously to oats contain more fungi than land cropped continuously to corn or wheat, suggesting a selective action by the oat plant. Qualitatively, the predominant fungus under oats is *Aspergillus fumigatus* while *Penicillium funiculosum* is the most numerous under corn (Figure 4.2). The selective action of the plant may be the result of a microbiological response either to specific root excretions or to chemical constituents of the sloughed-off tissues undergoing decomposition.

TAXONOMY

Considerable effort has been directed toward the establishment of the composition of the fungus flora of the soil. Mycologists in many countries have performed extensive studies of the genera and species that dominate in one or another ecological circumstance, and the results of their efforts have borne fruit. Although the entire dominant flora is not always well defined, a clear picture of the fungus inhabitants is now beginning to emerge.

Most isolates are placed in one of two classes, Hyphomycetes or Zygomycetes. The most frequently encountered group developing on agar media are strains belonging to the Hyphomycetes, fungi that produce spores only asexually. In species of this class, the mycelium is septate, and the conidial type of asexual spore is borne on specialized structures known as conidiophores.

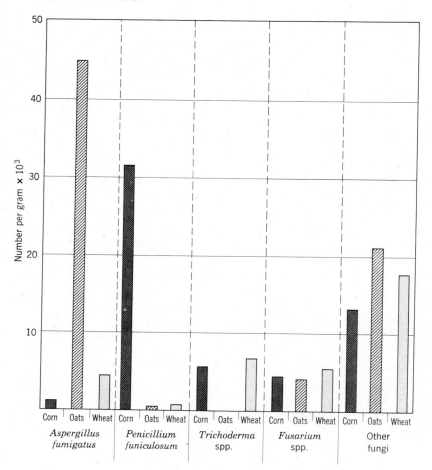

Figure 4.2. Frequency of fungi in soil under corn, oats, and wheat (18).

Zygomycetes and other fungi, on the other hand, produce spores by both sexual and asexual means.

The following list of classes is derived from the scheme of classification proposed by Ainsworth (1). Other classes than those given are present in soil, but they rarely appear on most agar media inoculated with soil samples. The genera tabulated are those most frequently encountered by conventional isolation procedures and, as such, they are undoubtedly of considerable importance in the biochemical transformations that take place. Several of the frequently encountered genera are shown in Figure 4.3.

    I. Hyphomycetes. Lack a sexual stage. Form a mycelium that has spores on special branches (the sporophores) or that bears no spores.

Alternaria, Aspergillus, Botryotrichum, Botrytis, Cladosporium, Curvularia, Cylindrocarpon, Epicoccum, Fusarium, Fusidium, Geotrichum, Gliocladium, Gliomastix, Graphium, Helminthosporium, Humicola, Metarrhizum, Monilia, Myrothecium, Paecilomyces, Penicillium, Rhizoctonia, Scopulariopsis, Stachybotrys, Stemphylium, Trichoderma, Trichothecium, Verticillium.

II. Coelomycetes. Lack a sexual stage. Spores in pycnidia or acervuli. *Coniothyrium, Phoma.*

III. Zygomycetes. The sexual resting spores are zygospores. *Absidia, Cunninghamella, Mortierella, Mucor, Rhizopus, Zygorhynchus.*

IV. Pyrenomycetes. Spores in the sexual stage are ascospores. *Chaetomium, Thielavia.*

V. Oomycetes. Possess biflagellate motile cells known as zoospores. Spores in the sexual stage are oospores. *Pythium.*

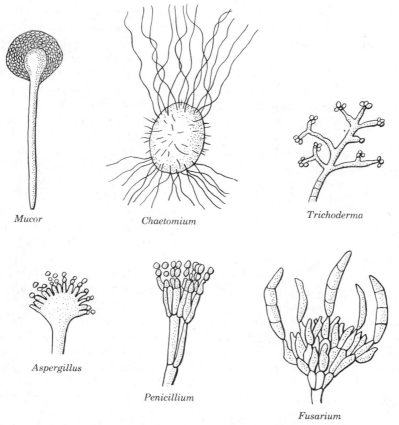

Mucor   *Chaetomium*   *Trichoderma*

Aspergillus   *Penicillium*   *Fusarium*

Figure 4.3. Common genera of soil fungi. (After J. C. Gilman.)

    VI. Chytridiomycetes. Possess uniflagellate zoospores and produce oospores. Commonly called chytrids.

    VII. Hymenomycetes. Sexual spores are basidiospores. Commonly called basidiomycetes.

    VIII. Acrasiomycetes. Have a free-living amebal phase, the amebae combining to form a pseudoplasmodium.

The media routinely used in microbiology laboratories are selective for individual nutritional types, and hence only certain fungi appear. A more complete nutrient substratum undoubtedly would show the presence of many other taxonomic groups. Conventional techniques, for example, fail to show the significance of the basidiomycetes classified as Hymenomycetes, despite their regular occurrence in forest and prairie soils. This omission is a serious shortcoming in characterization of a microbiological habitat since the basidiomycete size is enormous by comparison with the microscopic dimensions of those fungi that are better adapted to the artificial circumstances of agar media. Distribution of the Hymenomycetes must, therefore, be assessed by direct examination of the environment, particularly with a view toward observing the fruiting stages of the organisms. Such studies reveal that the mushroom fungi are common in forests and grasslands, and that their presence and activity are governed to no small extent by the availability of carbonaceous materials, moisture, and temperature. Difficulties in isolation have delayed intensive physiological investigations of these higher fungi, but their role in the initial rotting of woody tissue, in lignin decomposition, and in mycorrhizal associations has been well documented. Among the soil Hymenomycetes are found species of *Agrocybe, Ceratobasidium, Coniophora, Hyphodontia, Marasmius,* and *Pistillaria.*

    Chytridiomycetes are also rarely observed on the usual media, but by means of cellophane, chitin, or keratin baits, their presence is made known. These baits often favor the development of a limited number of chytrids that have differing affinities for the various substrates. Among the more frequently found genera are *Chytridium, Chytriomyces, Karlingiomyces, Nowakowskiella, Olpidium, Rhizophlyctis,* and *Rhizophydium.*

    Acrasiomycetes or cellular slime molds are a unique group found in soil. At one stage in their life cycle, these organisms exist as individual amebae that bear a remarkable resemblance to the true amebae. The amebae ultimately come together to form a fruiting body in which the organism produces its spores and, when conditions are suitable, the spores in turn give rise to the motile amebae. To demonstrate their presence, use is made of their habit of subsisting on bacteria; that is, the medium consists of a nonnutrient agar previously inoculated with an edible bacterium. Results obtained by this method show that the cellular slime molds are common in forest soils and reasonably infrequent in grasslands. In the forests of the temperate zone, moreover, they are more abundant in the layers of organic matter and leaf litter than in soil at some depth. The number of individuals, at least as detected by conventional counting methods, ranges from less than 100 to about 2500 per gram of soil, although occasionally values as high as 20,000 are recorded in forests (9, 10). The dominant genera tend to be *Acrasis, Acytostelium, Dictyostelium,* and *Polysphondylium.*

YEASTS

Little attention has been given to the yeasts, but their presence may be demonstrated in most soils. The term *yeast* has no taxonomic validity, but the group is commonly taken to include those fungi that exist primarily as unicellular organisms and that reproduce by budding or fission. Two broad categories may be differentiated, the sporogenous group that produces ascospores and those that do not form ascospores. The genera of soil yeasts most frequently isolated are *Candida, Cryptococcus, Debaryomyces, Hansenula, Lipomyces, Pichia, Pullularia, Rhodotorula, Saccharomyces, Schizoblastosporion, Sporobolomyces, Torula, Torulaspora, Torulopsis, Trichosporon,* and *Zygosaccharomyces.* Certain sugar-tolerant strains capable of carrying out an active fermentation of carbohydrates may also be demonstrated, but their scarcity in most field soils indicates that they are alien rather than native organisms.

The abundance of these organisms varies greatly with the locality under study, and counts from ca. 200 to 100,000 or more are not uncommon. Generally, populations of approximately 10^3 per gram are observed in temperate climates. Yeasts have been found in comparable numbers in soils of Antarctica, grasslands, cultivated fields, and forests, and they sometimes are particularly numerous on the roots of certain plants. It is not yet possible to correlate population size with environmental factors, nor is the role of yeasts in soil transformations defined.

GROWTH AND SURVIVAL

Fungi have many structures permitting survival of the population when environmental conditions are no longer favorable for active metabolism or when parasites become established around the organism. Sometimes it is the hypha itself that allows a particular species to endure adversity, but often specialized structures are associated with the persistence of the population. The specialized structures include conidia, chlamydospores, sclerotia, oospores, sporangia and sporangiospores, ascospores, and rhizomorphs. The longevity of some of these resistant bodies in soil is shown in Table 4.3. *Chlamydospores* are thick-walled cells appearing in some genera from preexisting cells in hyphae or from conidia. *Sclerotia* are hard, often large resting structures packed with mycelium. *Oospores* similarly are thick-walled structures associated with sexual stages of many genera. *Sporangiospores* are asexual spores borne in the *sporangia* of some fungi and contrast with the conidia, which are asexual spores produced at the ends or sides of hyphae. *Ascospores* are formed by meiosis in Pyrenomy-

TABLE 4.3

Persistence of Viable Fungal Structures in Soils (2)

Structure	Fungus	Persistence (years)
Chlamydospore	*Tilletia*	>5
Oospore	*Aphanomyces*	>10
Sclerotium	*Phymatotrichum*	>12
Microsclerotium	*Verticillium*	14

cetes in special saclike structures, and *rhizomorphs* are thick, frequently very long strands made up of hyphae that ultimately lose their separate identities. Because plant pathogenic fungi often owe their longevity in soil to such resting bodies and the frequency and severity of diseases these organisms cause frequently are related to the durability of the resistant stage, it is not surprising that particular attention has been given to the long-lived structures of the soil-borne pathogens.

The conidia of many genera are short-lived, and viability of these asexual spores is lost soon after they are formed in or enter soil. Thus, a fungus that sporulated profusely following a period of mycelial proliferation may be difficult to reisolate after several weeks. By contrast, the conidia of other genera or species persist in soil for some time. The durable conidia may remain inactive in dry soil, but some are stimulated to germinate as soon as moisture becomes available. On the other hand, conidia of many species do not germinate unless organic nutrients are in the surroundings, nutrients that may come from organic materials added to soil or from excretions of a root that grows to the site where the spore lies dormant.

Chlamydospore behavior has attracted widespread interest because the existence of various plant pathogens in soil, where they exist apart from their hosts, is attributable to the chlamydospores. Many of these species are not able to compete effectively with the soil residents for nutrients so that, were it not for the resistance of the chlamydospore, the population would be destroyed; for example, pathogenic species of *Fusarium* as well as *Phytophthora* and *Thielaviopsis* are able to survive as chlamydospores long enough following death of one host to infect new hosts planted some time later. Of considerable importance for the survival of the organism and its potential to invade plants is the fact that chlamydospores of many genera fail to germinate in soil, although many do so in water, but they do germinate when in proximity to roots or when organic compounds are added (27). This type of behavior allows a fungus that is a poor competitor for organic soil constituents to remain dormant as the chlamydospore, then to give rise to hyphae able to invade roots of susceptible hosts or to metabolize incoming nutrients for which it can effectively compete.

Sclerotia serve a similar role in allowing the populations of certain fungi to overcome deleterious environmental circumstances. These resistant masses of hyphae overlain by a hard coat are produced also by a number of pathogens, and they permit species of genera such as *Botrytis, Phymatotrichum, Rhizoctonia, Sclerotinia,* and *Sclerotium* to persist for months and often years, awaiting conditions to permit their germination and entry into suitable plant tissues. The survival ability of the tiny sclerotia of *Verticillium albo-atrum* is depicted in Figure 4.4. Sclerotia of some species remain viable and are not decomposed for extended periods whether moist or dry, but those of other species soon are subject to microbial attack if dried and then exposed to the community of moist soil (11).

Survival in the absence of host plants among species of *Aphanomyces, Phytophthora,* and *Pythium* has been attributed to their capacity to form oospores. For example, oospores of *Pythium aphanidermatum* retain viability in soil for at least 16 months and are able to germinate, even though the site in which they are deposited is exposed to extremes of temperature (29), and several years may elapse with at least some oospores being able to give rise to metabolically active hyphae. The oospores are probably produced only at certain times by the organisms, and those incorporated in the soil with plant tissues may then endure a few seasons until the surroundings are once again favorable for vegetative development. The sporangia of some fungi lose viability readily in soil, whereas the identical structures generated by different species persist for considerable periods. A surprising finding is that fungi not usually producing sporangia in culture may be induced to do so by soil bacteria (23). Indirect

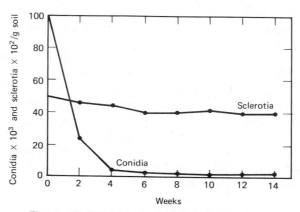

Figure 4.4. Survival of conidia and sclerotia of
***Verticillium albo-atrum* (14).**

evidence suggests, moreover, that a large percentage of the soil fungi emerging from dormant spores in fact develops from ascospores, and these ascospores do not germinate unless activated, as by heating (32).

Investigations of the behavior of the various spore types have revealed that soils have a substance or substances preventing the germination of conidia and other spore types. This inhibition is known as *fungistasis*, and soil fungistasis serves to prevent or inhibit germination of spores of numerous but not all genera. Fungistasis may be overcome with many of the organisms, and the spores then germinate when simple organic compounds, products of the decomposition of plant residues, or excretions of plant roots become available. Frequently, a spore may fail to germinate not because of the presence of fungistatic compounds but because germination requires energy derived from the metabolism of organic nutrients, which are not accessible to the spore (34). Nevertheless, many conidia germinate in distilled water, so that their inability to germinate in soil is the result of an actual toxicity and not merely the paucity of nutrients. The extent of inhibition of spore germination may sometimes be governed by their abundance, with the toxicity being absent at low spore densities (16). A few antifungal agents in soil have been characterized: ammonia in alkaline soils, possibly tannins from leaf litter, and aluminum. However, these substances do not account for the phenomenon of fungistasis in many sites where the level of these toxicants is not high.

Sensitivity to soil fungistasis is of great significance because the germ tubes emerging from many spores and the hyphae that may ultimately appear are frequently sensitive to lysis. The fungus thus survives in a resting and resistant stage and is not attacked by heterotrophs that could destroy the filaments, the fungus thus persisting until nutrients that would permit its extensive proliferation or roots that it could invade are in the immediate surroundings; the sensitive saprophytic species may then grow extensively on the organic materials, eventually to form again the spores influenced by fungistasis, whereas the sensitive parasite gains the opportunity to penetrate into a suitable host plant.

Lysis is readily in evidence among the fungi when they are observed microscopically. In lysis, the walls of cells or filaments are digested, probably usually as a result of enzymatic action by neighboring populations. In nature, fungi cannot remain viable without cell walls because the underlying membrane ruptures owing to the differences in osmotic pressure of the cell contents and the soil solution. Among many of the species sensitive to such attack, a germ tube emerging from a conidium or chlamydospore begins to grow, but lysis is soon initiated, growth terminates, and the germ tube is destroyed. Some species produce hyphae readily and the filaments spread through the soil or decaying carbonaceous materials, but with time they too succumb to the actions of lytic heterotrophs. On the other hand, not a few genera produce hyphae that withstand microbial digestion. Similarly, because they persist, sclerotia, chlamydospores, and many conidia also do not succumb readily. The durability of these resistant structures is frequently a result of the presence in the surface of either a dark pigment known as melanin (7, 19) or a polysaccharide composed of several sugars (5). Microorganisms that produce these lysis-resistant structures clearly have a unique advantage in that they will not be eliminated in regions where lytic bacteria and actinomycetes flourish.

FUNCTION AND ACTIVITY

The fungi contain no chlorophyll, and hence they must obtain carbon for cell synthesis from preformed organic molecules. In this regard, however, they are admirably suited since one or another strain can adapt itself to even the most complex of food materials. Among the carbon sources utilized are sugars, organic acids, disaccharides, starch, pectin, cellulose, fats, and the lignin molecule that is particularly resistant to bacterial degradation. Nitrogen frequently comes from ammonium or nitrate, but proteins, nucleic acids, or other organic nitrogenous complexes serve as well. Some species are nutritionally dependent, requiring B vitamins, amino acids, or other growth factors for active proliferation, but many develop fully in media containing only a sugar and inorganic salts. In the extreme condition of nutritional dependence are certain fungi that parasitize higher plants.

Predation is not rare among fungi. Various protozoa are especially susceptible to the active species. In the attack, the hypha penetrates the protozoan with a resulting decrease in motility of the animal and an eventual total cessation of movement. The fungus then slowly digests the cellular contents and assimilates the substances released. These predators seem to be obligate in their reliance on the protozoa, and no spore germination occurs unless the animal is present (28). Nematodes are also entrapped and devoured, frequently by means of specialized appendages or hyphal extensions (Figure 4.5). Among the more common *nematophagous*, or nematode-trapping, genera are *Arthrobotrys, Dactylaria, Dactylella*, and *Harposporium*. No definite function has been established for the predaceous species, but they may participate in the microbiological balance in soil, limiting the size and activity of the protozoan and nematode fauna.

Despite the limited value of quantitative estimates of numbers, considerable information is available on the function of the fungi. In the mycelial condition, one of the major activities is in the degradation of complex molecules, and the rapid fungal response following the addition of mature plant residues, green tissues, or animal manures bears witness to this capacity (Table 4.4). Upon the addition of organic matter, particularly to soils of low pH, the fungi become quite numerous. Representatives of all common classes can utilize and degrade the major plant constituents—cellulose, hemicelluloses, pectins, starch, and lignin. In woodland, the leaf debris becomes permeated with an extensive hyphal network that participates in the decomposition of the litter. The organic matter transformations brought about by filamentous fungi in well-aerated environments often may be more prominent than the reactions catalyzed by

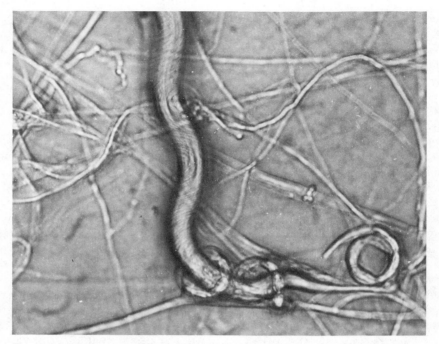

Figure 4.5. Nematode trapped by the fungus *Arthrobotrys conoides*. (Courtesy of D. Pramer.)

bacteria, but few attempts have been made to provide quantitative estimates of the relative activities of these two major microbial groups in the degradation of organic compounds. One method that has been proposed for this purpose involves an assessment of the metabolic activity of samples of soil amended with (*a*) an antibacterial chemical, (*b*) an antifungal substance, and (*c*) nothing. The results obtained by this procedure suggest that the fungi are dominant, at least in the decomposition of some simple sugars (3).

The utilization of proteinaceous substances is another common characteristic, and as a consequence the fungi are active in the formation of ammonium and simple nitrogen compounds. In the process of decomposition of the complex nitrogen-containing molecules, many genera and species participate. The microorganism benefits from the transformation since the proteinaceous material provides the organism with both nitrogen and carbon. Under certain conditions, however, the fungi will compete with higher plants for nitrate and ammonium and lead to a decrease in the soluble nitrogen content of the soil.

By the degradation of plant and animal remains, the fungi participate in the formation of humus from fresh organic residues. Species of *Alternaria, Aspergillus, Cladosporium, Dematium, Gliocladium, Helminthosporium, Humicola, Me-*

tarrhizum, and others synthesize substances resembling constituents of the soil organic fraction (26). Some fungi can also produce substances similar in chemical structure to several of the carbohydrates extracted from soil organic matter. In addition, this group carries out a number of inorganic transformations and also influences the formation of stable aggregates by means of hyphal penetration and the mechanical binding of particles.

Pathogenicity is another characteristic associated with several soil-borne fungi. Certain normally saprophytic species may, at the opportune occasion, invade living tissue and function as agents of plant disease. At one extreme are the facultative parasites that normally develop on inanimate materials but, for reasons not as yet fully understood, they do occasionally become concerned with the development of disease. At the opposite pole are the true parasites that are metabolically inactive in soil but that can persist for varying periods in an alien habitat when the host plant is no longer present. The former are indigenous to the environment and are capable of developing under the stress of intense microbiological competition. The latter are either allochthonous organisms or root inhabitants that may persist in soil, but they often find the environment inimical to their existence; these fungi have a unique advantage in that the host serves as a nutrient source not available to the soil inhabitants. Once the host plant is no longer present, the unique advantage is gone, and the fungus must cope with the residents of the soil ecosystem, where the parasite frequently fares poorly. Its abundance then may decline, and the population could even disappear completely. Only a very small portion of the fungi growing or surviving in soil is concerned with plant disease, and the more frequently encountered ones are generally classified in the genera *Armillaria, Fusarium, Helminthosporium, Ophiobolus, Phymatotrichum, Phytophthora, Plasmodiophora, Pythium, Rhizoctonia, Sclerotium, Thielaviopsis,* and *Verticillium.*

Fungi that may cause diseases among humans and animals are also

TABLE 4.4

Response of Fungi to Organic Matter Incorporation (21)

Soil Treatment	Fungi/g $\times 10^3$			
	7 days	21 days	35 days	49 days
None	7.90	7.55	4.06	4.74
Clover roots	70.0	68.0	64.4	43.2
Clover tops	—	—	48.0	43.0
Alfalfa roots	70.0	61.0	60.5	47.0
Alfalfa tops	—	—	72.5	36.8

encountered. Some of these organisms are dislodged from the soil surface by wind and are transmitted by air currents to be inhaled or alight on vulnerable portions of the body. Others are introduced with soil directly into injured areas of feet or other parts of the body, and infections associated with these pathogens are prominent in the tropics, where many people walk barefoot and are scantily dressed. A widespread pathogen is *Histoplasma capsulatum*, which is found in many regions of the world. In the United States alone, about 30 million people have been infected with *H. capsulatum*. This microorganism apparently colonizes soils contaminated with the fecal matter from chickens, starlings, pigeons, and other birds, and the fungus is also present in caves infested with bats. In soils containing *H. capsulatum*, the numbers range from 100 to 200,000 per gram (33). *Coccidioides immitis*, by contrast, is not cosmopolitan and is restricted to certain regions of North and South America. In these continents, moreover, it has a striking distribution for it is limited to only certain arid and semiarid localities where the soils are saline, receive little rainfall, and are exposed to high temperature (31). The basis for this dramatic biogeography is still unclear. The distribution of *Cryptococcus neoformans* is also associated with avian droppings, especially pigeon manure, but the organism is also found in soils around barns and near rabbit pens. Special techniques are required to demonstrate the presence of *H. capsulatum, C. immitis,* and *C. neoformans* because they do not appear on the agar media generally employed for the isolation of fungi. A highly selective procedure for isolating potential human pathogens from soil involves the use of keratin, a protein found in hair, nails, and animal horns. A common technique is to place sterilized hair in soil, and the fungi colonizing the hair and using its keratin are then isolated after suitable incubation periods. Many but not all the organisms that are thus obtained are found to be pathogenic; species of *Microsporum, Trichophyton,* and other genera are frequently found by means of these hair baits.

A unique fungus association with higher plants is found in the structure known as the *mycorrhiza* or fungus root, a two-membered relationship consisting of root tissue and a specialized mycorrhizal fungus. The microorganism is highly habitat-limited, and it usually is found only in the immediate vicinity of or directly within the roots. The fungus is not a soil microorganism in a strict sense, and its ecological niche is properly within the root association. The adaptation to root tissues may be linked with the complex nutrient demand of the microorganisms, many requiring mixtures of vitamins and amino acids and some having never been cultivated in artifical media.

Mycorrhizae are divided into ectotrophic and endotrophic categories. In the ectotrophic association, the fungus forms a mantle around the exterior of the roots, a network composed of a mass of hyphae entering into the spaces between individual plant cells. Many trees, including some of economic importance, bear this type of subterranean structure. The fungus in the endotrophic mycorrhiza, on the other hand, penetrates the cells of the host. The latter association is quite common among the Ericaceae and Orchidaceae as well as in fruit trees, citrus, coffee, and various legumes. Species of fungi rarely found in soil by dilution plating are capable of forming ectotrophic mycorrhizae. *Boletus*, *Lactarius*, *Amanita* and *Elaphomyces* are typical genera active on trees. Endotrophic mycorrhizal fungi include *Rhizoctonia*, *Phoma*, and *Armillaria*.

The formation of mycorrhizae is particularly pronounced in land low in phosphorus and nitrogen, and high nutrient levels are correlated with poor mycorrhizal development. Further, the production of these structures is most vigorous when the roots have a large reserve of available carbohydrates, especially following intensive photosynthesis. This may be an indication of the host supplying the invader with the carbohydrates necessary for its heterotrophic metabolism, but amino acids, B vitamins, or other growth factors cannot be excluded as factors provided by the root component. For many plants, the mycorrhiza exerts a beneficial influence; frequently no function can be attributed to the association, and occasionally detrimental effects ensue. Trees that bear ectotrophic mycorrhizae often develop well in the absence of the invader, but the existence of the fungal relationship may sometimes be advantageous or even essential. Mycorrhizae are thus important in forestry, involved as they are in problems of reforestation and in afforestation of new land.

The influence of mycorrhizae on the uptake of inorganic nutrients is often quite pronounced. Mycorrhizal roots frequently assimilate phosphate more readily than fungus-free roots, thus enabling the plant to grow well in phosphorus-deficient lands (24). The uptake of nitrogen, sulfur, zinc, and other essential elements is similarly promoted by the mycorrhizal fungus in many plant species. In addition, the fungus may protect the root against infection by a diverse array of soil-borne pathogens (22). Because of the many beneficial effects, extensive research has been conducted on inoculation of plant species of economic importance, and inoculation is sometimes practiced under field conditions.

REFERENCES

Reviews

Domsch, K. H. and W. Gams. 1973. *Fungi in agricultural soils.* Halsted Press (Wiley), New York.
Garrett, S. D. 1963. *Soil fungi and soil fertility.* Pergamon Press, Oxford.
Griffin, D. M. 1972. *Ecology of soil fungi.* Chapman and Hall, London.
Harley, J. L. 1969. *The biology of mycorrhiza.* Leonard Hill, London.
Tousson, T. A., R. V. Bega, and P. E. Nelson, eds., 1970. *Root diseases and soil-borne pathogens.* Univ. of Calif. Press, Berkeley.

Literature Cited

1. Ainsworth, G. C. 1973. In G. C. Ainsworth, F. K. Sparrow, and A. S. Sussman, eds., *The fungi*, vol. 4A. Academic Press, New York, pp. 1–7.

2. Alexander, M. 1975. *Microbial Ecol.*, 2:17–27.
3. Anderson, J. P. E. and K. H. Domsch. 1975. *Can. J. Microbiol.*, 21:314–322.
4. Apinis, A. E. 1972. *Mycopathol. Mycol. Applic.*, 48:63–74.
5. Ballesta, J.-P. G. and M. Alexander. 1971. *J. Bacteriol.*, 106:938–945.
6. Bisby, G. R., M. I. Timonin, and N. James. 1935. *Can. J. Res.*, C, 13:47–65.
7. Bloomfield, B. J. and M. Alexander. 1967. *J. Bacteriol.*, 93:1276–1280.
8. Burges, A. and E. Fenton. 1953. *Trans. Brit. Mycol. Soc.*, 36:104–108.
9. Cavender, J. C., 1972. *Can. J. Bot.*, 50:1497–1501.
10. Cavender, J. C. and K. B. Raper. 1964. *Amer. J. Bot.*, 52:297–302.
11. Coley-Smith, J. R., A. Ghaffar, and Z. U. R. Javed. 1974. *Soil Biol. Biochem.*, 6:307–312.
12. Eggleton, W. G. E. 1934. *J. Agr. Sci.*, 24:416–434.
13. Gams, W. and K. H. Domsch. 1969. *Trans. Brit. Mycol. Soc.*, 52:301–308.
14. Green, R. J. 1969. *Phytopathology,* 59:874–876.
15. Griffin, D. M. 1969. *Annu. Rev. Phytopathol.*, 7:289–310.
16. Griffin, G. J. and R. H. Ford. 1974. *Can. J. Microbiol.*, 20:751–754.
17. Gunner, H. B. and M. Alexander. 1964. *J. Bacteriol.*, 87:1309–1316.
18. Herr, L. J. 1957. *Ohio J. Sci.*, 57:203–211.
19. Kuo, M.-J. and M. Alexander. 1967. *J. Bacteriol.*, 94:624–629.
20. Martin, J. P. and D. G. Aldrich. 1954. *Soil Sci. Soc. Amer. Proc.*, 18:160–164.
21. Martin, T. L. 1929. *Soil Sci.*, 27:399–405.
22. Marx, D. H. 1972. *Annu. Rev. Phytopathol.*, 10:429–454.
23. Marx, D. H. and F. A. Haasis. 1965. *Nature,* 206:673–674.
24. Mosse, B., D. S. Hayman, and D. J. Arnold. 1973. *New Phytol.*, 72:809–815.
25. Newton, J. D., F. A. Wyatt, V. Ignatieff, and A. S. Ward. 1939. *Can. J. Res.*, C, 17:256–293.
26. Orpurt, P. A. and J. T. Curtis. 1957. *Ecology,* 38:628–637.
27. Papavizas, G. C. and M. F. Kovacs, Jr., 1972. *Phytopathology,* 62:688–694.
28. Peach, M. 1955. In D. K. M. Kevan, ed., *Soil zoology.* Butterworths Scientific Publications, London, pp. 302–310.
29. Stanghellini, M. E. and E. L. Nigh, Jr., 1972. *Plant Disease Reporter,* 56:507–510.
30. Stover, R. H., N. C. Thornton, and V. C. Dunlap. 1953. *Soil Sci.*, 76:225–238.
31. Swatek, F. E. 1970. *Mycopathol. Mycol. Applic.*, 41:3–12.
32. Warcup, J. H. and K. F. Baker. 1963. *Nature,* 197:1317–1318.
33. Weeks, R. J., F. E. Tosh, and T. D. Y. Chin. 1968. *Mycopathol. Mycol. Applic.*, 35:233–238.
34. Yoder, D. L. and J. L. Lockwood. 1973. *J. Gen. Microbiol.*, 74:107–117.

5
Algae

In almost every soil, in samples obtained from each continent, and from the most remote islands, one finds the presence of algae. These microorganisms are not as numerous as bacteria, actinomycetes, or fungi, and the lack of sufficient appreciation of this group can be attributed in part to the usually small numbers. Furthermore, since the algae are photosynthetic organisms that usually require access to sunlight, many early microbiologists felt that algal existence was too precarious for them to be of significance in soil. Recent work, however, has led to a more complete knowledge of the ecology and importance of the terrestrial algae.

The algae are abundant in habitats in which moisture is adequate and light accessible. Their development on the surface of cultivated or virgin land is frequently noted with the naked eye, but isolates can be obtained from lower depths as well. Their presence can be demonstrated readily by the addition of a small amount of soil to a medium containing nitrate, potassium phosphate, magnesium sulfate, calcium and iron salts, and traces of other inorganic nutrients. The resultant growth is visible macroscopically, usually as a green color, in the crude enrichments incubated in the light. The distinctive green pigmentation associated with these minute plants results from their possession of chlorophyll, but other pigments often mask the green color of the chlorophyll.

Morphologically, algae may be unicellular or they may occur in short filaments, but the soil strains as a group are characteristically smaller and structurally less complex than their aquatic counterparts. Several representative species are shown in Figure 5.1. The soil algae are divided into Chlorophyta or green algae, Cyanophyta or blue-greens, Bacillariophyta or diatoms, and Xanthophyta or yellow-greens. Other than the algae and certain algalike protozoa, the only photosynthetic microorganisms are a few genera of bacteria

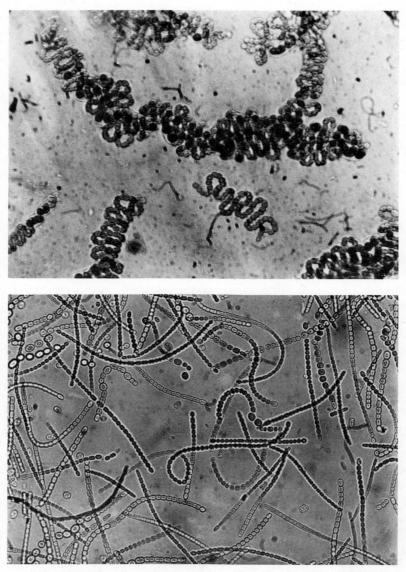

Figure 5.1a. Photomicrographs of some common algae. Top, *Anabaena spiroides*; bottom, *Anabaenopsis circularis* (22).

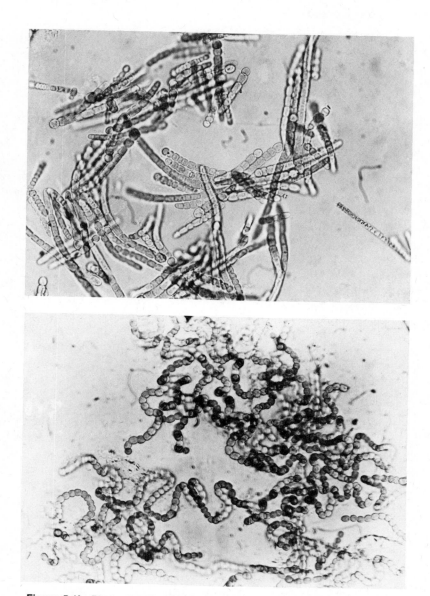

Figure 5.1b. Photomicrographs of some common algae. Top, *Tolypothrix tenuis*; bottom, *Nostoc* sp. (22).

whose habitat is usually aquatic rather than terrestrial. Because of their similarities to bacteria, the blue-green algae are sometimes classified together with the bacteria, and some microbiologists in fact prefer to call them bacteria rather than algae.

ECOLOGY

Algae are typified by the possession of a photoautotrophic nutrition that, through the agency of chlorophyll, endows them with the ability to use light as an energy source. The photosynthetic mechanism makes them independent of the preformed organic matter that limits the development of heterotrophic organisms in nature. For autotrophic development, the algae must obtain water, nitrogen, potassium, phosphorus, magnesium, sulfur, iron, and other micronutrients in minute quantities from the soil. The atmosphere provides carbon as CO_2 and energy in the form of light, but some species may make use of molecular nitrogen.

Algae found below the surface exist in complete darkness so that photoautotrophic life is impossible. Although many algae are obligate photoautotrophs and are, therefore, unable to grow in the absence of light, heterotrophy occurs in several species of Chlorophyta, Cyanophyta, and diatoms. These heterotrophic variants use the oxidation of organic carbon to replace the light in supplying energy for anabolic processes. Such species, properly classified as facultative photoautotrophs, metabolize a variety of carbohydrates including starch, sucrose, glucose, glycerol, and citric acid (21). Frequently, transfer of the organism to an inorganic medium and incubation in the light leads to an almost immediate resumption of photosynthesis, even following prolonged heterotrophic cultivation in the dark. Nevertheless, even with those species adapted to heterotrophy, the growth rate in the dark is less than during photoautotrophic development.

The occurrence of these photosynthetic microorganisms has been recorded in soils throughout the world, but no definitive geographical localization of families, genera, or species has yet been presented. As a group, the algae are moderately adaptable to environmental change, persisting in unfavorable circumstances such as in alkaline and desert soils. They are present also in alpine regions, in Antarctica, and on recent lava flows. Some species colonize the zone under the surface crusts of limestone and sandstone rocks in the desert, living at sites where the humidity is retained and where sufficient light penetrates to allow for photosynthetic activity (6).

The community tends to be concentrated immediately on and directly below the surface layer. Those organisms in the upper zones probably function as photosynthetic plants, using the sunlight that does penetrate. This locale is the site of dominant algal activity. On the other hand, isolates have been obtained from subterranean zones where light fails to penetrate, often to depths of 50 to 100 cm. The existence of algae at considerable depth poses a problem as to their mode of life: Do the subterranean forms have an active metabolism or do they exist passively? The latter alternative presupposes that the subsurface cells originate at the surface and are moved downward through water seepage, tillage practices, or by movement of the fauna. The finding of these organisms at depths of up to 1 m in sites undisturbed by seepage and cultivation suggests that at least some strains may proliferate within the profile itself. These subterranean forms might live as heterotrophs because the light needed for photosynthesis is unavailable. Yet, though many algae multiply heterotrophically in the dark, it is doubtful whether they can compete effectively with the dominant heterotrophs for the limited supply of available organic matter. The evidence at present argues against active growth below the sunlight zone. At the surface, on the other hand, algae are favored since they are not restricted by the organic matter level and need not compete for organic carbon.

Many estimates of community size have been carried out. Abundance is typically assessed by preparing tenfold dilutions of soil in sterile water and inoculating portions into a liquid medium or into sterilized sand containing suitable inorganic nutrients. After incubation of the medium or sand for 4 to 6 weeks in the light, the presence of algae is determined microscopically or by visual examination for the colored growth. Quantification is achieved by the most probable number procedure. Direct examination, either with an ordinary light microscope or by fluorescency microscopy, of soil dilutions allowed to dry on a measured area of a microscope slide is sometimes also employed to assess algal abundance. Viable counting procedures give values ranging usually from about 100 to 50,000 per gram for samples taken from immediatley below the surface of arable land, but results in excess of 10,000 per gram are uncommon. Under adverse conditions, the abundance of algae declines markedly. Surface samples frequently contain few organisms; alternatively, the population may be of the order of hundreds of thousands or even in the millions per gram where a distinct, visible bloom has developed.

The enumeration of algae in soil is of limited value because the significance of the observed numbers of algal units is difficult to interpret. Some species are filamentous and give low counts whereas others are unicellular, and each propagative unit represents an individual cell. At the same time, there are colonial forms in which the single colony yields many viable units.

The biomass of these organisms has been the subject of recent interest. Biomass can be estimated from the values obtained by counting procedures if one also measures cell volume and chooses a value, largely arbitrary, for the average specific gravity of the cell. The mass can also be estimated by extracting the chlorophyll pigments from soil with an organic solvent and determining the quantity of pigment by suitable analytical procedures. Such measurements give values ranging from as low as 7 to about 300 kg/ha. Values of 500 kg are sometimes encountered, and the biomass in highly localized areas showing surface blooms may have a mass equivalent to 1500 kg/ha (19).

   Qualitative approaches to ecological research have yielded fruitful dividends in defining the nature of the algal microflora (Figure 5.2). Such studies have demonstrated that only the green algae, diatoms, and blue-green algae are numerous in soil. Much less frequently noted are the Xanthophyta and certain of the chlorophyll-containing flagellates that are sometimes classified as protozoa. In temperate climates, the Chlorophyta are usually the predominant group, followed closely by the diatoms, while the Cyanophyta are the least numerous of the three major classes. The blue-greens are usually dominant to the Chlorophyta in tropical soils while the diatoms are the least numerous (16).

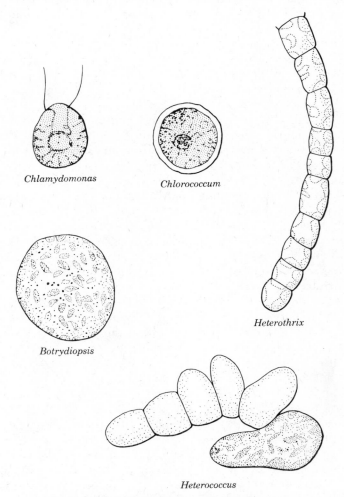

Chlamydomonas

Chlorococcum

Heterothrix

Botrydiopsis

Heterococcus

Figure 5.2. Several widespread algal genera (7).

Green algae, microorganisms classified as Chlorophyta, are characterized by the possession of chromatophores that impart to the organisms a grass-green color. In addition to chlorophyll, the cells contain xanthophyll and carotene pigments. In soil, these organisms are usually unicellular, but filamentous types are not unknown. Members of the class are found ubiquitously, entirely dominating the algal flora in acid soils but still numerous in neutral and alkaline environments. Species of *Ankistrodesmus, Characium, Chlamydomonas, Chlorella, Chlorococcum, Dactylococcus, Hormidium, Protococcus, Protosiphon, Scenedesmus, Spongiochloris, Stichococcus,* and *Ulothrix* are widely encountered.

Diatoms, on the other hand, are unicellular or colonial algae surrounded by a highly silicified outer layer. The cell wall consists of two separate halves, one overlapping the other. In comparison with aquatic types, the terrestrial diatoms tend to be of smaller size, and the difference in dimensions applies not only to species but to strains within the same species. Apparently, the environmental conditions in terrestrial habitats favor development of the smaller individuals. The small size may be advantageous since it permits greater water and salt absorption because of the greater surface:volume ratios of the cells. Diatoms are generally less frequent in acid soils, these algae faring best near neutrality or at slightly alkaline reactions; even in culture a number of species are limited to pH values greater than 6.0. The prominent genera in soil include *Achnanthes, Cymbella, Fragilaria, Hantzschia, Navicula, Nitzschia, Pinnularia, Surirella,* and *Synedra.*

Distinct from the other algal groups are the blue-green algae. In contrast to the aforementioned microorganisms, the Cyanophyta do not have their pigments localized in chromatophores but rather they are distributed throughout the cytoplasm. The nucleus, moreover, lacks a membrane and the clear morphological organization associated with other green plants, a point of similarity to the bacteria. Certain genera are unicellular and grow singly or in aggregates of individuals; *Anabaena* and others may be filamentous. The characteristic color of the group results from the presence, in addition to chlorophyll and the carotenoids, of a blue pigment known as phycocyanin. Studies in the field and in the laboratory indicate that the Cyanophyta prefer neutral to alkaline environments. This sensitivity to the hydrogen ion concentration is clearly evident from the work of Lund (14), who reported their absence in soils of pH below 5.2 and their frequent appearance in neutral and calcareous land. Many soil genera have been recorded, but the ones most frequently described are *Anabaena, Calothrix, Chroococcus, Cylindrospermum, Lyngbya, Microcoleus, Nodularia, Nostoc, Oscillatoria, Phormidium, Plectonema, Schizothrix, Scytonema,* and *Tolypothrix.* The prominent Cyanophyta in three tropical soils are shown in Table 5.1.

The yellow-green algae classified as Xanthophyta are relatively rare, but their isolation is not difficult. *Botrydiopsis, Bumilleria, Bumilleriopsis, Heterococcus,* and *Heterothrix* seem to be the most abundant and widespread, but even these

TABLE 5.1

Numbers of Named Species of Blue-Green Algae in Three Saline Soils (2)

	No. of Species Found		
Genus	Soil of pH 8.5	Soil of pH 8.7	Soil of pH 9.1
Anabaena	2	2	5
Lyngbya	10	4	9
Microcoleus	2	2	0
Nostoc	6	1	1
Oscillatoria	6	1	12
Phormidium	13	6	6
Plectonema	2	0	1

five are not prominent terrestrial organisms. Chlorophyll-bearing, unicellular flagellates of the genus *Euglena* and related genera are widely distributed but, although possessing a photosynthetic metabolism that relates them to the algae, these microorganisms also resemble the non-chlorophyll-containing protozoa. Frequently, the only significant cytological difference from the protozoa is the presence of the photosynthetic pigments. *Euglena* is classified as one of the genera of Euglenophyta.

In environments free of vegetation, the algae may play a critical pioneering role. Their early appearance in barren or denuded areas is especially noteworthy. For example, following volcanic eruptions that completely denude the surroundings of all higher forms of life, the algae are often primary colonizers (4). As the algae die and decay, the environment becomes more suitable for higher plants. Similar phenomena are not infrequently observed following burning. Large areas of eroded land in the United States bear algal crusts that initiate the plant succession cycle. In both volcanic and eroded areas, the Cyanophyta seem to be the pioneering invaders. Thus, because of its capacity to utilize simple inorganic compounds, the photoautotrophic microflora seems to be among the earliest living forms in surroundings where life has been eliminated by natural or artificial agencies.

These organisms sometimes also flourish to an extent that a distinctive visible bloom appears at the surface of arable land. Such growths require that the moisture level be high, and the algal abundance may then attain levels in excess of 10^6 cells/sq cm of surface and biomasses in the zone of the bloom equivalent to 700 to 750 kg/ha (13). In some desert soils receiving occasional precipitation, the organisms appear at the surface following a period of rain and cause an increase in the tensile strength of the upper crust of the soils. Even in regions where rainfall levels are high, extensive blooms probably bind small soil particles together into larger aggregates.

Another environment in which the algae could have a great agronomic significance is in flooded paddy fields. The microbiological action may be associated with the utilization of atmospheric nitrogen, the release of O_2, or the excretion of products stimulating plant development. During the extended periods in which rice soils are waterlogged, an algal film forms at the liquid surface, eventually making up an appreciable mass. Providing the pH is above about 6.0 and the phosphorus level is high, the algal bloom consists largely of blue-greens. Among the more common inhabitants of these waterlogged areas are species of *Anabaena, Calothrix, Nostoc, Oscillatoria,* and *Tolypothrix.* The large area of land cropped to lowland rice makes potential microbial contributions to the nitrogen and oxygen status of the paddy field or an algal synthesis of stimulatory products important to food production for vast numbers of people.

ENVIRONMENTAL INFLUENCES

One of the major environmental factors governing the activity of the hetero-trophic microflora, the content of readily available organic carbon, has no appreciable bearing on algal distribution because of their photosynthetic metabolism. On the other hand, this same photosynthetic attribute imposes on the algae the need for sunlight and CO_2. Obtaining an adequate supply of the latter rarely poses a problem as CO_2 and bicarbonate usually are produced in excess of the autotrophic demand whereas light accessibility is a dominant factor governing the distribution of photoautotrophic microorganisms. The need for sunlight is reflected particularly clearly in the vertical distribution of the algae. Thus, the population is most dense in the upper 5 to 10 cm and falls off dramatically with depth (12). Often the point of greatest concentration is in the surface centimeter, but sometimes a layer immediately below the topmost stratum has the most organisms; generally, however, the trend is for a decline in numbers with depth. It cannot be doubted that algae are present far below the zone of light penetration, and counts up to 10^3 per gram have been recorded from the C horizon. However, as pointed out above, because of their feeble competitive powers in the heterotrophic state, a large proportion of these organisms undoubtedly is dislocated and carried down through mechanical cultivation, the burrowing habits of earthworms and other lower animals, and by the movement of water. Therefore, these cells probably exist in a dormant condition as aliens in a foreign environment.

Acidity determines to a large extent the qualitative composition of the photoautotrophic microflora. Each individual strain has an optimum pH and a range outside of which the organism fails to multiply. More than any other broad taxonomic group of microorganisms, the algae exist independently of other organisms. The reason for the relative independence lies in the fact that the chief nutritional limitation to heterotrophic development is the supply of organic matter, a restriction that has no bearing on autotrophic proliferation. In contrast is the sparser fungal community in many neutral than in acid soils, the result not of the inability of fungi to grow at neutral reaction but of the expropriation of much of the organic matter by rapidly growing bacteria; that is, the distribution of fungi is affected markedly by other members of the microflora. One may expect, therefore, that the optimum for algal development in nature would agree more closely to that defined in vitro than for any other broad microbial group with the possible exception of the chemoautotrophs—which likewise are independent of carbonaceous materials. For example, blue-green algae generally develop best in pure culture from pH 7 to 10, and their occurrence in soil is similarly limited to neutral or alkaline conditions. They are often absent at pH values below 5 and uncommon below 6 (Figure 5.3). Diatoms are similarly less frequently encountered in acid soils whereas they are numerous in calcareous areas. In marked contrast, species of Chlorophyta are not appreciably limited by reaction, and they appear in regions with a diversity

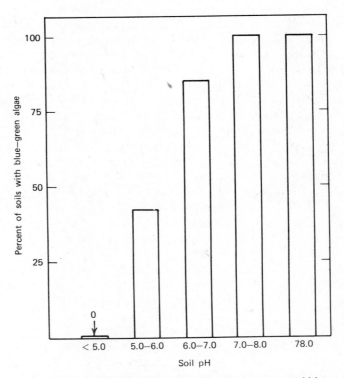

Figure 5.3. Relation between soil pH and occurrence of blue-green algae in Swedish soils (8).

of hydrogen ion concentrations; as a consequence, the Chlorophyta dominate the algal flora of acid habitats because of the absence of other forms.

Moisture is apparently a common limitation to growth since algal development is usually enhanced by increasing the supply of available water. In agricultural land, the quantity of water is often insufficient for algal development, and the extreme moisture variations in the surface layer play havoc with the metabolism of these microorganisms. Because of the dependence on moist conditions, the community responds greatly following periods of precipitation or, in regions having a wet and dry season, to irrigation. In times of drought, when the soil becomes desiccated, the organism density falls drastically. Of the three dominant groups, the diatoms are most sensitive to drying while the Chlorophyta and Cyanophyta exhibit a greater persistence and may endure in a resting stage for several years even in sun-baked, tropical regions. Laboratory studies show, moreover, that various species survive at least for 10 years in dry soil (20).

The requirement for moisture and adequate sunlight defines in part the seasonal influence. The water status is most favorable in spring and autumn in the temperate zone, periods of the year when algae show maximum vigor. Freezing is highly detrimental as evidenced by the rapid population decline during the winter months. This stage is followed by a spurt in activity at the onset of the spring thaw. In the driest portion of the summer, the floral status is poor as the low water and intense sunlight take their toll (17). By and large, algal proliferation is associated with wet, cool seasons during which the light intensity is not excessively high. Persistence under conditions of adversity is probably linked with the formation of dormant stages that provide the organism with a means of survival until the time that the environment becomes more favorable.

Herbicides are widely and frequently applied in order to control weeds that might have a serious effect on crop production. These chemicals have a degree of selectivity and kill only a portion of the plant species in the treated area. The toxicity of many herbicides is not limited to higher plants, however, and often a herbicide applied to a field to control undesirable rooted plants will also have a devastating impact on individual algal species or a major segment of the algal community.

Many species are also quite susceptible to attack by other subterranean inhabitants, both macroscopic and microscopic. Bacteria, fungi, and *Streptomyces* can destroy the integrity of the cells and filaments of diverse species of blue-green and green algae, although some populations are notably resistant to such attack. This type of attack is generally initiated by the heterotroph producing enzymes that cause the disintegration of the cell wall, the underlying part of the alga then being unable to maintain its structural integrity without the essential surface structure (9). The extensive digestion of the cells is of great significance because the algae assimilate many inorganic nutrients and thus make them unavailable to other organisms, but the decomposition leads to the release of the nutrient elements to the surroundings where they are once again available for use by members of the community. Nitrogen and phosphorus assimilation and their ultimate release during decomposition are of especial importance (Figure 5.4). The algae that resist destruction by their heterotrophic neighbors obviously must have structural features able to withstand the organisms that eliminate the susceptible species, and the unique structural characteristics appear to be components of the cell wall: ligninlike substances, sporopollenin, and special types of polysaccharides (10).

Among the animals that consume algae are protozoa, nematodes, mites, and earthworms, and some visible growths of the photosynthetic organisms may be rich in animals attacking them. The feeding is selective in that some species serve as good nutrient sources, whereas others are either poor nutrients or are

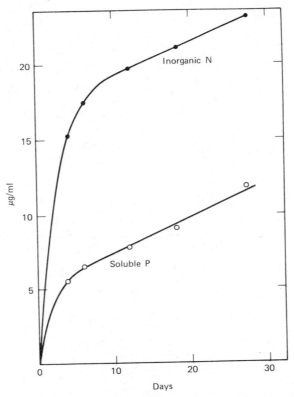

Figure 5.4. Formation of inorganic nitrogen and soluble phosphorus during the decomposition of _Chlamydomonas oblonga_ (15).

entirely rejected. The grazing by earthworms has been proposed as one means by which the algal biomass is reduced (3), and it is quite likely that both the macro- and microfauna contribute to the diminution in biomass of the surface or subsurface photosynthetic populations. An interesting illustration of the effect of the invertebrates became evident when it was noted that an insecticide stimulated algal populations in flooded rice soils; studies designed to explain this anomalous enhancement disclosed that the chemical suppressed the small animals that consumed the indigenous algae, so that the latter could then flourish (18).

SIGNIFICANCE

In a general sense, the algae cannot be considered as contributing appreciably to the many biochemical transformations necessary for soil fertility except in flooded soils planted to rice. Under the stress of competition from the bacteria, fungi, and actinomycetes, particularly below the surface, a group poorly adapted to heterotrophy could make only a small impression on the many biological reactions. Yet, the photosynthetic microflora is capable of exerting a definite influence in certain environments in which it occupies a unique position.

One of the major algal functions in terrestrial habitats is an outcome of their photoautotrophic nutrition. This function is in the generation of organic matter from inorganic substances. Those algae living at the soil surface convert CO_2 to carbonaceous materials; consequently, the photosynthetic microflora in some habitats is responsible for increases in the total quantity of organic carbon. The magnitude of these additions in agricultural land has not been accurately estimated, but the algal role in creating organic carbon *de novo* by colonizing denuded, barren, or eroded areas is beyond dispute.

Coincidental with the colonization of barren surfaces is the ability of the algae to corrode and weather rocks. A thick layer of algal cells is often found covering the surfaces of rocks, and the organic matter in these cells, on their death, supports the growth of bacteria and occasionally fungi that appear as secondary colonizers. The weathering of rocks through biological agencies of this type may be the result of carbonic acid formation from the respiratory CO_2 of the algae or it may be associated with the products of the bacterial and fungal utilization of the organic matter supplied by the algal protoplasm. In addition, lichens—which contain an algal and a fungal symbiont—release compounds that bring about weathering.

Algae also are conspicuous through their contribution to soil structure and erosion control. Thus, the surface blooms reduce erosion losses, probably by means of the binding together of soil particles. In the rain crust of deserts, the algal community that develops following periods of precipitation tends to increase the tensile strength of the crust by an analogous mechanism; this too affects the physical structure of the surface soil.

The photosynthetic microflora of flooded paddy soils has a special sphere of influence. In algae as well as in higher plants, the photosynthetic process liberates molecular oxygen. Through the evolution of this gas, the algae can beneficially affect the growth of rice by providing part of the O_2 required by the submerged roots. The abundance of algae in flooded fields in the tropics is of such magnitude that the O_2 contribution cannot be ignored. Under controlled conditions, growth of tobacco is enhanced by the presence in unaerated culture solution of green algae (5) so that extension of the oxygenation phenomenon to the wet paddy field is not difficult.

Recent years have seen an upsurge of interest in one major agronomic contribution of certain of the algae, their capacity to utilize N_2 as nitrogen source for growth. The process ultimately leads to the enrichment of the environment with combined forms of nitrogen since the protoplasmic constituents formed from N_2 are released and decomposed upon decay of the cell. The capacity for N_2 assimilation is associated only with the class Cyanophyta, but not all of the blue-greens can utilize molecular nitrogen. *Anabaena, Calothrix, Chroococcus, Nostoc, Oscillatoria, Scytonema,* and *Tolypothrix* are prominant genera concerned in nitrogen-enriching activities. Not only free-living cells but also those blue-greens participating in the lichen symbiosis are capable of using N_2. Such N_2-assimilating activities have been observed in temperate grasslands, the tropics, tundra regions, desert crusts following a rainfall, and on rocks colonized by lichens.

In vast areas of Asia, rice has been produced each year for centuries with no known addition of nitrogen in the form of manure or chemical fertilizers. The nitrogen would thus seem to be derived from the air over the paddy field and, as rice itself cannot utilize N_2, the nitrogen gains were assumed to be associated with free-living microorganisms. The results of careful experimentation have revealed that an increase in bound nitrogen frequently occurs in waterlogged soils containing an abundant blue-green algal bloom (24), and

TABLE 5.2

Increase in Rice Growth Resulting from Inoculation with Algae' (23)

Alga Used	Location	Yield Increase, %
Tolypothrix tenuis	Japan	2–20[a]
Aulosira fertilissima	India	114
Mixture	India	30
Anabaena azotica	Asia	24

[a] Value increased in each of the first four years.

there exists little doubt of the significance of the blue-greens in the nitrogen economy of paddy soils.

Because of the relatively high cost of nitrogen fertilizers in many of the countries where rice is grown, flooded soils have sometimes been inoculated with N_2-utilizing species of blue-green algae. The use of such inocula has often proved to be of practical value because of the resulting increase in rice yield (Table 5.2). On the other hand, indirect evidence suggests that the benefit of such inoculation may sometimes result from the algal synthesis of metabolites stimulatory to the rice plant rather than by their conversion of N_2 to usable products (1).

Some species of Chlorophyta parasitize cultivated plants as well as an array of noncultivated ones. Diseases of tea, citrus trees, cacao, and nutmeg result from the parasitic relationship (11). The occurrence of these algae in soil, however, has not yet been explored.

REFERENCES

Reviews

Fogg, G. E., W. D. P. Stewart, P. Fay, and A. E. Walsby. 1973. *The blue-green algae*. Academic Press, New York.

Lund, J. W. G. 1967. Soil algae. In A. Burges and F. Raw, eds., *Soil biology*. Academic Press, New York, pp. 129–147.

Round, F. E. 1973. *The biology of the algae*. Edward Arnold, London.

Literature Cited

1. Aiyer, R. S., S. Salahudeen, and G. S. Venkataraman. 1972. *Indian J. Agr. Sci.*, 42:380–383.
2. Ali, S. and G. R. Sandhu. 1972. *Oikos*, 23:268–272.
3. Atlavinyte, O. and C. Pociene. 1973. *Pedobiologia*, 13:445–455.
4. Brock, T. D. 1973. *Oikos*, 24:239–243.
5. Engle, H. B. and J. E. McMurtrey. 1940. *J. Agr. Res.*, 60:487–502.
6. Friedman, E. I. 1971. *Phycologia*, 10:411–428.
7. Fritsch, F. E. and R. P. John. 1942. *Ann. Botany*, 6:371–395.
8. Granhall, U. and E. Henriksson, 1969. *Oikos*, 20:175–178.
9. Gunnison, D. and M. Alexander. 1975. *Can. J. Microbiol.*, 21:619–628.
10. Gunnison, D. and M. Alexander. 1975. *Appl. Microbiol.*, 29:729–738.
11. Joubert, J. J. and F. H. J. Rijkenberg. 1971. *Annu. Rev. Phytopathol.*, 9:45–64.
12. Jurgensen, M. F. and C. B. Davey. 1968. *Can. J. Microbiol.*, 14:1179–1183.
13. Kulikova, R. M. 1965. *Soviet Soil Sci.*, pp. 166–169.
14. Lund, J. W. G. 1947. *New Phytol.*, 46:35–60.
15. Mills, A. L. and M. Alexander. 1974. *J. Environ. Qual.*, 3:423–428.
16. Mitra, A. K. 1951. *Indian J. Agr. Sci.*, 21:357–373.
17. Pomelova, G. I. 1970. *Pochvovedenie*, No. 8, pp. 70–74.
18. Raghu, K., and I. C. MacRae. 1967. *Can. J. Microbiol.*, 13:173–180.
19. Shtina, E. A. 1974. *Geoderma*, 12:151–156.
20. Trainor, F. R. 1970. *Phycologia*, 9:111–113.

21. van Baalen, C., and W. M. Pulich. 1973. *CRC Crit. Rev. Microbiol.*, 2:229–255.
22. Watanabe, A. 1959. *J. Gen. Appl. Microbiol.*, 5:21–29.
23. Watanabe, A. and Y. Yamamoto. 1971. In T. A. Lie and E. G. Mulder, eds., *Biological nitrogen fixation in natural and agricultural habitats.* M. Nijhoff, The Hague, pp. 403–413.
24. Yoshida, T., R. A. Roncal, and E. M. Bautista. 1973. *Soil Sci. Plant Nutr.,* 19:117–123.

6
Protozoa

Many representatives of the animal kingdom spend part or all of their life underground. The subterranean fauna contains protozoa, earthworms, nematodes, insects, and a variety of mammals. Invariably, however, the most abundant of the invertebrates found are the protozoa, the simplest forms of animal life. The phylum Protozoa contains primitive, unicellular organisms ranging in size from several micrometers up to one or more centimeters. The terrestrial species are all microscopic, however, and they are characteristically smaller than their aquatic relatives. These animal cells are typically devoid of chlorophyll, but transitional genera resemble the algae and possess chloroplasts containing chlorophyll pigments.

The life cycle of many protozoa consists of an active or *trophozoite* phase where the animal feeds and multiplies and a resting or *cyst* stage where the cell produces a thick coating about itself. In its encysted condition, many species can withstand deleterious environmental influences and persist for many years. Reproduction of protozoa is usually asexual, taking place by fission of the mother into two daughter cells, a process that occurs by either longitudinal or transverse division. Only a few of the protozoa reproduce sexually. Here, two cells similar in appearance fuse, their nuclei unite with an exchange of genetic material, and two new individuals ultimately emerge.

The distribution of protozoa has been investigated intensively by microbiologists throughout the world. The presence of the unicellular animals has been noted in equatorial, subtropical, and temperate regions and in the arctic and antarctic as well. No arable soil examined to date has been entirely devoid of protozoa although some localities yield but a single species while others contain a great diversity of types. From the ecological and the agronomic viewpoints, however, geographical studies have not been of profound importance because

89

the reasons for the dominance of individual species or genera in given localities have not been established.

TAXONOMY

Soil protozoa are classified on the basis of their means of locomotion. Some move about by virtue of one or more long *flagella* or whips, others by means of short, hairlike *cilia*, and a third group by temporary organelles known as *pseudopodia*. On the criterion of locomotion, subterranean representatives of the phylum Protozoa are divided into three groups: (*a*) Mastigophora or flagellates, which are motile by means of flagella; (*b*) Sarcodina, sometimes termed rhizopods or amebae, which possess pseudopodia; and (*c*) Ciliophora or ciliates, which bear cilia through the entire active stage of life.

Species of Mastigophora usually are endowed with one to four flagella, but occasional groups possess more than four. The strains found in soil characteristically are small, 5 to 20 μm in length. Protozoologists often separate this group into Phytomastigophora and Zoomastigophora, the former containing chlorophyll and growing photosynthetically and the latter being devoid of the green pigment and thereby limited to a heterotrophic existence. There is little question that the flagellates dominate the microfauna of terrestrial habitats. Among the many genera described, special mention may be made of *Allantion, Bodo, Cercobodo, Cercomonas, Entosiphon, Heteromita, Monas, Oikomonas, Sainouran, Spiromonas, Spongomonas,* and *Tetramitus*. The algalike flagellates are best represented by *Euglena*. Some typical genera are shown in Figure 6.1.

Members of the class Sarcodina move by means of temporary protoplasmic extrusions from the cell body. Because there is no rigid external surface, the shape of the animal body changes frequently as the organism sends forth or withdraws its pseudopodia. In this way, the Sarcodina differ markedly from the flagellates and ciliates in which the organelles of locomotion are essentially permanent structures. The soil amebae or Sarcodina are of two types: some have a shell-like (or *test*) structure, others possess none. When the shell is present, the pseudopodia extend through distinct openings. The most frequently encountered representatives of the class are *Acanthamoeba, Amoeba, Biomyxa, Difflugia, Euglypha, Hartmanella, Lecythium, Naegleria, Nuclearia,* and *Trinema* (Figure 6.1).

Movement among the Ciliophora results from the action of the vibrating hairs situated around the protozoan cell. The hairs are short and numerous, and several thousand may be found on a single individual. In size, the terrestrial ciliates are often as small as 10 μm yet they range up to 80 μm in length, but aquatic species are distinctly larger and some attain a size of 2 mm. The typical soil forms include *Balantiophorus, Colpidium, Colpoda, Enchelys, Gastrostyla, Halteria, Oxytricha, Pleurotricha, Uroleptus,* and *Vorticella* (7, 8, 14, 16).

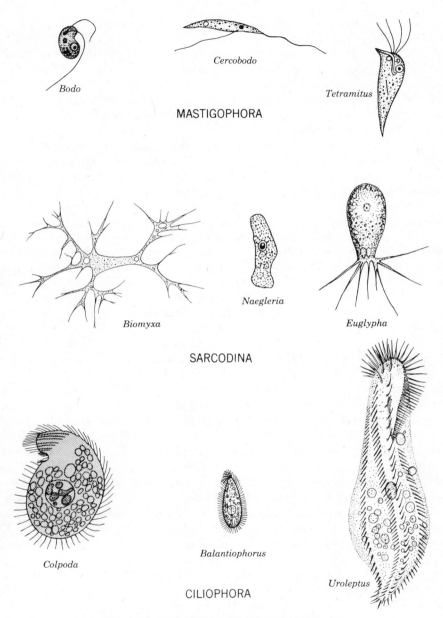

Bodo

Cercobodo

Tetramitus

MASTIGOPHORA

Biomyxa

Naegleria

Euglypha

SARCODINA

Colpoda

Balantiophorus

Uroleptus

CILIOPHORA

Figure 6.1. Typical soil protozoa (14).

NUTRITION

The energy for protozoan growth is obtained in several ways. At the extreme of nutritional independence are the photosynthetic protozoa, those algalike flagellates that synthesize their protoplasm from CO_2 using energy derived from sunlight. Photoautotrophy, however, is rare in the animal kingdom, the chlorophyll-containing phytoflagellates being the sole animals to possess the capacity for photosynthesis.

The vast majority of protozoa are dependent on preformed organic matter either as *saprobic* feeders, obtaining their nutriment from soluble organic and inorganic substances, or by a *phagotrophic* nutrition characterized by a direct feeding upon microbial cells or other particulate matter. The significance in soil of saprobic feeding is unknown, and the dominant mode of nutrition is generally considered to be phagotrophic. Available to the predaceous protozoa are bacteria and other kinds of microorganisms. Some protozoa are even cannibalistic, feeding on cells of their own species. In phagotrophic nutrition, the ingested particle of food is surrounded by a vacuole where digestion takes place. Since such food particles contain proteins, polysaccharides, sugars, and lipids, the individual protozoan cell must be able to form all the enzymes necessary to mediate the decomposition. Ultimately, undigested portions are released back into the external environment.

The preying on bacteria can be demonstrated by simultaneous inoculation of pure cultures of protozoa and bacteria into sterile soil and noting the change in abundance of the latter group. Initially, the bacterial population rises, reaching a maximum size by about the end of one week. The active protozoa are scarce up to this point, but their subsequent numerical increase is accompanied by a drastic fall in bacterial density as a result of the animals' activities. For any one predatory protozoan, many bacterial cells must be ingested to generate enough protoplasm to permit a single cell division. For example, certain species of Sarcodina may consume several thousand bacteria per protozoan cell division (4). Results obtained from the few estimates made of growth rates in soil suggest that small amebae and ciliates may divide once or twice each day (17) so that the bacteria in a soil with many of such protozoa must reproduce at a rapid rate merely to keep pace with their predators. On the other hand, some of the Sarcodina bearing shells may have generation times of about a week (12).

The micropredators are quite selective in the prey they consume. Investigations of individual populations of protozoa reveal that they consume many different species of rod-shaped bacteria and cocci, but often only a few species of *Mycobacterium* or algae can be eaten. Spores of *Clostridium* and the hyphae and often spores of fungi are generally resistant.

The feeding of amebae on various prey species illustrates the range of organisms potentially consumed, at least in culture. Thus, isolates of *Acantham-*

oeba, Hartmanella, and *Mayorella* feed on many bacterial species and yeasts, they consume certain encapsulated bacteria only with difficulty, they are often poor predators on *Streptomyces* and algae, and they are unable to consume the hyphae .of many fungi (10). By contrast, isolates of *Amoeba* and *Tetramitus* readily prey on many but not all algae provided to them (11). The feeding on an alga by an amebal cell is depicted in Figure 6.2. Protozoa that devour cells of other protozoan species are similarly not a rarity.

Prey may be divided into three types: readily digested organisms, slowly eaten strains, and those that are attacked only on rare occasions or not at all. Any single species may be a food source for one protozoan but be entirely inedible by another. The preference goes beyond species boundaries because some strains of one species may be susceptible while others are resistant. Attempts to ascertain the precise reasons that organisms are rejected as food sources have met with little success. By and large, inedible bacteria excrete no toxins inhibitory to protozoa, a fact supported by the observation that suscepti-ble strains are commonly eaten in the presence of resistant bacteria. Some bacteria appear to be protected by toxic cellular constituents, however. Large size and filamentous habits of growth frequently are associated with resistance to predation, but exceptions to this generalization are not difficult to find. Hence, much research must yet be done before the bases for selective feeding can be established. Nevertheless, regardless of the explanations for the selectiv-ity, it seems plausible to suggest that the composition of the bacterial community is influenced by protozoa since strains of lesser edibility would have a greater persistence and would not be subject to the vagaries of the microfauna.

When edible prey cells are no longer available or when the environment becomes in some way unfavorable, the active protozoan enters the cyst stage. *Encystment* allows the organism to persist in conditions unsuitable for develop-ment of the biochemically active stage of the protozoan. These specialized bodies are more tolerant to harmful chemicals, acids, and high temperatures than are the vegetative cells, and they often remain viable through long periods of drought. Consequently, the cysts usually serve a protective rather than a reproductive function. Although cysts of the common soil species survive drying, the cysts of many other protozoa fail to remain viable during a drought. Once nutrients are again accessible or the harmful influences are dissipated, the microorganism will *excyst* and enter into its active stage of life in which it feeds, reproduces, and moves from place to place. The return to the metabolically active form is related to the bacterial species in the vicinity of the cyst. Thus, some bacteria allow for rapid excystment, others permit only a slow return, and certain bacterial groups favor quick but incomplete excystment.

DISTRIBUTION AND ABUNDANCE

Examination of cultivated and virgin soils of all continents and from a variety of land management practices has revealed the presence of a rich and heteroge-

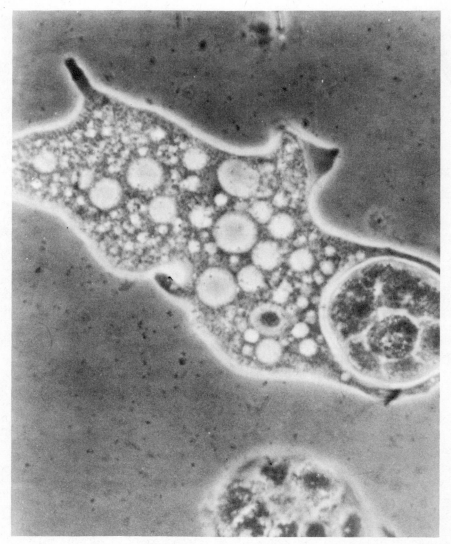

Figure 6.2. *Amoeba discoides* feeding on an engulfed cell of *Pandorina morum* (11).

neous protozoan fauna. Populations are sometimes as large as 100,000 to 300,000 cells per gram although values between 10,000 and 100,000 are more typical. Yet, only a small percentage of the individuals constituting the subterranean community are protozoa. Counts made of bacteria in comparable areas are always severalfold higher. On the other hand, the protozoan cell has a mass appreciably greater than that of the bacterium.

For the development, enrichment, or enumeration of protozoa, the use of liquid or agar media containing soil extract is often recommended; these media are especially suitable for amebae and ciliates. Hay or manure infusions have also proved to be of value. A nutrient medium for culture of soil protozoa should not permit overgrowth by the bacteria which are numerically in excess in soil dilutions. One technique based upon the principle of utilizing deficient media employs a nonnutrient agar to which is added an edible bacterial strain for the protozoa to feed on. The nutrient source is commonly a short, gram negative, non-spore-forming bacterium. Inherent to this technique is the problem of choosing a bacterium that is digestible by the greatest proportion of the protozoan fauna. At best, the estimates of community size will not be absolute since the results will vary with the bacterium chosen as food source, that is, its relative edibility. If the indigenous bacteria in the soil dilutions are required to play the role of prey on the agar plates, the results will be far more variable.

For the enumeration of protozoa, a common procedure is the standard dilution technique, in which serial tenfold dilutions are made from the soil sample and portions inoculated into a medium containing the selected bacterium. Because protozoa do not form distinct colonies, it is necessary to rely on the criterion of growth on the dilution plates. Thus, the absence or presence of protozoa can be ascertained by microscopic examination of each plate and the record of positives at each dilution used to estimate the most probable number of protozoa in the original sample. Measurements of protozoan density can thereby be made with a reasonable degree of accuracy.

By the techniques described, it has been established that the flagellates usually are more abundant than the small amebae whereas the ciliates are relatively uncommon. Occasionally, the amebae are the rare type and the ciliates numerous. With certain media, the amebae may seem as populous as the flagellates, but as a rule the flagellates tend to dominate in absolute numbers and in the variety of species while the ciliate population is sparse. Providing that the environment is not deleterious, there may be from 3000 to 200,000 flagellates, a similar range for the amebae, and usually less than 1000 ciliates per gram. Forest soils may sometimes contain more than 1000 ciliates per gram. Such values, although only the grossest of approximations, indicate that the flagellates and amebae are well adapted to the physical and chemical environment of the soil, possibly because of their small size, while the limited moisture supply and physical barriers make the ciliates poorly suited to the habitat. The usual dilution techniques often fail to provide reliable figures for the abundance of the shell-bearing amebae, but estimates of their numbers suggest that soils may contain only a few or as many as 8000 per gram (1). In Table 6.1 are given the numbers of several genera in a typical study.

An almost invariable feature of the estimates is the marked degree of variability in samples taken at daily intervals. Fluctuations on a day-to-day basis

TABLE 6.1
Abundance of Protozoa in a Scottish Soil (7)

Protozoan		No. per Gram
Ciliophora:	*Colpoda*	140
	Pleurotricha	1,830
Mastigophora:	*Heteromita*	20,700
	Oikomonas	2,600
	Cercomonas	1,830
	Phalansterium	266

are observed in total numbers, in abundance of the various species, and in percentage of active forms. In a one-day period, the numbers may rise from hundreds to hundreds of thousands. Some evidence has been obtained that the daily changes in protozoa are related inversely to the size of the bacterial community, one increasing as the other decreases. The idea of a natural equilibrium between predator and prey is highly tempting, but the experimental results are still equivocal.

Counting procedures of the types outlined above estimate the active cells as well as cysts. The number of cysts, however, indicates the protozoan potential rather than the biochemically important fauna. For the differentiation of cysts from active cells, soil is treated overnight with 2 percent hydrochloric acid, a procedure that destroys vegetative but not resting forms. The difference in counts made prior to and following acid exposure represents the number of active animal cells. Data obtained by this method have established that the cyst stage predominates at low soil moisture levels while excessive water permits emergence of the active individual. This change of stage is demonstrated most dramatically in the population shift that takes place when a dry soil is moistened. But even in daily examinations, sometimes the cysts predominate, sometimes the active protozoa. On some days, no active individuals can be found, and the entire microfauna exists in the encysted condition, yet, within 24 hours, the metabolically active cells emerge and reach numbers in the vicinity of 10,000 per gram. The diurnal fluctuations are probably related to moisture, temperature, and food supply. Some typical data for active and encysted amebae are presented in Table 6.2.

These organisms do not constitute a large portion of the biomass of the microbial community. Sometimes in forests or in grasslands in cool regions, the protozoan biomass may reach 20 g/m^2 of surface area of soil, but commonly their mass is less than 5 g/m^2 in the temperate zone (17).

TABLE 6.2
Effect of Manure on Number of Amebae in Barnfield Soil (15)

| | No./g | | | |
| | Untreated | | Manured | |
Date	Active	Cystic	Active	Cystic
April 27	530	1,790	16,040	2,060
May 13	4,040	4,840	29,210	9,590
May 28	13,900	4,100	49,550	3,550
June 20	6,540	1,940	34,130	5,870
July 8	4,040	4,020	52,500	11,000
August 27	8,770	3,730	22,300	10,500

ENVIRONMENTAL INFLUENCES

The presence of an adequate food supply is critical to the well-being of the soil protozoa, and the size and activity of the microfauna are seemingly interrelated with the bacterial density. Although this relationship is not firmly established, nevertheless the general rule that environmental circumstances favoring bacteria also tend to affect protozoa still holds within certain limits.

Protozoa are found in greatest abundance near the surface of the soil, particularly in the upper 15 cm. They are scarce in subsoils, but occasional isolates may be obtained from depths of a meter or more. The population is thus most dense where the bacteria are especially numerous in the profile. A similar explanation may account for the greater protozoan numbers in manured plots than in parallel plots receiving no animal manure (Table 6.2); that is, the applied organic matter permits the development of a larger microflora, which then serves as nutriment for the micropredators. Alternatively, the beneficial effect of manure may be partly the result of a direct microbiological stimulation by the extensive root system developed as a consequence of the improved fertility status. Should the saprobic habit be of greater importance in soil than present concepts suggest, then the influence of depth may result from the availability of organic matter in the A horizon.

Moisture level is of significance both qualitatively and quantitatively. It has often been stated, with considerable justification, that the water content of the environment is a major limitation to protozoan proliferation. An adequate water supply is essential for physiological activity and lateral or vertical movement. The flagellates are tolerant of low moisture, and they can develop in drier conditions than the other microfaunal types. Indeed, the flagellates are dominant in regions of the Sahara desert (18). Ciliates, on the other hand, are abundant only if the moisture level is high. When the supply of water is too low for life processes, the protozoa encyst and remain in the cyst form until the environment becomes more conducive to their growth, at which time the trophozoites again appear (12).

Unequivocal conclusions regarding the role of aeration, hydrogen ion concentration, and temperature are not possible because of inadequate study of the influence of these environmental factors upon protozoa. Aerobic metabolism is the rule for these microorganisms, but occasional species grow at low partial pressures of O_2 or under complete anaerobiosis. The survival of the obligate aerobes at low O_2 tensions is probably rather short and their functions sluggish at best. With regard to acidity, most protozoa exhibit no marked sensitivity to pH although an optimum can always be established. Certain species can, in pure culture, proliferate at pH 3.5 and others at values above 9.0, observations that are in accord with the abundance of protozoa in soils of the same range of acidities. On the other hand, many strains will not tolerate extreme acidity or alkalinity and fare poorly outside the range from pH 6 to 8. Certain of the Sarcodina are favored by high acidity such as that found in acid peats, but they are infrequent in arable land of neutral to alkaline reaction. Temperature is another important ecological determinant, the most favorable environments being both cool and damp. Excessive warmth is detrimental. The influences of moisture, aeration, pH, and temperature are highly complex and cannot be explained entirely on the basis of the supply of bacterial cells.

The significance of parasites and predators to the ecology of these animals is not yet clear. Filamentous fungi, for example, are known to feed on the amebae, including those surrounded by shells, and these fungi have been observed in soil and in decomposing plant remains. Other microorganisms are able to attack and destroy amebal cysts, leading to extensive degradation of the resting stage, in culture at least (19). Such organisms may be important in killing the trophozoites, which frequently die readily in soil, and the cysts, but too little information is at hand to indicate the causes of protozoan decline under natural conditions.

SIGNIFICANCE IN SOIL

Despite the ubiquity and abundance of protozoa, little is known of their function in soil. Techniques for direct experimentation are still inadequate, and the attributes of the protozoa in the dynamic biological equilibrium of terrestrial habitats therefore often must be inferred on the basis of their activities in culture solution. The lack of knowledge is especially disturbing because their great number and large cell size indicate that these organisms must be important members of the microscopic community.

The chief role postulated for these organisms, based on their feeding habits in enrichment cultures, is that they serve to regulate the size of the bacterial community. Evidence for a direct influence on bacteria has been obtained in investigations in which large numbers of bacterial cells are added to soil. Under such conditions, the population density of the introduced species falls rapidly and abruptly, and the only group of inhabitants that replicates coincidental with the decline of the added bacterium is the protozoa (Figure 6.3). Moreover, the marked decline in bacteria does not occur when a chemical suppressing indigenous protozoa is added to the soil. Such experiments have been done with species of *Xanthomonas* and *Rhizobium* as prey (5, 9). If these animals do in fact consume bacteria and other microbial prey, they probably are more important in localities where the species serving as food sources are actively reproducing than at sites where the populations under attack are not extensively proliferating.

If it is assumed that protozoa consume bacteria at a rapid rate, it is necessary to explain why the predators do not eliminate their prey, particularly because so many of the bacteria are readily consumed when provided in culture to flagellates, amebae, and ciliates. For example, thousands or even tens of

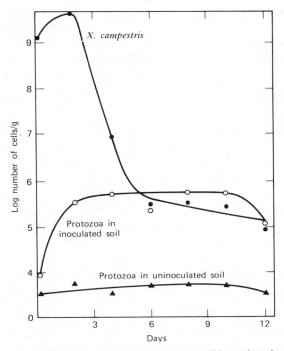

Figure 6.3. Numbers of protozoa in soil inoculated with *Xanthomonas campestris* and in uninoculated soil (9).

thousands of bacteria may be consumed by each protozoan prior to cell division or 10 to 150 yeasts may be eaten per day. Several hypotheses have been proposed to explain the apparent coexistence of protozoa and prey: (*a*) the proportion of cells in the prey population that is consumed declines as the prey density falls; (*b*) the small cell of the prey is protected in soil pores not penetrated by the large protozoa; and (*c*) the predator is itself under attack by parasites or other predators. That the animals do not eliminate bacteria is evident in soils to which a particular bacterium is added: although the indigenous protozoa multiply and attain high densities, the population of the introduced prey species, after falling appreciably in abundance, remains in reasonable numbers. Individual protozoan species similarly do not eliminate the introduced bacteria if both organisms are added to sterilized soil or liquid media. Hence, though organisms keeping the predators under control and refuges for bacteria may be important, the protozoa are incapable of totally eliminating the populations on which they feed, at least among the bacteria so far examined (4, 13). The failure to destroy entirely the prey also could result from the protozoa being unable to get enough energy from the few surviving cells they capture to continue hunting or from the bacteria being able to replicate at a rate just rapid enough to replace the cells that are devoured.

The unicellular animals also may function to allow different competing bacteria to coexist in a soil in which one bacterial species might otherwise have eliminated its neighbor. Typically, the better competitors outgrow the less active competitors and attain higher numbers. However, the more abundant bacterial species may in turn be more heavily fed on because more of its cells are available to the animal. This would then lead to a decline in the population of the good competitor, and the second bacterial species in its turn would not be eliminated by competition, although it would be under strong attack if it subsequently became preponderant. Such feeding on species that might other-wise dominate the bacterial community could explain why so many different organisms live and grow in the same environment.

Because many flagellates, amebae, and ciliates can grow in media free of other microorganisms, it may be that they participate in the decomposition of plant remains rather than living in nature solely by feeding on viable or dead cells. Particles of organic materials may indeed be engulfed and support growth, but in view of the intense competition for organic compounds by bacteria and fungi, it seems unlikely that protozoa would be too active in utilizing soluble organic substances or many of the polysaccharides readily metabolized by other groups of microorganisms. Furthermore, the protozoa have complex nutrient requirements, and these probably can be met more readily by ingesting entire cells of prey species than by competing with less fastidious bacteria or fungi for growth factors in solution.

Some evidence also exists that protozoa may actually enhance certain bacterial transformations, such as the utilization of N_2 or the degradation of phosphorus-containing organic materials (2, 6). It is not yet clear whether such an enhancement takes place in soil itself or what is the mechanism of the stimulation.

Soil may be a reservoir for pathogenic amebae. Of particular interest is the causative agent of amebic dysentery, *Entamoeba histolytica*. This pathogen may enter soil with fecal matter derived from an infected person, and it becomes of public health concern when the viable cysts then alight on vegetables grown in such polluted fields. Fortunately, the cysts of *E. histolytica* do not survive for more than a few days. On the other hand, strains of *Naegleria* can cause meningoencephalitis, and strains of *Hartmanella* produce infections in experimental animals (3); the significance of soil as a reservoir for the infectious strains remains uncertain. Not only may mammals be parasitized by soil protozoa but earthworms, larvae of insects, and other invertebrates may also be infected.

NONPROTOZOAN FAUNA

Many higher animals spend a large part of their life cycle in the soil. Various of these organisms are permanent inhabitants of the subterranean habitat; others are merely transients. Some consideration needs to be given to the nonprotozoan fauna although the subject falls outside of the scope of microbiology. These soil animals include nematodes, earthworms, flatworms, slugs, snails, centipedes, millipedes, wood lice, certain arachnids, and many insects. Such organisms feed on other animals, animal excreta, plants, or on inanimate materials. As a rule, the development of the macrofauna requires well-aerated environments, adequate moisture, and warm temperatures. Manuring tends to be beneficial.

Each hectare of soil contains up to several hundred kilograms of animal tissue. The dominant groups are the earthworms, insects, nematodes, and millipedes. The earthworms are of considerable agronomic importance because of their burrowing and channeling habits, the result of this activity being reflected in the improvement of soil aeration, drainage, and structure. By the channeling, a considerable quantity of soil material is translocated. These segmented worms are sensitive to environmental change, and they are benefited by high organic matter levels, good drainage, and nonacid conditions.

Nematodes have attained prominence because of their role in attacking higher plants, but free-living types are similarly found in soil. The latter use organic debris, microorganisms, or other nematodes as food sources. The presence of numerous species of insects has been demonstrated. Especially common are ants, termites, springtails, and the larvae of flies and beetles. As with the nematodes, the insects may be free-living or, alternatively, they may feed on plant roots. Snails and slugs spend part of their life cycle underground, but they are frequently seen beneath rocks, plant debris, and in shady areas. Spiders, mites, and ticks are among the other common invertebrates. In forests and prairies are seen animals that actively burrow into the ground, and their channels may occasionally be quite prominent.

The agricultural importance of the macrofauna rests upon its contributions to fertility, soil structure, and plant disease. Nematodes are of special concern to the pathologist, but physical injury to plant roots by other animals is not uncommon. By the burrowing of the various animals and by the contribution of earthworm casts to aggregate formation, the macrofauna exerts a beneficial action on drainage, aeration, and soil structure. In addition, the fauna serves as an adjunct to the microflora by direct participation in organic matter decomposition and, indirectly, by the physical intermixing of crop debris and forest litter with the underlying soil so as to permit more rapid microbiological action.

REFERENCES

Reviews

Darbyshire, J. F. 1975. Soil protozoa: Animalcules of the subterranean environment. In N. Walker, ed., *Soil microbiology: A critical review*. Halsted Press (Wiley), New York, pp. 147–163.

Grell, K. G. 1973. *Protozoology*. Springer-Verlag, New York.

Sleigh, M. A. 1973. *The biology of protozoa*. Elsevier Publishing Co., New York.

Stout, J. D. and O. W. Heal. 1967. Protozoa. In A. Burges and F. Raw, eds., *Soil biology*. Academic Press, New York, pp. 149–195.

Literature Cited

1. Bamforth, S. S. 1971. *J. Protozool.*, 18:24–28.
2. Barsdate, R. J., R. T. Prentki, and T. Fenchel. 1974. *Oikos*, 25:239–251.
3. Culbertson, C. G. 1971. *Annu. Rev. Microbiol.*, 25:231–254.
4. Danso, S. K. A. and M. Alexander. 1975. *Appl. Microbiol.*, 29:515–521.
5. Danso, S. K. A., S. O. Keya, and M. Alexander. 1975. *Can. J. Microbiol.*, 21:884–895.
6. Darbyshire, J. F. 1972. *Soil Biol. Biochem.*, 4:359–369.
7. Darbyshire, J. F., R. E. Wheatley, M. P. Greaves, and R. H. E. Inkson. 1974. *Rev. D'Ecol. Biol. Sol*, 11:465–475.
8. Flint, E. A., M. E. Di Menna, and J. D. Stout. 1973. *Rev. D'Ecol. Biol. Sol*, 10:475–493.
9. Habte, M. and M. Alexander. 1975. *Appl. Microbiol.*, 29:159–164.
10. Heal, O. W. and M. J. Felton. 1970. In A. Watson, ed., *Animal populations in relation to their food resources*. Brit. Ecol. Soc. Symp. no. 10, pp. 145–162.
11. Ho, T.-S. and M. Alexander. 1974. *J. Phycol.*, 10:95–100.
12. Lousier, J. D. 1974. *Soil Biol. Biochem.*, 6:19–26.
13. Luckinbill, L. S. 1973. *Ecology*, 54:1320–1327.
14. Sandon, H. 1927. *The composition and distribution of the protozoan fauna of the soil*. Oliver and Boyd, London.
15. Singh, B. N. 1949. *J. Gen. Microbiol.*, 3:204–210.
16. Stout, J. D. 1970. New Zeal. Soil Bureau Publ. 456.
17. Stout, J. D. and O. W. Heal. 1967. In A. Burges and F. Raw, eds., *Soil biology*. Academic Press, New York, pp. 149–195.
18. Varga, L. 1936. *Ann. Inst. Pasteur*, 56:101–123.
19. Verma, A. K., M. K. Raizada, O. P. Shukla, and C. R. Krishna Murti. 1974. *J. Gen. Microbiol.*, 80:307–309.

7
Viruses

The individual cells or filaments of bacteria, actinomycetes, fungi, algae, and protozoa are consistently small, but all are visible by light microscopy. Beyond the resolution of the light microscope, however, is a group of unique organisms dependent for their development on a suitable host. These submicroscopic agents, the viruses, are of considerable economic and medical importance because of the diseases of plants, animals, and humans for which they are responsible. Yet, not until recent years has appreciable attention been focused on the response and behavior of these infective agents in soil.

Each viral particle requires for its reproduction the presence of a viable, metabolizing organism. In the absence of the host, little activity and no reproduction or duplication are possible. Moreover, viruses are limited in their host ranges; that is, they parasitize only specific plants, animals, or microorganisms. This specificity has led to a categorization on the basis of the type of host. There is thus a group of viruses pathogenic to plants, a second to animals, and another to microorganisms. The infective agents of the last group include viruses parasitizing bacteria (*bacteriophages*), actinomycetes (sometimes termed *actinophages*), fungi, yeasts, algae, and protozoa. Among certain of the fungi, yeasts, algae, and protozoa, evidence for the existence of submicroscopic parasites rests solely on the finding in the cell of structural entities resembling viruses, and these structures are usually designated as viral-like particles. Whether these particles are indeed viruses and whether they occur in soil remain still to be established.

Even among those genera and species of microorganisms which have representatives susceptible to viral attack, a remarkable degree of host specificity is apparent. It is not uncommon for a virus capable of infecting representatives of one genus to have no such action on a phylogenetically related genus. Furthermore, among the bacteriophages, a single type may find several but not all species of the genus suitable for invasion, and susceptibility to a specific bacteriophage is often associated with only certain of the strains of a single species (Table 7.1).

More is known about the ecology of bacteriophages than about the other kinds of viruses parasitizing soil residents. Like their counterparts which affect higher plants and animals, the bacteriophages have a minute body that can pass without difficulty through ultrafine filters designed to retain bacterial cells. Morphologically, the bacterial viruses usually possess head- and taillike structures (Figure 7.1). The diameter of the bacteriophage rarely exceeds 0.05 to 0.10 μm. The tail is somewhat larger, about 0.2 μm in length, but it is quite narrow.

Following the entry of the bacteriophage into the bacterium, the tail being the point of attachment, *lysis* of the host cell takes place. If the bacterium is growing on agar, the lytic area is seen as a zone of clearing known as a *plaque*; in like fashion, infection in liquid media is evidenced by a decrease in turbidity of the bacterial suspension as lysis proceeds. The cellular dissolution is accompanied by a marked increase in the number of bacteriophage particles and a sharp decline in the number of viable bacteria in the medium (Figure 7.2). Purification of the viruses can be accomplished by passing the suspension of lysed bacteria through a filter designed to retain residual, intact cells. Such enrichments are not difficult to achieve since the rupture of a single, infected cell often leads to the release of several hundred virus particles, each of which is potentially invasive. Thus, by use of high bacterial concentrations, counts in excess of 10^{10} bacteriophage particles per milliliter of culture medium can be attained.

TABLE 7.1

Effect of Bacteriophages on Strains of *Azotobacter* (8)

Bacterium Tested for Lysis	Bacteriophage Strain			
	PCan	P18	P	PBulg
Azotobacter chroococcum				
strain C12	−	+	+	−
strain C18	+	+	+	−
Azotobacter vinelandii				
strain V5	−	+		+
strain P3	−	−		−
Azotobacter beijerinckii				
strain B2-e	−	+		±

+ Lysis. − No lysis.

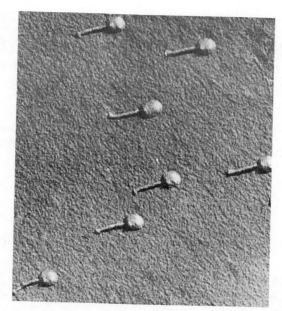

Figure 7.1. Bacteriophage particles viewed under the electron microscope.

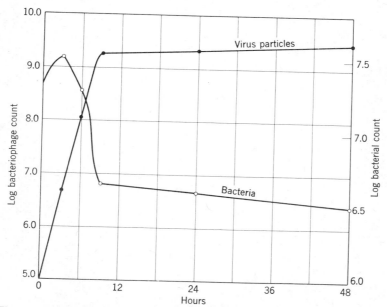

Figure 7.2. Changes in populations of bacteria and bacteriophages during infection (13).

Viruses are usually first recognized through the symptoms they induce in the host. With bacterial and actinomycete viruses, the initial abnormality is the appearance on solid media of plaques or, in liquid media, the clearing of turbid cell suspensions. The bacteriophage may then be propagated in cultures of a susceptible microorganism from which it can be ultimately separated and purified by filtration and high-speed centrifugation. When necessary, the enumeration of infective units is performed on agar in a fashion analogous to the conventional plate count. After dilutions are made, the bacteriophage suspension is pipetted onto a plate previously inoculated with a suitable microbial species and, following an appropriate incubation period, the clear plaques counted (Figure 7.3). Under such conditions, each plaque originates from a single bacteriophage particle deposited upon the plate, there to multiply at the expense of the adjacent bacterial growth.

In order to demonstrate the presence in soil of a specific bacteriophage, a

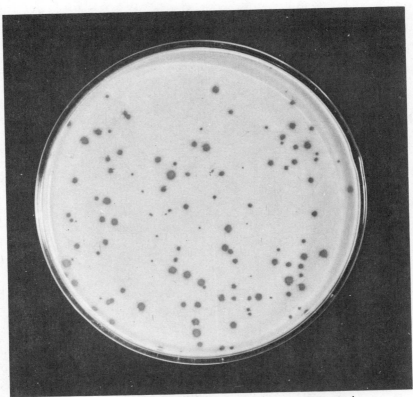

Figure 7.3. Plaques on an agar plate seeded with bacteria.

sample of soil is incubated with the host bacterium to allow for an increase in the population of the virus in question. A small quantity of the treated sample is then added to a nutrient medium previously inoculated with the host. After 24 to 48 hours, the lysed suspension is passed through a sterile bacteriological filter and the filtrate tested for its capacity to lyse a fresh, rapidly growing culture of the microorganism. By such methods, viruses specific for *Agrobacterium, Arthrobacter, Azotobacter, Bacillus, Bdellovibrio, Clostridium, Corynebacterium, Erwinia, Mycobacterium, Pseudomonas, Rhizobium,* and *Xanthomonas* as well as *Nocardia* and *Streptomyces* have been demonstrated in a variety of soils. In addition, bacteriophagelike particles have been found in *Nitrobacter* (3).

Current evidence indicates that bacteriophages for any given bacterial species are not usually numerous in soil so that deliberate addition of the host to soil facilitates isolation of the virus. On the other hand, enhancing the reproduction of the host cells already in the soil in a manner that provides more individuals to parasitize frequently leads to increases in bacteriophage abundance. This enhancement is most readily accomplished by amending the soil with a carbon source for the bacterium, which then will proliferate and be more accessible to the virus. Thus, addition of simple sugars, by stimulating *Arthrobacter globiformis*, ultimately leads to large populations of bacteriophages lysing that bacterium (5). In a similar fashion, adding nutrients to a soil and incubating it at high temperatures not only favors the growth of thermophilic bacteria but also results in many bacteriophages able to lyse the thermophiles (15).

One group of bacteriophages potentially important in agriculture is capable of bringing about the lysis of the bacteria that form nodules on the roots of legumes. These bacteria, members of the genus *Rhizobium*, function in symbiosis with legumes to convert atmospheric nitrogen to a form utilized by the plant. Should the bacteria within the nodule be parasitized by the virus, material economic losses could ensue. Bacteriophages specific for *Rhizobium* spp. can be isolated directly from soil or from roots of a number of leguminous species. For example, Katznelson and Wilson (12) found the bacteriophage for the alfalfa rhizobium in every alfalfa field examined, but in only occasional sites not cropped to alfalfa was the virus encountered (Table 7.2). It has been suggested that the poor yields attendant on continuous cropping of alfalfa and clovers are caused by a buildup of the bacteriophage population with the resultant destruction of the plant-*Rhizobium* symbiotic association (6). Despite the fact that no final conclusion can yet be advanced to rule out this possibility, there is little definitive evidence in its behalf, and the hypothesis is now deemed to be probably invalid.

Mention has been made only of those bacterial viruses that vigorously lyse the host and cause it to release large numbers of free bacteriophage particles. This is the *lytic* or virulent type of bacteriophage. Not infrequently, however, the host bacterium carries the virus within itself, transferring the infecting unit to its daughter cells without apparent lysis but with an occasional release of free

TABLE 7.2
Presence of Bacteriophage for *Rhizobium meliloti* in New York Soils (12)

Soil	pH	Age of Stand yr	Bacteriophage Incidence
Forest soil	4.90		−
Forest soil	4.99		−
Pasture	5.27		−
Alfalfa field	5.27	6	+
Alfalfa field	5.85	3	+
Alfalfa field	6.10	8	+
Wheat field	6.11		+
Pasture	6.20		−
Alfalfa field	7.39	1	+

viral particles. Such bacteriophages are *temperate*, the phenomenon is termed *lysogenicity*, and the carrier cell is known as a lysogenic bacterium. Many actinomycetes are also lysogenic. Detection of lysogenic bacteria is difficult since lysis is not immediately apparent. For purposes of identification, the lysogenic strain is mixed with an indicator or susceptible strain that will be attacked by the viruses borne by the former. Lysogenicity may have associated with it a unique phenomenon, *transduction*, in which the bacteriophage serves to transmit certain genetic characteristics from one host to a newly infected bacterium. Transduction serves essentially as a unidirectional transfer of a portion of the nuclear apparatus of the microbial architecture. The characters thus transferred, traits that may alter the nutrition or physiology of the newly infected cells, usually are passed between strains of the same species, but the transmission of genetic material may occasionally take place between closely related genera.

Bacteriophages have been thoroughly and extensively investigated, but only in the last few years has it been recognized that viruses or viruslike particles can occur within the cells and sometimes cause lysis of individual genera of fungi, yeasts, algae, and protozoa. Viruses acting on fungi first attracted attention because they were identified as the causal agents of a disease of the commercial mushroom, *Agaricus bisporus*. The disease was found to be transferred from the affected mycelium to normal hyphae, and the infective particles so transferred were then characterized as a new type of virus (9). Subsequent research has shown that either viruses or bodies morphologically akin to viruses are present in many genera of the fungi. Among the recognized hosts are species of *Aspergillus, Boletus, Cephalosporium, Fusarium, Gliocladium, Mucor, Ophiobolus, Poly-*

porus, Penicillium, Rhizopus, and many other genera. The taillike structures common to bacteriophages are typically absent from these viruses so that they are morphologically distinct from the bacterial parasites. The influence of the viruses on their hosts, in culture at least, ranges from those that cause no or little detectable harm to others that bring about abnormal growth of hyphae and induce the development of abnormal fruiting bodies. The degeneration and loss of viability of some fungi in media sometimes appear to result from damage done by these submicroscopic agents, although most of the viruses are reasonably or entirely avirulent and do not induce lysis.

Viruses infecting blue-green algae are similar to those associated with the bacteria. They produce plaques on agar containing visible growth of the host algae, cause extensive lysis of suitable algal cultures, and pass through bacteriological filters. They resemble bacteriophages also in morphology and in the process of infection. These viruses—often called *cyanophages* but sometimes phycoviruses or algophages—have been noted in rivers, lakes, ponds, and brackish waters, but except for a report of their presence in flooded rice field (16), their distribution in soils is largely undefined. Their hosts in culture include *Anabaena, Anacystis, Cylindrospermum, Microcystis, Nostoc, Oscillatoria, Phormidium*, and *Synechococcus*, and though the viruses usually are virulent and lyse the host culture, lysogenic viruses have likewise been observed. Viruslike structures have also been found by microscopic means in the cells of a few aquatic green, red, and brown algae, but whether they are indeed viruses, can be transmitted to uninfected populations, and occur in soil remain to be established.

Infectious agents similar and possibly identical to viruses have been observed in cells of one ameba that finds its habitat in soil (7), and viruslike particles have also been noted recently inside of a few protozoa inhabiting other environments. Because of the paucity of information, an assessment of their distribution, host ranges, and significance is not yet feasible.

Despite the fact that viruses cause diseases of many agronomic and horticultural crops, it is the rare plant-infecting virus that can persist or overwinter in soil. These uncommon forms are often considered to be soil-borne, but they are soil-borne only in the sense that they retain their infective capacity for some time after crop removal. Such viruses do not multiply in soil; instead, they persist for varying periods in a condition resistant to inactivation. By this criterion, the viruses responsible for the mosaic diseases of wheat, oats, and tobacco and the pathogens inducing the big vein disease of lettuce and the corky ringspot of potatoes are soil-borne. A noteworthy characteristic of many of the soil-borne plant-infecting viruses is their reliance on either nematodes or fungi for transmission. For example, the arabis mosaic, raspberry ringspot, and tobacco rattle viruses are spread through soil by nematodes of the genera *Longidorus, Trichodorus*, and *Xiphinema*, whereas the tobacco necrosis and lettuce big vein viruses are disseminated by the fungus, *Olpidium brassicae* (4). The transmission is illustrated in one study in which it was shown that tobacco necrosis virus is released from roots of infected tomato plants in the absence of the fungus, but the liberated viruses then become associated with and are transmitted by the motile zoospores of *O. brassicae* (17).

The movement, persistence, and inactivation of viruses affecting humans, livestock, and even insects are of importance because of the diseases associated with the infective agents. Viruses, like those causing infectious hepatitis, may be discharged with the effluents from septic tanks, and should the pathogens enter ground water, they may move to nearby wells in suburban or rural communities and become a substantial hazard to people drinking the water. Infectious hepatitis has thus been spread from home to home. The growing interest in water reuse and disposal of sewage on land has prompted an evaluation of the capacity of soils to remove viruses of public health importance from percolating water. As a rule, viruses are readily removed or inactivated when passed through soil in moving waters, although it is not yet clear whether the mechanism is by sorption, biological inactivation, or other means. On the other hand, some soils do not readily remove or inactivate viruses present in percolating waters (18), and hence the parasites may enter ground water. This danger is particularly prominent in the case of infectious hepatitis. Viruses of some insects appear in soil in connection with outbreaks of disease among the insects, and the submicroscopic agents probably are derived from the bodies of insects that died of the disease. The abundance of the insect parasites in soil appears to be related to the severity of the outbreak (10).

Viruses of several types persist in soil, and instances of prolonged survival have been documented for certain bacteriophages, actinophages, and also plant-, animal-, and insect-infecting viruses. The tobacco mosaic virus, as a case in point, may sometimes endure in plant remains buried at lower soil depths for more than a year (2), and other viruses may remain active for some time in moist but not in dry soil (Table 7.3). Although many of the plant viruses

TABLE 7.3

Survival of Plant Viruses in Soil (17)

Virus	Days of Survival	
	Moist Soil	Dry Soil
Tobacco necrosis virus	8	<1
Southern bean mosaic virus	>25	3
Petunia asteroid mosaic virus	>25	<1
Cucumber necrosis virus	>25	3

transmitted by nematodes are inactivated readily when soil dries, the plant viruses borne by fungi may remain active in dry soil for many years. Certain animal and human viruses likewise will remain infective in soil for periods ranging up to several months. The persistence of the mouse encephalomyelitis virus, for example, is greatest in neutral environments, negligible in acid conditions, and of intermediate duration in alkaline soils (14). Some insect viruses, moreover, are capable of resisting inactivation for more than five years after their introduction into soil (10).

Of potential significance is the capacity of organic residues, humus, and clay to adsorb viruses. Substances known to remove bacteriophages from solution include nonsusceptible bacterial cells, fungi, clay, and plant materials. Soils rich in clay generally adsorb more of the virus particles than sandy soils, but the organic matter similarly seems to be of consequence (1, 11). Factors such as clay and organic matter content of soil undoubtedly influence the action and spread of bacterial viruses, and they may at the same time be influential in the spread of plant diseases by soil-borne viruses.

The practical or ecological significance of viruses in soil is difficult to assess. The overwintering of agents of plant disease no doubt has a bearing on crop production because the pathogen can initiate infection of susceptible crops the following season. Viruses associated with human infections may pose a hazard if they move through soil into ground waters, but they are really transients rather than soil residents. The insect parasites may serve to regulate populations of some insects, including those of economic significance, but convincing studies have yet to be conducted of this possibility. Similarly, no role can be assigned to the viruses or viruslike particles of algae, fungi, protozoa, or yeasts. Little is known of the agronomic importance of the bacteriophages, moreover, and the extensive investigations of the viruses lysing *Rhizobium* spp. have not yielded sufficiently strong evidence to warrant the conclusion that the infective agents have any relationship to legume yields. Despite the abundance of suitable bacteria of many genera in soil, the viruses attacking them are never numerous. This fact suggests that there is some check on the spread of the submicroscopic particles. The presence of such a check can be demonstrated indirectly by showing the sensitivity in vitro of many soil bacteria to soil-derived bacteriophages; that is, the bacteria are vigorously lysed in culture solution, yet they exist and multiply in nature.

It is often assumed that the importance of the bacteriophage to the ecology of its host is solely a result of lytic influences, the phage presumably acting by the destruction of susceptible bacterial strains with which it comes in contact. Alternatively, it is possible that the virus may participate in the transmission of genetic material from one bacterium to another through transduction. This transfer could give to the receptor cell physiological properties that might be of competitive advantage, such as new nutritional or biochemical attributes. Experimental verification of transduction as an event in nature, however, is lacking, nor is there evidence that viruses infecting other microbial groups are associated with the transfer of genetic information. Nevertheless, such means of genetic exchange may prove to be of importance in the evolution and variation of microorganisms.

REFERENCES

Reviews
Grogan, R. G. and R. N. Campbell. 1966. Fungi as vectors and hosts of viruses. *Annu. Rev. Phytopathol.,* 4:29–52.
Lemke, P. A. and C. H. Nash. 1974. Fungal viruses. *Bacteriol. Rev.,* 38:29–56.
Luria, S. E. and J. E. Darnell, Jr. 1967. *General virology.* Wiley, New York.
Padan, E. and M. Shilo. 1973. Cyanophages—Viruses attacking blue-green algae. *Bacteriol. Rev.,* 37:343–370.

Literature Cited
1. Bershova, O. 1938. *Mikrobiol. Zh.,* 5:161–180.
2. Broadbent, L., W. H. Read, and F. T. Last. 1965. *Ann. Appl. Biol.,* 55:471–483.
3. Bock, E., D. Düvel, and K.-R. Peters. 1974. *Arch. Microbiol.,* 97:115–127.
4. Cadman, C. H. 1963. *Annu. Rev. Phytopathol.,* 1:143–172.
5. Casida, L. E., Jr. and K.-C. Liu. 1974. *Appl. Microbiol.,* 28:951–959.
6. Demolon, A. and A. Dunez. 1935. *Trans. 3rd Intl. Cong. Soil Sci.,* Oxford, 1:156–157.
7. Dunnebacke, T. H. and F. L. Schuster. 1974. *J. Protozool.,* 21:327–329.
8. Hegazi, N. A. and V. Jensen. 1973. *Soil Biol. Biochem.,* 5:231–243.
9. Hollings, M., D. G. Gandy, and F. T. Last. 1963. *Endeavour,* 22:112–117.
10. Jaques R. P. and D. G. Harcourt. 1971. *Can. Entomol.,* 103:1285–1290.
11. Katznelson, H. 1939. *Trans. 3rd Comm., Intl. Soc. Soil Sci.,* New Brunswick, A:43–48.
12. Katznelson, H. and J. K. Wilson. 1941. *Soil Sci.,* 51:59–63.
13. Kleczkowska, J. 1945. *J. Bacteriol.,* 50:81–94.
14. Murphy, W. H., O. R. Eylar, E. L. Schmidt, and J. T. Syverton. 1958. *Virology,* 6:612–622.
15. Reanney, D. C. and S. C. N. Marsh. 1973. *Soil Biol. Biochem.,* 5:399–408.
16. Singh, P. K. 1973. *Arch. Mikrobiol.,* 89:169–172.
17. Smith, P. R., R. N. Campbell, and P. R. Fry. 1969. *Phytopathology,* 59:1678–1687.
18. Young, R. H. F. and N. C. Burbank, Jr. 1973. *J. Amer. Water Works Assoc.,* 65:598–604.

THE CARBON CYCLE

INTRODUCTION

The most important single element in the biological realm and the substance that serves as the cornerstone of cell structure is carbon. Plant tissues and microbial cells contain large quantities of carbon, approximately 40 to 50 percent on a dry weight basis, yet the ultimate source is the CO_2 that exists in a perennially short supply, only some 0.03 percent of the earth's atmosphere. Carbon dioxide is converted to organic carbon largely by the action of photoautotrophic organisms—the higher green plants on land, the algae in aquatic habitats. These photoautotrophs supply the organic nutrients needed for heterotrophic animals and the non-chlorophyll-containing microorganisms.

Carbon is constantly being fixed into organic form by photosynthetic organisms under the influence of light and, once bound, the carbon becomes unavailable for use in the generation of new plant life. It is thus essential for the carbonaceous materials to be decomposed and returned to the atmosphere in order for higher organisms to continue to thrive. It has been estimated that the vegetation of the earth's surface consumes some 1.3×10^{14} kg CO_2 per annum, about one twentieth of the total supply of the atmosphere or one-thousandth of that in the oceans. With the conversion of so much of the plant-available carbon to organic form each year and the limited supply in the air and seas, it is manifestly apparent that the major plant nutrient element would become exhausted in the absence of microbial transformations.

In its barest outlines, the carbon cycle revolves about CO_2 and its fixation and regeneration. Chlorophyll-containing plants utilize the gas as their sole carbon source, and the carbonaceous matter thus synthesized serves to supply the animal world with preformed organic carbon. Upon the death of the plant or animal, microbial metabolism assumes the dominant role in the cyclic sequence. The dead tissues undergo decay and are transformed into microbial cells and a vast, heterogeneous body of carbonaceous compounds known collectively as humus or as the soil organic fraction. The cycle is completed and carbon made available with the final decomposition and production of CO_2 from humus and the rotting tissues.

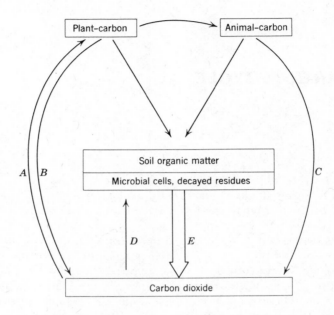

A. Photosynthesis   C. Respiration, animal
B. Respiration, plant   D. Autotrophic microorganisms
E. Respiration, microbial

The carbon cycle.

8
Some Aspects of Microbial Physiology

Prior to a consideration of the various transformations brought about by the microflora, it is necessary to review some of the details of microbial nutrition and physiology. The capacity to grow in a given habitat is determined by an organism's ability to utilize the nutrients in its surroundings. At the same time, the organism exists in an environmental complex, and its nutritional and physiological characteristics will determine to a great extent its ability to get along with its neighbors. Hence, not only the function but the very existence of a species in the soil habitat is conditioned by its nutritional and biochemical versatility.

Many of the points made in the following discussion are presented in considerable detail in textbooks on microbial physiology and biochemistry. However, a brief discussion will serve to set the stage for an understanding of the specific processes that are brought about by the microflora of natural environments.

NUTRITION

Nutrients serve three separate functions: providing the materials required for protoplasmic synthesis, supplying the energy necessary for cell growth and biosynthetic reactions, and serving as acceptors for the electrons released in the reaction that yields energy to the organism (Table 8.1). In aerobes, O_2 serves the last function. In strict or facultative anaerobes, either an organic product of metabolism or some inorganic substance replaces the O_2. Among the energy sources for soil heterotrophs are found cellulose, hemicelluloses, lignin, starch, pectic substances, inulin, chitin, hydrocarbons, sugars, proteins, amino acids, and organic acids. The conversion of these organic substances to more oxidized products releases energy, and a portion of the energy released is used in the synthesis of protoplasmic constituents. However, in order to be made use of, the

TABLE 8.1

Nutrients Required by Microorganisms

1.	Energy source	Organic compounds
		Inorganic compounds
		Sunlight
2.	Electron acceptor	O_2
		Organic compounds
		NO_3^-, NO_2^-, N_2O, $SO_4^=$, CO_2
3.	Carbon source	CO_2, HCO_3^-
		Organic compounds
4.	Minerals	N, P, K, Mg, S, Fe, Ca, Mn, Zn, Cu, Co, Mo
5.	Growth factors[a]	
	a. Amino acids	Alanine, aspartic acid, glutamic acid, etc.
	b. Vitamins	Thiamine, biotin, pyridoxine, riboflavin, nicotinic acid, pantothenic acid, p-aminobenzoic acid, folic acid, lipoic acid, B_{12}, etc.
	c. Others	Purine bases, pyrimidine bases, choline, inositol, peptides, etc.

[a] Where growth proceeds in the absence of growth factors, the compounds are presumably synthesized by the organism.

foodstuff must penetrate into the organism. Often, the energy source enters with no difficulty, but microbial cells are impermeable to many complex molecules; these compounds must be first solubilized and simplified prior to their serving within the cell's confines as energy sources.

The elemental composition of microorganisms is remarkably similar, and bacteria, fungi, actinomycetes, algae, and protozoa generally contain the same elements. Nitrogen, phosphorus, potassium, magnesium, sulfur, iron, and probably calcium, manganese, zinc, copper, cobalt, and molybdenum are integral parts of the protoplasmic structure. From these essential nutrients plus carbon, hydrogen, and oxygen is built the microbial cell. The differences in nutrition among members of the microflora are not linked with the essential mineral substances but rather with molecules containing carbon and nitrogen. In several instances, an element is required for a specific reaction in decomposition or in cell synthesis. A number of these inorganic substances are assimilated in extremely small quantities, so small that the requirement is not always detected. For example, probably most microorganisms contain vitamin B_{12}, a cobalt-containing molecule, yet the need for cobalt is difficult to establish because the optimum cobalt concentration is appreciably less than 0.001 ppm. Ingredients of the culture solution usually contain quantities of cobalt in excess of the need.

Organic constituents make up a large fraction of the total protoplasmic material. Carbohydrates, proteins, amino acids, vitamins, nucleic acids, purines, pyrimidines, and other substances constitute the working apparatus of the organism. The complexity of the cell's interior is in marked contrast to the simplicity of its surroundings, and only a few of the materials from the habitat—or culture medium—are assimilated without further modification. Nevertheless, microorganisms differ greatly in their synthetic abilities. Frequently, an organism is unable of itself to synthesize one or more of its structural building blocks. Where this occurs, the organism must be provided with the necessary materials. These substances, the *growth factors*, are organic molecules required in trace quantities for growth. The need for one or several growth factors—amino acids, vitamins, or other structural units—has considerable ecological importance because a species needing them will only grow in habitats where the organic molecules are present. On the other hand, a growth factor need not be essential. Many species are stimulated by growth factors, but they grow, slowly to be sure, in the absence of individual substances; in these instances, the chemical is stimulatory rather than essential. Metabolically, the position of a growth factor as stimulatory rather than essential may be taken to reflect a metabolic condition in which the organism produces the substance in question but at a rate insufficient to meet the demands for rapid proliferation.

At the simplest nutritional extreme are the chemoautotrophs (chemolithotrophs), organisms that synthesize all protoplasmic constituents from inorganic salts—CO_2, O_2, and water. Somewhat more exacting are the heterotrophic (chemoorganotrophic) bacteria, actinomycetes, and fungi that require some simple carbon source—often a sugar or organic acid suffices—and inorganic nutrients. In this group are found bacteria such as *Pseudomonas* and fungi like *Aspergillus*. A large proportion of the soil microflora develops at least to some extent in a medium containing a simple source of organic carbon, nitrogen in the form of ammonium or nitrate salts, and a number of inorganic substances. More exacting yet are the microorganisms that, because they are unable to form all of their own vitamins or amino acids, require one or more of these growth factors. Development of these more exacting species depends on a supply of the appropriate factor or factors. *Bacillus* spp., for example, often lack the ability to synthesize one or several amino acids or vitamins. At the extreme of nutritional dependence are microorganisms requiring a host of essential substances, a nutritional complexity associated with the inability of the organism to synthesize many of the requisite building blocks of the cell's constituents (Table 8.2). In soil, the existence of species with complex requirements is tenuous since the growth factors must be obtained from the environment; this group contains organisms whose activity is closely regulated by other species in the environment.

Carbon dioxide, a product of both aerobic and anaerobic metabolism, is important not only because it completes the carbon cycle but also because of its direct influence on growth. Chemoautotrophic and photoautotrophic microorganisms must have CO_2 as it is their sole carbonaceous nutrient. However, the

TABLE 8.2

Composition of Culture Media for Several Aerobic Bacteria

	Nitrobacter winogradskyi	Pseudomonas sp.	Arthrobacter sp.	Bacillus subtilis
Energy source	KNO_2	Glucose	Sucrose	Glucose
Carbon source	$KHCO_3$	Glucose	Sucrose	Glucose
Minerals	KNO_2	NH_4Cl	$(NH_4)_2SO_4$	NH_4Cl
	K_2HPO_4	K_2HPO_4	KH_2PO_4	K_2HPO_4
	$MgSO_4$	$MgSO_4$	$MgSO_4$	KH_2PO_4
	$FeSO_4$	$FeSO_4$	$FeSO_4$	$MgSO_4$
		$CaCl_2$	$CaCO_3$	Na_2SO_4
			$MnSO_4$	$FeSO_4$
			$ZnSO_4$	$MnSO_4$
			$CuSO_4$	$CaCl_2$
Growth factors	—	—	Biotin	Glutamic acid
			Thiamine	Cysteine
			B_{12}	

gas is stimulatory to and often required by many heterotrophs, and growth of some, and possibly most, species will not proceed in its absence. A part of the CO_2 supplied, even to heterotrophs, is incorporated into the cell structure. The requirement for this gas rarely presents a problem in soil because of its continual evolution from decaying organic matter. On the other hand, CO_2 at certain concentrations is toxic to a number of fungi, and the inhibition may have a considerable biological influence since individual fungus species vary in their sensitivity to the gas. Hence, its release during decomposition may alter the composition of the subterranean community.

Oxygen, a requirement for all aerobes, may also serve as an ecological determinant. Thus, although most fungi are obligately aerobic, there is considerable variation among microorganisms in the capacity to grow at low partial pressures of O_2. Certain strains develop to some extent using O_2 dissolved in solution when there is none in the gas phase; others exhibit no such capacity. Furthermore, the rate and extent of growth and the ability to sporulate are affected by the aeration status of the environment.

GROWTH

In any discussion of the rate of microbiological processes, some attention must be given to the growth of the dominant microorganisms. Most investigations of the growth patterns have been concerned with the true bacteria for which the characteristics of proliferation are now well established. Consider the development in optimal conditions of a bacterial population of size a. As bacteria reproduce by binary fission, one mother cell dividing into two daughters in a period known as the *generation time*, the final population, b, is

after one generation	$2 \times a$
after two generations	$2 \times 2 \times a$
and after three generations	$2 \times 2 \times 2 \times a$

Therefore, the final population, b, after n generations is

$$b = a \times 2^n$$

Hence, bacterial growth is ideally exponential because the size of the population is related to an exponential function of the base 2. As logarithms are similarly exponents of some base figure, bacterial growth is logarithmic in character, and a plot of the logarithm of the number of bacteria as a function of time yields a straight line, at least during the active period of growth. This stage is therefore known as the *logarithmic phase* of growth. Should each daughter have the same biochemical activity as its mother cell, then a plot against time of the logarithm of some change brought about by the culture would be linear. This can be shown for a number of bacterial transformations; for example, the assimilation of N_2 by *Azotobacter* spp. or the oxidation of ammonium by *Nitrosomonas* spp. (Figure 8.1).

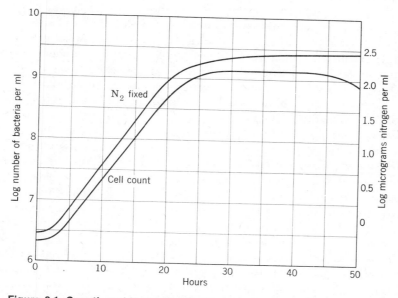

Figure 8.1. Growth and N_2 fixation by *Azotobacter* sp. in culture medium.

It is simple to calculate the generation time, g, from the relationship between the initial and final population in the logarithmic phase.

$$b = a \times 2^n$$

$$\log b = \log a + n \log 2$$

$$0.301n = \log b - \log a$$

$$n = \frac{\log b - \log a}{0.301}$$

However, the generation time is the number of generations in the selected time interval, t. Hence

$$g = \frac{t}{n}$$

$$g = \frac{0.301t}{\log b - \log a}$$

The generation time varies considerably, depending on the organism, the temperature, and the medium. For example, the period required for one generation—equivalent to the time in which the population exactly doubles—may be 20 minutes for *Bacillus cereus,* 35 minutes for *Pseudomonas fluorescens,* 2 hours for *Rhizobium leguminosarum*, and 11 hours for *Nitrosomonas europaea.* The generation time is similarly shortest at the optimum temperature and pH and increases as the environment becomes progressively more unfavorable.

Logarithmic growth occurs only during a portion of the culture cycle. As the supply of nutrients diminishes or as metabolic wastes accumulate, the rate of development declines. Frequently, the factor responsible for the decline is difficult to establish, but nutrient deficiencies or waste products are the usual causes. Ultimately, the culture reaches its peak population density, and the dying of the individual cells offsets the appearance of new bacteria; this is reflected by a diminution in the population size.

In nature, bacteria do not exist in pure culture, and the logarithmic phase is not frequently encountered in a heterogeneous community. However, the addition to soil of certain substances, ammonium or thiosulfate salts for example, may result in a logarithmic transformation rate; that is, the logarithm of the quantity of ammonium or thiosulfate oxidized is linear. Under such circumstances, the demonstration of a logarithmic conversion serves to indicate that bacteria are the responsible agents.

The growth patterns of filamentous microorganisms have not received as much attention as those of the bacteria. Among the filamentous microorganisms, however, there is an active stage during which time the rate of increase of cell mass and cell activity is highest. In some fungi and actinomycetes, there is evidence that the increase is cubic so that a plot of the cube root of mycelium weight or activity against time yields a straight line. With other filamentous species growing in well-agitated media, the increase in cell mass appears to be logarithmic. The period of optimal development is maintained until some factor in the environment becomes limiting. When the supply of O_2 or nutrients is insufficient to meet the demand, the increase in protoplasmic mass of many of the filamentous microorganisms appears to be linear. In the later stages, there is no further gain in mass; finally, *autolysis* or self-decomposition leads to the slow digestion of the protoplasm with a release of soluble substances into the medium.

Many factors affect microbial development. Two of the more important variables to which frequent references are made are temperature and pH. In soil, the influence of temperature and pH on organic matter decomposition, pesticide persistence, and on other processes has considerable agronomic importance. The effect of temperature and acidity is exerted in two ways, by altering the composition of the microflora and by directly influencing the individuals making up the community. It might be assumed that a specific microorganism would be most prominent in soils having a pH or maintained at a temperature near the organism's optimum. That this is not the case is a result of the interactions between individual populations. A species that proliferates readily at the warmer or colder temperature extremes or at the extremes of acidity or alkalinity will predominate in these circumstances because of the lack of competition, despite the fact that the environment may not be particularly favorable for the dominant species. The same can be stated for moisture.

BIOCHEMICAL CONSIDERATIONS

Chemical reactions may take place with the liberation or utilization of energy. The energy released by one reaction may be used to do work or to drive a second reaction that will not proceed on its own. In an isolated system, the transfer of energy from one process to another is complete; that is, there is neither a net loss nor a net gain. In natural processes, however, a portion of the energy is dissipated to the surroundings in the form of heat.

Growth of microorganisms requires an energy input. This is accomplished by the biological oxidation of organic or inorganic compounds. In aerobic heterotrophs, the oxidation may be visualized as

$$C_6H_{12}O_6 + 6O_2 \rightarrow 6CO_2 + 6H_2O + \text{energy} \tag{I}$$

For an aerobic autotroph, a typical reaction is

$$2NH_4Cl + 3O_2 \rightarrow 2HNO_2 + 2H_2O + 2HCl + \text{energy} \tag{II}$$

The conversion of glucose to CO_2 or ammonium to nitrite releases considerable energy, but not all of that released is captured by the microorganism. The ratio of the amount captured by the biological system to the amount released in the oxidation is known as the *free energy efficiency*. Thus, when it is stated that the oxidation of ammonium to nitrite releases 66 kcal, the quantity actually used by *Nitrosomonas* is the product of the energy yield × free energy efficiency.

In equations I and II, the oxidant appears to be O_2; that is, the energy seems to be liberated when O_2 acts on the sugar or on ammonium. In reality, biological oxidations usually proceed not by the addition of O_2 but by the removal of hydrogen (dehydrogenation) or of electrons. Thus, the following two equations entail oxidations although O_2 is not involved.

$$Cu \rightarrow Cu^{++} + 2e^- \tag{III}$$
$$RH_2 \rightarrow R + 2H \tag{IV}$$

The electrons or hydrogens must now be disposed of. This is usually accomplished by a reaction with O_2, the oxygen thereby acting as an *electron (or hydrogen) acceptor*. Consequently, O_2 is not the immediate cause of the oxidation but rather the acceptor in aerobic microorganisms of the electrons liberated. In the absence of this gas, a number of other substances may be electron acceptors, for example, nitrate for strains of *Pseudomonas*, sulfate for *Desulfovibrio*, and CO_2 for *Methanobacterium*. Instead of water being produced through the reduction of O_2 by the electrons (or H), the products are N_2, H_2S, and CH_4 for nitrate, sulfate, and CO_2, respectively. These substances do not serve as sources of oxygen, an early but incorrect concept, but as acceptors of electrons.

With anaerobic bacteria, the metabolic system is basically the same. *Clostridium*, *Lactobacillus*, and other anaerobes grow in the absence of O_2 yet no inorganic electron acceptors are used. In these bacteria, hydrogens removed from the organic compound are dissipated by reaction with one of the products of carbohydrate breakdown. Thus, in the lactic acid fermentation of glucose:

$$\underset{\text{glucose}}{C_6H_{12}O_6} \rightarrow \underset{\text{pyruvic acid}}{2C_3H_4O_3} + 4H \tag{V}$$

$$4H + 2C_3H_4O_3 \rightarrow 2C_3H_6O_3 \tag{VI}$$

Not all reactions produce or consume energy. Many complex compounds must be transformed to simpler forms prior to use by the organism. For example, in those microorganisms which utilize cellulose, the long-chain carbohydrate is converted to simple sugars.

$$\underset{\text{cellulose}}{(C_6H_{10}O_5)_n} + nH_2O \rightarrow \underset{\text{glucose}}{nC_6H_{12}O_6} \tag{VII}$$

The stage depicted in equation VII provides no useful energy to the active species, and it is only the subsequent metabolism of glucose that yields energy for cell synthesis. Yet, the initial attack is necessary in order to convert the cellulose into the sugars that are assimilated. The same holds for hemicelluloses, chitin, pectin, and other structurally complex carbohydrates.

The mere carrying out of an oxidation, either in the presence or in the absence of O_2, is not sufficient for the acquisition of biologically useful energy. In order for the energy to be applied effectively in growing organisms, its storage and release must be carefully regulated. This is effected by means of adenosine diphosphate (ADP) and adenosine triphosphate (ATP). When the microorganism is releasing energy by oxidation, a portion of that liberated is used to convert adenosine diphosphate and inorganic phosphate into adenosine triphosphate.

$$ADP + phosphate + energy \rightarrow ATP \qquad\qquad (VIII)$$

When there is a demand for energy in cell synthesis or for reductive reactions, adenosine triphosphate is converted back to adenosine diphosphate with the controlled release of the energy.

$$ATP \rightarrow ADP + phosphate + energy \qquad\qquad (IX)$$

Consider once again the case of the anaerobes. These microorganisms accumulate certain incompletely oxidized products during growth; that is, not all of the potential energy in the carbon source is released. This incomplete oxidation and incomplete energy yield is reflected in the few molecules of ATP produced for each molecule of carbonaceous nutrient metabolized. In the anaerobic decomposition of glucose by yeast, to cite a single example, the net yield is only two ATP molecules for each glucose molecule degraded. In those aerobic organisms that convert simple sugars to CO_2 and water, getting the full value of the oxidation, a total of about 38 ATP molecules may be formed for each molecule of glucose transformed. Thus, aerobic processes liberate far more energy than anaerobic reactions. Furthermore, since the oxidation of carbohydrates is used for the formation of protoplasm, the greater energy release in air is associated with a greater microbial cell synthesis per unit of organic nutrient.

The various reactions concerned in microbial metabolism require the presence of enzymes. An *enzyme* may be defined as a protein produced by a living cell that functions in catalyzing a chemical reaction. Being proteins, enzymes are denatured at high temperatures, and their activity is also affected by pH. The compound that is changed by enzymatic action, the *substrate*, serves as the basis for enzyme nomenclature. Thus, cellulase, chitinase, and xylanase are the catalysts concerned in the degradation of cellulose, chitin, and xylan, respectively. Some enzymes are named on the basis of both the substrate and the type of reaction; for example, the enzyme removing hydrogen from succinic acid is succinic dehydrogenase while the catalyst concerned in reducing nitrate is designated as nitrate reductase. A formal scheme of enzyme nomenclature has been devised, but the names used here will be those commonly employed by researchers.

Enzymes exhibit a marked specificity for individual substrates or individual processes; that is, they usually function in catalyzing a single type of transformation of one or of a few closely related substrates. This implies that a multitude of enzymes is required in the metabolism of even a unicellular organism. In addition to specificity, another attribute to be borne in mind is the cellular site of enzyme action. Some catalysts, the *intracellular* enzymes, perform their function within the confines of the cell. Others, the *extracellular* enzymes, are concerned with reactions outside of the organism that synthesized the catalyst. The latter are essential in decomposition of polysaccharides such as cellulose and the hemicelluloses because the microbial cell is impermeable to the large polysaccharide molecule. Without the extracellular enzymes, polysaccharide decomposition could not take place. Another useful distinction in enzymology is made between those catalysts always produced by the cell, the *constitutive* enzymes, and those formed only in the presence of the specific substrate. The latter, the *inducible* enzymes, are also of significance in the decomposition of polysaccharides since many of the extracellular catalysts implicated in polysaccharide decomposition are only excreted when a polysaccharide or related compound is available.

ENZYMATIC ACTIVITY IN SOIL

Enzymatic activity is measured by incubating a sample containing the enzyme in question with its substrate, and after an appropriate period, one or more products of the reaction or the amount of substrate lost is measured. Enzymatic activity in soil can be measured in a similar fashion except for the problem of the existence of viable cells. Such organisms would affect the assay in two ways: (a) they would grow during the incubation period at the expense of the substrate so that the activity would increase with time and (b) they would assimilate or further degrade products of the reaction so that little or none would remain.

In considering interpretations of the meaning of measurements of enzymatic activity of soil, it is necessary to remember that much or all of the activity and that many of the enzymes may not exist apart from the cell because many of the enzymes are intracellular; nevertheless, their activity can be evaluated. Only those enzymes that are excreted by members of the community and those that may be released to the surroundings when cells are lysed—that is, digested by autolysis or as a result of activities of other populations—are found free in the soil. Enzymes may also be derived from subterranean animals, roots, or plant remains. The relative contribution of intracellular enzymes to an observed activity cannot as yet be assessed, although enzymes acting on high-molecular-weight substrates such as polysaccharides are probably largely free in the soil because they are characteristically extracellular.

The problems of cell growth or further metabolism of products have been partially overcome in several ways. Sometimes toluene is added to the sample of soil as a germicide to inactivate the cells during the test period, but use of this antimicrobial agent has been criticized because it does not kill all organisms and also induces lysis of sensitive populations so that intracellular enzymes are released. Sometimes the soil is sterilized by high-energy radiation, a treatment that leads to loss of cell viability but, if excessively high levels of irradiation are avoided, at least allows certain enzymes to function. For this purpose, an electron beam, gamma rays from a ^{60}Co source, or X rays are used. Neither of these techniques is ideal for assaying activity of free enzymes that exist in soil: the perfect procedure would entail a sterilization method (*a*) that prevents the biosynthesis of additional enzymes and assimilation or degradation of products because of cell replication, (*b*) that does not lead to rupture of the cell surface so that the substrate is accessible to intracellular enzymes, and (*c*) that does not affect the extracellular enzymes in the test sample.

The list of enzymatic activities in soil that have been investigated is long. A few of the reactions and the common names of the enzymes are presented in Table 8.3. Many of those enzymes whose activities are frequently determined, however, are rarely free in the soil since they are chiefly or wholly intracellular and only would appear outside of the organism on lysis. Nevertheless, a few enzymes have been extracted from soil, and these presumably existed outside the organism that synthesized them.

Because enzymes are proteins, they can be substrates for protein-destroying enzymes and thus would be inactivated and destroyed readily. A striking feature of a few enzymes that have been extracted from soil is their resistance to such attack (1, 2), a useful property in an environment containing a highly diverse community, some members of which probably could destroy proteins that are not in some way made less susceptible to degradation.

Protection of enzymes from decomposition may sometimes result from the enzyme being bound to clay or humus constituents. These catalysts, like other proteins, are frequently adsorbed by clay minerals, but such sorption may not

TABLE 8.3
Enzymatic Activities Observed in Soils

Enzyme	Reaction
Amylase	Hydrolysis of starch
Lipase	Lipid $\rightarrow$ glycerol + fatty acids
Cellulase	Hydrolysis of cellulose
Proteinases	Conversion of protein to amino acids
Urease	Urea $\rightarrow$ NH_3 + CO_2
Dextranase	Dextran $\rightarrow$ glucose
Sulfatase	Sulfate ester $\rightarrow$ inorganic sulfate
Catalase	$2H_2O_2 \rightarrow 2H_2O$ + O_2
Invertase	Sucrose $\rightarrow$ glucose + fructose

only afford protection to the catalyst but also may reduce or occasionally increase its activity. Furthermore, the apparent optimal pH for the reaction that is catalyzed may be altered if the catalyst is complexed with a clay mineral (3).

A number of environmental factors affect the rate of reactions brought about by these enzymes. Each enzyme has a range of pH values and temperatures and an optimum pH and temperature for activity. Similar ranges and optima can be established for the actions of the community itself. Some reactions are most rapid at low, others at medium, and still others at high pH values or temperature. Almost invariably, the activity declines with depth, as is shown in Figure 8.2 for the enzyme arylsulfatase. The rate of substrate disappearance or product formation also is governed by soil type, season of year, and kind of vegetation, and because most and possibly nearly all of the

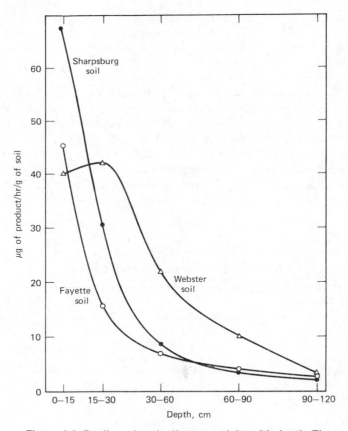

Figure 8.2. Decline of arylsulfatase activity with depth. The activity is expressed as μg of product formed/g of soil/hr (4).

enzymes are microbial in origin, it is not surprising that other factors governing community size influence enzymatic activity.

REFERENCES

Reviews

Ainsworth, G. C. and A. S. Sussman, eds. 1965. *The fungi: Vol. 1. The fungal cell.* Academic Press, New York.

Alexander, M. 1971. Biochemical ecology of microorganisms. *Annu. Rev. Microbiol.*, 25:361–392.

Doelle, H. W. 1975. *Bacterial metabolism.* Academic Press, New York.

Kiss, S., M. Dragan-Bularda, and D. Radulescu. 1975. Biological significance of enzymes accumulated in soil. *Advan. Agron.*, 27:25–87.

Kuprevich, V. F. and T. A. Shcherbakova. 1971. *Soil enzymes.* National Technical Information Service, U.S. Dept. of Commerce, Springfield, Va.

Mahler, H. R. and E. H. Cordes. 1971. *Biological chemistry.* Harper and Row, New York.

Stewart, W. D. P., ed. 1974. *Algal physiology and biochemistry.* Blackwell Scientific Publications, Oxford.

Literature Cited

1. Burns, R. G., A. H. Pukite, and A. D. McLaren. 1972. *Soil Sci. Soc. Amer. Proc.*, 36:308–311.
2. Satyanarayana, T. and L. W. Getzin. 1973. *Biochemistry*, 12:1566–1572.
3. Skujins, J., A. Pukite, and A. D. McLaren. 1974. *Soil Biol. Biochem.*, 6:179–182.
4. Tabatabai, M. A. and J. M. Bremner. 1970. *Soil Sci. Soc. Amer. Proc.*, 34:427–429.

9
Organic Matter Decomposition

The organic matter subjected to microbial decay in soil comes from several sources. Vast quantities of plant remains and forest litter decompose above the surface. Subterranean portions of the plant and the above-ground tissues that are mechanically incorporated into the soil body become food for the microflora. Animal tissues and excretory products are also subjected to attack. In addition, the cells of the microorganisms serve as a source of carbon for succeeding generations of the microscopic community. The chemistry of the organic matter is clearly very complex, and investigations of the transformations and the responsible organisms have therefore been extremely interesting but not without problems arising from the heterogeneity of the natural substrates.

The diversity of plant materials that enter the soil presents to the microflora a variety of substances which are both physically and chemically heterogeneous. The organic constituents of plants are commonly divided into six broad categories: (*a*) the most abundant chemical constituent, cellulose, varying in quantity from 15 to 60 percent of the dry weight; (*b*) hemicelluloses, commonly making up 10 to 30 percent of the weight; (*c*) lignin, which usually makes up 5 to 30 percent of the plant; (*d*) the water-soluble fraction, in which is included simple sugars, amino acids, and aliphatic acids, these contributing 5 to 30 percent of the tissue weight; (*e*) ether- and alcohol-soluble constituents, a fraction containing fats, oils, waxes, resins, and a number of pigments; and (*f*) proteins that have in their structure much of the plant nitrogen and sulfur. The mineral constituents, usually estimated by ashing, vary from 1 to 13 percent of the total tissue.

As the plant ages, the content of water-soluble constituents, proteins, and minerals decreases, and the percentage abundance of cellulose, hemicelluloses, and lignin rises. On a weight basis, the bulk of the plant is accounted for by cellulose, the hemicelluloses, and lignin. In wood, there are particularly large amounts of cellulose, lignin, and also hemicelluloses while the water- and solvent-soluble materials occur in small quantities. These substances constitute the mixed and highly diverse substrates utilized by the community in the decomposition and mineralization of carbon.

CARBON ASSIMILATION

Organic matter decomposition serves two functions for the microflora: providing energy for growth and supplying carbon for the formation of new cell material. Carbon dioxide, methane, organic acids, and alcohol are merely waste products as far as microbial development is concerned, metabolic wastes released in the acquisition of energy. The essential feature for the soil inhabitants themselves is the capture of energy and carbon for cell synthesis.

The cells of most microorganisms commonly contain approximately 50 percent carbon. The source of the element is the substrate being utilized. The process of converting substrate to protoplasmic carbon is known as *assimilation*. Under aerobic conditions, frequently from 20 to 40 percent of the substrate carbon is assimilated; the remainder is released as CO_2 or accumulates as waste products. The extent of assimilation can be estimated roughly by adding known quantities of various simple organic compounds to soil and determining the percent of the added substrate carbon that is retained; for example, about 40 percent of the radioactive carbon from ^{14}C-glucose added to one soil was retained after 104 days (18). The chemical composition of the organic material has a bearing on the magnitude of assimilation, but ultimately the carbon incorporated into newly generated microbial tissues will in turn be decomposed. With heterogeneous organic materials and plant remains, carbon assimilation is not easily measured because of difficulties in determining whether the carbon left represents microbial cells or a portion of the added organic matter that has not yet been degraded.

The fungal flora generally releases less CO_2 for each unit of carbon transformed aerobically than the other microbial groups because the fungi are more efficient in their metabolism. Efficiency is here considered as the effectiveness in converting substrate carbon into cell carbon and is commonly calculated from the ratio of cell carbon formed to carbon source consumed, expressed as a percentage. The more efficient the organism, the smaller the quantity of organic products and CO_2 released. Inefficient cultures, by contrast, lose most of the carbon as wastes and form little cell substance. By and large, filamentous fungi and actinomycetes exhibit a greater efficiency than aerobic bacteria although individual species vary greatly. Anaerobic bacteria utilize carbohydrates very inefficiently, leaving considerable carbonaceous products. Much of the energy in the original substance is not released by the anaerobes, and the incompletely oxidized compounds that are excreted may still be utilized for growth of other species when air reenters the habitat. During decomposition by fungi, some 30 to 40 percent of the carbon metabolized is used to form new mycelium. Populations of many aerobic bacteria, less efficient organisms, assimilate 5 to 10 percent while anaerobic bacteria incorporate only about 2 to 5 percent of the substrate carbon into new cells. These values should only be taken as approximations because some aerobic bacteria are notably efficient, and certain fungi exhibit low efficiencies.

At the same time as carbon is assimilated for the generation of new protoplasm, there is a concomitant uptake of nitrogen, phosphorus, potassium, and sulfur. Assimilation of inorganic substances can be of great practical significance because, agronomically, nutrient assimilation is an important means of *immobilization*, that is, a mechanism by which microorganisms reduce the quantity of plant-available nutrients in soil. Because microbiological immobilization is determined by the utilization of nutrient elements for cell synthesis, the magnitude of immobilization is proportional to the net quantity of microbial cells and filaments formed and is related to carbon assimilation by a factor governed by the $C:N$, $C:P$, $C:K$, or $C:S$ ratio of the newly generated protoplasm. For example, if the average cell composition of the microflora is taken as 50 percent carbon and 5 percent nitrogen, the nitrogen immobilized would be equal to one-tenth of the carbon going into the production of microbial cells.

The efficiency of cell synthesis is governed by environmental conditions, and it may vary over a considerable range. Organisms under one set of circumstances may liberate an end product not produced in another situation; for example, acid or alkaline reactions frequently alter the type of products. At the low level of available nutrients associated with soil, one might expect that a microorganism must be efficient if it is to compete well, particularly if it happens to be a slow grower. Among the rapidly growing species as typified by many bacteria, on the other hand, inefficiency may not be a serious handicap.

DECOMPOSITION AND CARBON DIOXIDE EVOLUTION

The most important function of the microbial flora is usually considered to be the breakdown of organic materials, a process by which the limited supply of CO_2 available for photosynthesis is replenished. The number and diversity of compounds suitable for microbiological decay are enormous. A host of organic acids, polysaccharides, lignins, aromatic and aliphatic hydrocarbons, sugars, alcohols, amino acids, purines, pyrimidines, proteins, lipids, and nucleic acids undergo attack by one or another population. Any compound that is synthesized biologically is subject to destruction by the soil inhabitants; otherwise these compounds would have accumulated in vast amounts on the earth's surface. In addition to biological products, many of the compounds synthesized by the organic chemist are readily decomposed.

Since organic matter degradation is a property of all heterotrophs, it is commonly used to indicate the level of microbial activity. Several techniques have been developed to measure decomposition rates. These include (*a*) measurement of CO_2 evolution or O_2 uptake, (*b*) determination of the decrease in organic matter either chemically or by weight loss, and (*c*) observation of the disappearance of a specific constituent such as cellulose, lignin, or hemicellulose. The results of a study utilizing two of these techniques are shown in Figure 9.1. The evolution of CO_2 is usually measured by passing CO_2-free air over the surface of a soil sample maintained at constant temperature. The liberated CO_2 enters the flowing air stream and can be estimated by gravimetric or volumetric means following absorption. Manometric procedures have also been adapted to the assay of organic matter decomposition. In the manometric technique, gas exchange is measured in two respirometer flasks in the presence and absence of alkali. The first flask detects O_2 uptake, the second O_2 uptake plus CO_2 evolution; the difference between the manometers attached to the two flasks gives the CO_2 evolution rate. In both the manometric and the flowing-air

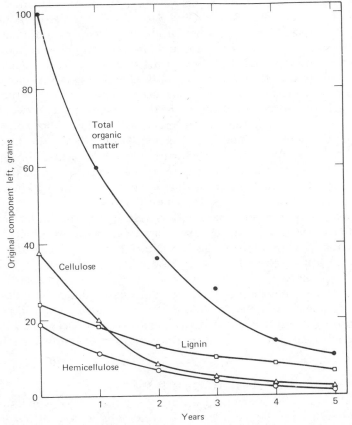

Figure 9.1. Decomposition of *Miscanthus sinensis* leaf litter (30).

procedures, decomposition can be measured at regular time intervals without physically disturbing the soil.

Three separate, simultaneous processes can be distinguished during organic matter transformations. First, plant and animal tissue constituents disappear under the influence of microbial enzymes. At the same time new microbial cells are synthesized so that the proteins, polysaccharides, and nucleic acids typical of bacteria and fungi appear. Third, certain end products of the breakdown are excreted into the surroundings, there to accumulate or to be further metabolized.

The diversity of substrates and their chemical heterogeneity are staggering, but certain biochemical phenomena are universal in microbial metabolism. An organism gets energy for growth only from reactions occurring within the confines of the cell so that, should the substrate be too large or complex to penetrate the cell surface, the compound first must be transformed into simpler molecules to allow the organism to derive energy from the oxidation. Insoluble polysaccharides are commonly hydrolyzed to soluble, simple compounds. And, although the polysaccharides, proteins, aromatic substances, and other nutrients are quite dissimilar in their chemical and physical properties, following their initial degradation, the metabolic sequences concerned in the decomposition within the cell consist of the same general biochemical pathways. Regardless of the structural peculiarities of the starting material, the carbon in the substrate will ultimately be metabolized through the same steps and via the same intermediates. With molecules as different as cellulose, the hemicelluloses, proteins, pectin, starch, chitin, and aromatic hydrocarbons, the final steps in metabolism involve only a few simple sugars and organic acids. The initial stages alone differ, the steps transforming the original compounds into the common intermediates. This is the basis of the doctrine of comparative biochemistry, that there is a certain underlying unity in metabolic reactions.

Two decomposition processes are of significance to the present discussion: the decomposition of soil organic matter and the decay of added substrates. The decomposition of native organic matter (humus) reflects the biological availability of soil carbon while the release of CO_2 following the addition of relatively simple substrates is an estimation of the biodegradability of the test compound. *Mineralization* is a convenient term used to designate the conversion of organic complexes of an element to the inorganic state. The two processes— the breakdown of humus and the decay of added carbonaceous materials—will be considered separately although the characteristics of the two are frequently similar.

DECOMPOSITION OF SOIL ORGANIC MATTER

The rate at which CO_2 is released during the mineralization of humus varies greatly with soil type. Under controlled conditions in the laboratory and at temperatures maintained in the mesophilic ranges, 20 to 30°C, the rate of CO_2

production is commonly from 5 to 50 mg of CO_2 per kg of soil per day, but figures of 300 mg or more are occasionally encountered. In the field, the rate of CO_2 evolution may be as low as 0.5 or up to 10 g of CO_2 per square meter per day, and values as high as 25 g are sometimes encountered. Field estimates, however, may include CO_2 from the respiration of roots and soil animals as well as from microbial activity, and the values obtained depend on soil temperature and water content as well as time of day and season of year. From field data, it is possible to demonstrate that about 2 to 5 percent of the carbon present in humus can be mineralized per annum, but the figures vary appreciably in different localities. Thus, a significant portion of the organic matter is mineralized each year, but the amount that is lost is usually compensated for by the return of organic matter from the vegetation. As a rule, the quantity of CO_2 carbon lost through heterotrophic action is about equal to that introduced into soil by roots, leaf litter, and other plant remains, although some years the loss may exceed and some years be less than that reintroduced (6).

The major factors governing humus decomposition are the organic matter level of the soil, cultivation, temperature, moisture, pH, depth, and aeration. It is evident that those environmental influences that affect microbial growth and metabolism will modify the rate at which either native organic matter or added compounds are transformed. The magnitude of carbon mineralization is directly related to the organic carbon content of the soil; that is, the release of CO_2 is proportional to the organic matter level. A similarly high correlation is noted between the percentage of humus and the O_2 uptake. The production of CO_2 is enhanced by the addition of organic materials. To ascertain whether the additional CO_2 arises largely from the decomposition of added carbonaceous materials or from humus, the added substance is labeled with radioactive ^{14}C, thereby permitting the investigator to distinguish between the two organic matter sources. By such means, it has been found that fresh substrates sometimes accelerate and sometimes reduce the rate of humus decomposition (20). The enhancement is known as *priming*. Whether it occurs or not depends on the soil and the type of organic substrates or plant remains added. The effect of one substrate on the metabolism of another is not restricted to effects on humus degradation since analogous enhancements or reductions in rates of decomposition can be shown with two added materials. For example, ^{14}C-tagged plant tissue can be allowed to decompose in soil and then nonradioactive glucose is added. Assessment of the rate of release of $^{14}CO_2$ reveals whether the sugar promoted or diminished the degradation of the residual tissue. As shown in Table 9.1, sometimes glucose accelerates and sometimes it reduces the mineralization process. Priming and the reduced activity associated with a second substrate have not yet been adequately explained.

Cultivation enhances organic matter destruction. For example, after 25 or more years of cropping, the mean organic matter content of 28 soils of Georgia had decreased from 3.29 to 1.43 percent, a loss of more than half. After a rapid

TABLE 9.1

**Effect of Glucose Addition on the Rate of Decomposition of
Organic Matter in Soil (25)**

Days	Alfalfa Meal[a]		Cellulose-Lignin Preparation[a]	
	Glucose Added	Water Added	Glucose Added	Water Added
		mg $^{14}CO_2$-carbon evolved		
7	33	11	18.7	34.8
15	46	20	41.7	65.2
21	52	24	54.7	77.6

[a] The ^{14}C-labeled material was allowed to decompose in soil for 42 days before ^{12}C-glucose or water was added.

decline of the organic carbon level in the first few years, the decrease becomes more gradual with further cultivation. In one sandy loam, for example, a virgin forest soil contained 2.30 percent organic matter, but the concentration had fallen to 1.59 percent after 3 years of cultivation (7).

Temperature, moisture, and pH are also critical environmental variables. Humus decomposition can proceed at temperatures down to the freezing point, but it is accelerated by increasing temperatures. Freezing, however, brings about changes in soil such that the rate of release of CO_2 is greater than in a soil not exposed to a freezing phase (11). Moisture level also affects soil respiration, and the environment must contain sufficient water for maximum microbiological action. The rate of CO_2 evolution is enhanced when soil is exposed to a cycle of drying and wetting, and several such cycles appreciably stimulate microbial activity as compared to soils that are moist continuously (21). Other factors being equal, carbon mineralization is most rapid in neutral to slightly alkaline soils. As expected, therefore, liming acid soils enhances carbon volatilization.

Heterotrophic activity in soil is limited not by the supply of nitrogen, phosphorus, or other inorganic nutrients but by the paucity of readily utilizable organic nutrients. This is easily demonstrated experimentally by adding a simple organic compound to one sample of soil and combinations of inorganic nutrients to others: CO_2 evolution is enhanced just by the added carbon source. However, responses of the community to nitrogen and phosphorus become evident when sufficient readily degradable carbonaceous matter is provided to satisfy the microbial demand.

The greatest rate of CO_2 evolution occurs near the surface of the profile where the highest concentration of plant remains is found. At greater depths, the rate of CO_2 production diminishes, and little is usually volatilized at depths of 50 cm or more. This decrease in activity parallels the drop in the organic carbon level.

One of the major microbiological changes in organic soils is the phenomenon of subsidence, where the soil itself shrinks through biological decomposition. The subsidence is of great practical importance since it leads to loss of agricultural productivity and to problems in road construction and maintenance. Subsidence may range from 0.2 to more than 7 cm per year. Although the major cause appears to be biological, wind erosion and physical shrinkage contribute to the effect.

Most intermediates in the decomposition of the soil organic fraction are probably metabolized as quickly as they are produced since the rate-limiting step in the breakdown is undoubtedly the attack on the complex molecules of humus. In well-drained soils, acids and alcohols are formed, but they rarely accumulate in appreciable amounts because they are readily metabolized by aerobic bacteria, actinomycetes, and fungi. The chief organic acids that accumulate are formic and acetic (3), and the common alcohols are methyl, ethyl, *n*-propyl, and *n*-butyl (29). The yield of the organic acids found in a typical study is given in Table 9.2.

Simple substrates that are added to soil are readily metabolized but almost always with an apparent lag period prior to the maximal oxidation rate. The lag represents the time necessary for the active populations to increase to an extent sufficient to cause rapid organic matter turnover. Ethanol, however, is oxidized readily with no preliminary lag. This anomalous characteristic of ethanol decomposition has been reported in a number of soil types. A break in the rate of ethanol oxidation does occur when the amount of gas consumed is equivalent

TABLE 9.2

Organic Acids in Two Belgian Soils (3)

	Quantity found, μmol/g	
Acid	Meerdael Soil	Heverle Soil
Formic	6.69	4.48
Acetic	1.19	7.65
Propionic	0.85	0.36
3-Hydroxybutyric	0.27	0.17
Vanillic	0.96	0.88
Syringic	0.50	0.56
Coumaric	0.44	0.52

to 1 mole O_2 for each mole of ethanol, a ratio that suggests the accumulation of acetic acid.

$$CH_3CH_2OH + O_2 \rightarrow CH_3COOH + H_2O \qquad\qquad (I)$$

Acetate is also oxidized without a lag period, but the activity quickly declines (23). Since ethanol and acetate are metabolized immediately, it seems likely that the community has repeated and frequent encounters with these substrates.

BREAKDOWN OF ADDED CARBONACEOUS MATERIALS

A number of factors affect the mineralization of added organic materials. The rapidity with which a given substrate is oxidized will depend on its chemical composition and the physical and chemical conditions in the surrounding environment. Temperature, O_2 supply, moisture, pH, inorganic nutrients, and the C :N ratio of the plant residue are the chief environmental influences. The age of the plant, its lignin content, and the degree of disintegration of the substrate presented to the microflora also govern the decomposition. As with humus breakdown, those factors that affect microbial growth and metabolism will alter the rate of decay of added plant or animal remains.

Temperature is one of the most important environmental conditions determining how rapidly natural materials are metabolized. A change in temperature will alter the species composition of the active flora and at the same time have a direct influence on each organism within the community. Microbial metabolism and hence carbon mineralization are slower at low than at elevated temperatures, and warming is associated with greater CO_2 release. Appreciable organic matter breakdown occurs at 5°C and probably at cooler values, but plant tissue degradation is increased with progressively warmer conditions; the individual constituents of the plant also disappear more rapidly.

Each individual microbial species and the biochemical capacities of the community as a whole have temperature optima. Because the composition of the flora varies from locality to locality and is altered even in a single site treated with different plant residues, a single optimum for organic matter breakdown cannot be found. Thus, there are reports that the maximum rates of decay of carbonaceous nutrients take place at 30 to 35, at 37, and at 40°C. In the vicinity of the optimum, taken at about 30 to 40°C, temperature fluctuation has little effect on decomposition. In the range below the optimum, generally from 5 to 30°C, rising temperature accelerates plant residue destruction. Above about 40°C, the rapidity of decomposition declines except in those special circumstances where thermophilic decay is initiated.

Air supply also governs the extent and rate of dissimilation of added substrates. This effect is a consequence of the role of O_2 in microbial metabolism. Carbon dioxide is released from completely anaerobic environments through the activities of the obligate and facultative anaerobes, but aeration invariably stimulates carbon mineralization. The decay of the major plant constituents is similarly depressed as the supply of O_2 diminishes.

Moisture, too, must be adequate for decomposition to proceed. Microorganisms grow readily in liquid media provided the O_2 supply is ample; in soil, on the other hand, high moisture levels reduce microbial activities not as a result of the water itself but rather indirectly, by hindering the movement of air and thus reducing the O_2 supply. Hence, when an increase in moisture is observed to stimulate CO_2 release, water is limiting; however, if additional water reduces the rate of transformation, then there is a deficiency of O_2. At low moisture, supplemental water has a profound influence on decay while similar additions at moisture levels near the optimum result in little change. Respiration of the soil microflora developing at the expense of simple or complex organic nutrients is commonly greatest at about 60 to 80 percent of the water-holding capacity of the soil. The effect of moisture tension on carbon mineralization is shown in Figure 9.2.

Another major factor determining the rate of carbon turnover is the hydrogen ion concentration. Each bacterium, fungus, and actinomycete has an optimum pH for growth and a range outside of which no cell proliferation takes place. In addition, individual enzymes elaborated by a single microbial strain are affected by pH. The pH also governs the type of microorganisms concerned in the carbon cycle of any habitat. Decomposition typically proceeds more readily in neutral than in acid soils. Consequently, the treatment of acid soils with lime accelerates the decay of plant tissues, simple carbonaceous compounds, or native soil organic matter.

Nitrogen is a key nutrient substance for microbial growth and hence for organic matter breakdown. Plant and animal tissues always contain some nitrogen, but its availability and amount vary greatly. If the nitrogen content of the substrate is high and the element is readily utilized, the microflora satisfies its needs from this source, and additional quantities are unnecessary. If the substrate is poor in the element, decomposition is slow, and carbon mineralization will be stimulated by supplemental nitrogen. In the latter circumstances, nitrogenous amendments cause an increase in CO_2 evolution and a greater loss of cellulose, hemicelluloses, and other plant polysaccharides. Nitrogen-rich materials such as legume tissues are metabolized very rapidly, and the microflora responds little if at all to supplemental nitrogen while the addition of ammonium or nitrate to straw or other nitrogen-deficient substrates greatly enhances decomposition. Differing from mineral soils where the level of available nitrogen is usually too low to allow for maximum rates of carbohydrate breakdown, application of inorganic nitrogen salts to peats does not stimulate glucose decomposition, suggesting a large reserve in the organic soils (24).

Despite the greater carbon loss from the soil as a result of nitrogen treatment of protein-poor crop residues, humus formation is benefited. The explanation for this observation rests on the fact that plant residues remain

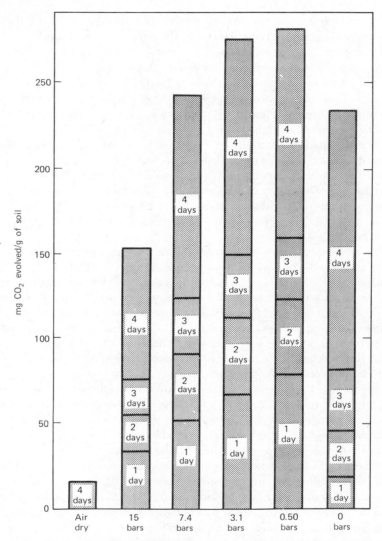

Figure 9.2. The influence of moisture tension on CO₂ evolution from soil (14). *Bar* is a term to designate water tension; the higher the value, the lower the soil water content.

partly decomposed for long periods of time if nitrogen is lacking, and they do not become converted to humus. Yet, although applied nitrogen commonly stimulates the rate of residue breakdown, the total quantity of CO_2 ultimately liberated is the same with or without the supplement. The limited inorganic nutrient supply is merely recirculated through successive populations.

A number of investigators have reported that the rate of decomposition of plant materials depends on the nitrogen content of the tissues, protein-rich substrates being metabolized most readily (13). This can be seen if plant residues are arranged in order of decreasing rates of mineralization: sweet clover (3.14 percent nitrogen); alfalfa (3.07 percent); a group containing red clover (2.20 percent), soybeans (1.85 percent), millet (1.17 percent), and flax (1.73 percent); another group decayed even less quickly containing hemp (0.88 percent), corn stalks (1.20 percent), and sudan grass (1.06 percent); and lastly wheat and oat straw with 0.50 and 0.61 percent nitrogen. Such observations are not unexpected in view of the high nitrogen demands of the community. Because crop plants generally contain about the same amount of carbon, usually about 40 percent of the dry weight, their nitrogen contents can be compared by use of the C:N ratio. Thus, a low nitrogen content or a wide C:N ratio is associated with slow decay.

Generalizations of this sort must nevertheless be accepted with some reservation since it is not easy to determine the precise causal relationship of the enhanced decay. Other factors are operating in addition to nitrogen. For example, the report that tissues of young plants are metabolized faster than mature tissue apparently substantiates the nitrogen or C:N ratio hypothesis for velocity of decomposition because the immature plants have a higher nitrogen content. But a complete chemical investigation shows changes in other plant constituents as well; for example, maturation is accompanied by lignification and related alterations. The nitrogen content or C:N ratio of plant residues thus frequently is a convenient tool for predicting the rate of decomposition, yet it is not the sole determinant.

During the mineralization of materials containing little nitrogen, the C:N ratio tends to decrease with time (Figure 9.3). This results from the gaseous loss of carbon while the nitrogen remains in organic combination for as long as the C:N ratio is wide. Therefore, the percentage of nitrogen in the residual substance continuously rises as decomposition progresses. The narrowing of the ratio in the decay of nitrogen-poor substrates is not linear, the curve approaching a ratio of approximately 10:1 asymptotically.

The C:N ratio of soil is one of its characteristic equilibrium values, the figure for humus being roughly 10:1 although values from 5:1 to 15:1 are not uncommon. This critical ratio is a reflection of the dynamic equilibrium that results from the dominating presence of a microbial community, the ratio being similar to the average chemical composition of microbial cells. As a rule, microbial cells contain 5 to 15 parts of carbon to 1 part of nitrogen, but 10:1 is a reasonable average for the predominant aerobic flora. A change in the population brought about by anaerobiosis or the accumulation of fractions resistant to further decay can modify the C:N equilibrium value of humus.

Consider the incorporation into soil of a residue having a wide C:N ratio. The microflora carrying out the decomposition will develop to the extent of the

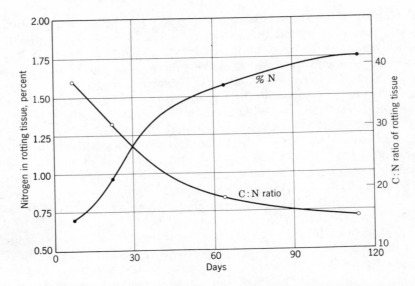

Figure 9.3. Changes in the nitrogen content of decomposing barley straw (8).

supply of available nitrogen and other inorganic nutrients, and all the immediately available nitrogen will be assimilated and bound in organic complexes. Assuming that the aerobic community contains 50 percent carbon and 5 percent nitrogen and assimilates one-third of the substrate carbon, then 1 unit of available nitrogen incorporated into cell material will allow for the assimilation of 10 units of cell carbon but will be accompanied by the volatilization of 20 units of CO_2 carbon. In this first stage, no nitrogen is lost because the demand exceeds the supply, but CO_2 is released and the C:N ratio narrows. As the populations in the initial community die and are decomposed, the nitrogen liberated will be assimilated by later populations, which then synthesize 10 times more microbial carbon and volatilize 20 times more CO_2 carbon than nitrogen made available. The C:N ratio is narrowed further. This process is repeated until the equilibrium C:N ratio is attained, ca. 10:1. At this point, the organic nitrogen that becomes mineralized is no longer necessary for microbial growth, and it remains in the inorganic form. Henceforth, nitrogen and carbon mineralization run parallel, and the humus C:N ratio has at this stage reached the value determined largely by the chemistry of the microbial cell.

Natural materials rich in lignin are less readily utilized by microorganisms than lignin-poor products. It is not uncommon to find that the rates of decay of plant debris are proportional to their content of lignin, and the quantity of lignin in plant residues may be of greater importance in predicting decomposition velocity than the C:N ratio. The resistance of wood and sawdust to microbial attack is probably linked to the abundance of lignin in such materials.

Young, succulent tissues are metabolized more readily than residues of mature plants. As the plant ages, its chemical composition changes; the content of nitrogen, proteins, and water-soluble substances falls, and the proportion of cellulose, lignin, and hemicelluloses rises. Although aging makes the vegetation more resistant to decay, the changes with age of many individual constituents have prevented a final explanation of the precise reason for the effect. A large part of the resistance associated with aging probably is a consequence of the abundance of lignin, but other factors may also be operative.

The rate of decomposition is also governed by size of the organic particles subject to attack. As a rule, the small particulate materials are more readily degraded than are the large particles (19).

Before the widespread use of chromatographic, and especially gas chromatographic, procedures in soil microbiology, few of the products generated in the microbial decay of organic materials or simple compounds added to soil were known. Such methods, however, have revealed an astonishing array of metabolites that appear and persist for short or sometimes long periods of time. Amino acids and gluconic acid thus are produced from glucose (28), phenolic compounds are formed in the decomposition of plant remains (10), and a host of simple or occasional novel metabolites are excreted by microorganisms utilizing pesticides and other synthetic molecules used in food production. The list of metabolites that have recently been characterized, given the availability of sophisticated analytical methods, is growing rapidly.

CHANGES DURING ORGANIC MATTER DECOMPOSITION

In studies of decomposition, the entire plant residue, extracted tissue constituents, or pure organic compounds may be utilized. The various techniques each have their values since every material is metabolized in a different way and often by dissimilar populations.

As a result of the development of a mixed flora on chemically complex natural products, some components quickly disappear while others are less susceptible to microbial enzymes and persist. The water-soluble fraction contains the least resistant plant components and is thus the first to be metabolized. As a result, in those tissues in which 20 to 30 percent of the dry matter is water-soluble, decomposition proceeds rapidly. Cellulose and hemicelluloses, on the other hand, disappear not as quickly as the water-soluble substances, but their persistence usually is not too great. The lignins are highly resistant and consequently become relatively more abundant in the residual, decaying organic matter.

In succulent tissues, the bulk of the organic matter lost during decay is derived from the cellulosic, hemicellulosic, and water-soluble constituents. In contrast, the major part of the weight loss in woody materials results from the disappearance of cellulose. The magnitude of dry matter loss is reduced under anaerobiosis, but here too the percentage of sugars, water-soluble constituents, and cellulose declines and the percentage of lignin rises with time. The metabolism of the highly available components of the plant residue is accompanied by a qualitative alteration in the chemical composition of the remaining portion since the character of the organic matter is now dominated by the newly formed microbial cells and by those plant fractions exhibiting the greatest resistance to attack, for example, aromatic substances related to and possibly derived from lignin. The change in the chemistry of the soil from one reflecting the presence of recently added organic molecules to one dominated by microbial cells and products is well illustrated by investigations in which ^{14}C-labeled chemicals are applied to soil; for example, following the addition of ^{14}C-acetate and its metabolism, radioactive carbon appears in amino acids, carbohydrates, other cellular components, and polymers characteristic of humus (12, 22). Similar changes are evident when plant tissues undergo decay.

Other modifications take place in the organic matter as it undergoes decomposition. The data presented in Table 9.3 demonstrate that the hydroxyl content of the remaining residue declines while the carboxyl content and cation exchange capacity rise as rotting progresses. Residues remaining after prolonged decomposition of cellulose or glucose contain little lignified carbon whereas tissues rich in lignin yield a decayed fraction containing a high concentration of ligninlike substances (17).

When carbonaceous substrates are incorporated into the soil, there is an immediate and marked drop in the O_2 and an increase in the CO_2 content of

TABLE 9.3
Changes in Properties of Oat Straw During Decomposition (5)

Days of Incubation	Lignin %	Hydroxyl %	Chemical Properties, meq/100 g	
			Carboxyl Content	Cation Exchange Capacity
0	19.3	7.40	28	25
14	21.8	5.94	24	26
40	28.0	8.03	81	42
88	30.8	5.29	95	47
135	34.3	5.48	113	58
180	39.4	5.59	142	60
244	38.3	4.69	139	81
355	37.6	4.62	139	82

the soil air; at the same time, the oxidation-reduction potential (E_h) is shifted to a more reduced condition. The rate and magnitude of the increase in reducing power varies with the substrate added. A similar fall in oxidation-reduction potential and disappearance of dissolved O_2 takes place in flooded soil. If a readily available carbohydrate is added to the waterlogged field, the drop in oxidation-reduction potential is accelerated, yet, should the sample receiving the organic compounds be sterilized immediately, no difference in potential would be detected between treated and control samples, regardless of flooding. Consequently, microorganisms cause the change in E_h through the consumption of O_2 and the liberation of reduced products.

The quantity and type of clay in a soil have a bearing upon carbon mineralization because clays adsorb many organic substrates, extracellular carbohydrate-splitting enzymes produced by microorganisms, and even bacterial cells. Clays have marked carbon-retaining capacity, and decomposition is suppressed in their presence. Furthermore, the addition of certain clays to culture media inoculated with soil enrichments retards the degradation of a variety of substrates.

Not only clays but also sand and silt may influence decomposition. These structures may serve as mechanical barriers to microbial movement to particulate organic nutrients or prevent contact between the potentially active cells or their enzymes and a substrate deposited at a microsite shielded by noncarbonaceous particles (16).

ANAEROBIC CARBON MINERALIZATION

The main products of aerobic carbon mineralization are CO_2, water, cells, and humus components. In the absence of O_2, organic carbon is incompletely metabolized, intermediary substances accumulate, and abundant quantities of CH_4 and smaller amounts of H_2 are evolved. At the same time, the energy yield during anaerobic fermentation is low, resulting in the formation of fewer microbial cells per unit of organic carbon degraded. Consequently, organic matter breakdown is consistently slower under total anaerobiosis than in environments containing adequate O_2; the rate in waterlogged soils is intermediate between the two extremes.

When a soil is waterlogged or flooded, there is a shift from aerobic to anaerobic transformations. As a result of this shift, a variety of products accumulate, some in appreciable quantity (Figure 9.4). As O_2 is metabolized and lost, CH_4 and H_2 appear in addition to CO_2. The amount of CH_4 is frequently great, but the quantity of H_2 is invariably small. In flooded paddy fields and in other waterlogged environments, much of the carbon mineralization is anaerobic although O_2 is present in the very top portion of the soil or at the soil-water interface. In addition, O_2 may be formed biologically by the algae developing in the liquid phase. Where sufficient available carbohydrates are present, most of

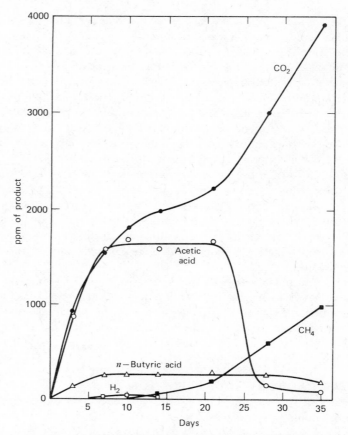

Figure 9.4. Products of anaerobic breakdown of corn residues (15).

the O_2 is utilized before it penetrates too deeply in the liquid-mud layer, and the transformation at the lower depths is almost entirely anaerobic.

Organic acids accumulate because of the fermentative character of the microflora of wet soils. In flooded fields, the dominant acids often are acetic, formic, and butyric, but lactic and succinic acids appear too. Of particular importance is the fact that metabolites such as these are frequently detrimental to root development (26, 27). Also accumulating when soils devoid of O_2 receive simple organic molecules are simple alcohols and a number of carbonyl compounds (1). The organic acids, alcohols, H_2, and possibly other simple organic molecules serve as energy sources for the bacteria that produce the CH_4 in these habitats. Anaerobic carbon transformations are thus characterized by the formation of organic acids, alcohols, CH_4, and CO_2 as major end products.

FLORA

The amount, type, and availability of organic matter will determine the size and composition of the heterotrophic community that a soil will contain. The nature of the flora will vary with the chemical composition of the added substrates, certain microbial groups predominating for a few days, others maintaining high population levels for long periods. Each individual organism has a complex of enzymes that permits it to oxidize a fixed array of chemical compounds, but no others. If the proper substances are present in an accessible state, then the microorganism will proliferate, providing that it can cope with the competition of other organisms having similar enzymatic potentials and the harm done by predators and parasites.

The microorganisms preferentially stimulated by the components of added carbonaceous substances make up the primary flora. A secondary flora also develops, one growing on compounds produced by the primary agents or growing on the dead or living cells of the initial flora. This succeeding group of organisms has a different biochemical makeup from those appearing initially. The population responding to organic carbon amendments thus feeds on (a) the organic substrates added, (b) intermediates formed during decomposition, and (c) the protoplasm of microorganisms active in the degradation of a or b.

When succulent plant tissues are incorporated into the soil, the abundance of bacteria around and within the buried materials increases rapidly. A rise in bacterial numbers only occurs directly on the plant substance, the populations here reaching 10^{10} per gram in the first week while the viable counts of bacteria in the adjacent soil are not markedly altered. By the seventh day, the bacterial numbers begin to decline, falling to a point where the counts are essentially the same as in unamended soil. There is a concomitant rise followed by a subsequent diminution in the numbers of protozoa, the changes paralleling the bacterial fluctuations. Plate counts of fungi and actinomycetes, however, are often not appreciably affected by turning under young plants. Mature crop residues, having a distinctly different chemical composition from succulent tissues, support a flora better adapted to utilize resistant carbonaceous compounds. This population is largely fungal although bacteria and actinomycetes also are stimulated to some extent.

The addition of simple sugars to some soils prompts a rapid rise in the abundance of bacteria, the apparent generation time of the species bringing about the degradation sometimes being about two hours (4). In other soils, addition of the same substrate, although stimulating the bacteria, leads to marked enhancement in fungal activity, and the latter organisms are chiefly responsible for the decomposition process (2). Substrates rich in amino acids such as peptone stimulate the spore-forming bacilli (9). The flora concerned in humus decomposition differs from that concerned with the breakdown of freshly added plant materials.

The relationship between microbial numbers and CO_2 evolution has still not been fully resolved. Because the abundance of microorganisms depends on the presence of available carbonaceous and energy materials, a correlation between microbial numbers and CO_2 release might be expected and is sometimes observed. Yet, reports to the contrary are not lacking. If the carbon source were homogeneous and the community composed of a single species, a definite relationship might be clear. But, with the diversity of microbial types and the variety of carbon sources, a poor correlation between numbers and CO_2 formation is not surprising. Moreover, even in the development of bacteria in pure culture, there is no clear relation between population size and activity in the late stages of growth. In unamended soils, the bacteria are largely not in the active phases of growth, and CO_2 production is thus rarely proportional to the size of the community. Only with rapid increases in microbial numbers is a clear association to be expected.

REFERENCES

Reviews

Burges, A. 1967. The decomposition of organic matter in the soil. In A. Burges and F. Raw, eds., *Soil biology*. Academic Press, New York, pp. 479–492.

Dickinson, C. H. and G. J. F. Pugh, eds. 1974. *Biology of plant litter decomposition*. Academic Press, London.

Wagner, G. H. 1975. Microbial growth and carbon turnover. In E. A. Paul and A. D. McLaren, eds., *Soil biochemistry*. Marcel Dekker, New York, vol. 3, pp. 269–305.

Literature Cited

1. Adamson, J. A., A. J. Francis, J. M. Duxbury, and M. Alexander. 1975. *Soil Biol. Biochem.*, 7:45–50.
2. Anderson, J. P. E. and K. H. Domsch. 1975. *Can. J. Microbiol.*, 21:314–322.
3. Batistic, L. 1974. *Plant Soil*, 41:73–80.
4. Behera, B. and G. H. Wagner. 1974. *Soil Sci. Soc. Amer. Proc.*, 38:591–594.
5. Broadbent, F. E. 1954. *Soil Sci. Soc. Amer. Proc.*, 18:165–169.
6. deJong, E., H. J. V. Schappert, and K. B. MacDonald. 1974. *Can. J. Soil Sci.*, 54:299–307.
7. Giddens, J. 1957. *Soil Sci. Soc. Amer. Proc.*, 21:513–515.
8. Hende, A. van den, A. Cottenie, and R. de Vlieghere. 1952. *Trans. Intl. Soc. Soil Sci., Comm. II and IV*, 2:37–47.
9. Hirte, W. F. 1972. *Symp. Biol. Hungar.*, 11:221–227.
10. Kuwatsuka, S., and H. Shindo. 1973. *Soil Sci. Plant Nutr.*, 19:219–227.
11. Mack, A. R. 1963. *Can. J. Soil Sci.*, 43:316–324.
12. McGill, W. B., J. A. Shields, and E. A. Paul. 1975. *Soil Biol. Biochem.*, 7:57–63.
13. Millar, H. C., F. B. Smith, and P. E. Brown. 1936. *J. Amer. Soc. Agron.*, 28:914–923.
14. Miller, R. D. and D. D. Johnson. 1964. *Soil Sci. Soc. Amer. Proc.*, 28:644–647.
15. Moraghan, J. T. and K. A. Ayotade. 1968. *Trans. 9th Intl. Cong. Soil Sci.*, 4:699–707.
16. Ou, L.-T. and M. Alexander. 1974. *Soil Sci.*, 118:164–167.
17. Peevy, W. J. and A. G. Norman. 1948. *Soil Sci.*, 65:209–226.

18. Shields, J. A., E. A. Paul, W. E. Lowe and D. Parkinson. 1973. *Soil Biol. Biochem.*, 5:753–764.
19. Sims, J. L. and L. R. Frederick. 1970. *Soil Sci.*, 109:355–361.
20. Smith, J. H. 1966. In R. A. Silow, ed., *The use of isotopes in soil organic matter studies.* Pergamon Press, New York, pp. 223–233.
21. Sorensen, L. H. 1975. *Soil Biol. Biochem.*, 6:287–292.
22. Sorensen, L. H. and E. A. Paul. 1971. *Soil Biol. Biochem.*, 3:173–180.
23. Stevenson, I. L. and H. Katznelson. 1958. *Can. J. Microbiol.*, 4:73–79.
24. Stotzky, G. and J. L. Mortensen. 1957. *Soil Sci.*, 83:165–174.
25. Szolnoki, J., F. Kunc, J. Macura, and V. Vancura. 1963. *Folia Microbiol.*, 8:356–361.
26. Takijima, Y. 1964. *Soil Sci. Plant Nutr.*, 10:204–211.
27. Takijima, Y. 1964. *Soil Sci. Plant Nutr.*, 10:212–219.
28. Vancura, V., J. Macura, and J. Szolnoki. 1967. *Trans. 8th Intl. Cong. Soil Sci.*, Bucharest, 3:779–789.
29. Wang, T. S. C. and T.-T. Chuang. 1967. *Soil Sci.*, 104:40–45.
30. Yamane, I. 1974. *Sci. Rep. Res. Inst. Tokoku Univ., Ser. D.*, 25:25–30.

10
Microbiology of Cellulose

A prominent carbonaceous constituent of higher plants and probably the most abundant organic compound in nature is cellulose. Because a large part of the vegetation added to soil is cellulosic, the decomposition of this carbohydrate has a special significance in the biological cycle of carbon. As a result, considerable attention has been given to the microorganisms participating in the decomposition of this substance.

In structure, cellulose is a carbohydrate composed of glucose units bound together in a long, linear chain by β-linkages at carbon atoms 1 and 4 of the sugar molecule. Most evidence suggests that there are between 2000 and 10,000 and on occasion as many as 15,000 glucose units in the molecule, but the number of sugar units per chain and the molecular weight of cellulose vary with the plant species. Molecular weight determinations give values ranging from 200,000 to about 2.4 million.

Cellulose occurs in seed-bearing plants, in the algae, in many of the fungi, and in cysts of at least some protozoa. The polysaccharide is localized in the cell wall where it is found not as simple chains but as submicroscopic rodshaped units known as micelles. The micelles then are further arranged into a larger structure, the microfibril. In the cell wall, the cellulose probably is organized into discrete units separated by a space, which in mature plant tissue is often filled with lignin. A number of other polysaccharides are also associated with the cellulose of the plant cell wall. Cellulose has also been found in soil organic matter (6).

The cellulose content of higher plants is never fixed, and the concentration changes with age and type of plant. The carbohydrate is especially prominent in woody substances and in straw, stubble, and leaves. Succulent tissues are commonly poor in cellulose, but the concentration increases as the plant matures. In young grasses and legumes, for example, cellulose may account for as little as 15 percent of the dry weight, but the figure may be greater than 50 percent in woody materials. A concentration range of 15 to 45 percent includes most of the common crop species, the lower extreme being typical of younger plants.

Both starch and cellulose are polymers of the same building block, glucose, but the individual peculiarities of the two molecules permit ready microbial attack of the former substance while the latter is far more resistant to microbiological and enzymatic breakdown. Furthermore, because of their structural differences, the two carbohydrates stimulate entirely different populations.

FACTORS GOVERNING DECOMPOSITION

The rate at which cellulose is metabolized is governed by a number of environmental influences, and soils varying in their physical and chemical characteristics possess markedly different cellulolytic capacities. The major environmental factors affecting the transformation are the available nitrogen level, temperature, aeration, moisture, pH, the presence of other carbohydrates, and the relative proportion of lignin in the plant residue. It is evident from Figure 10.1 that plant constituents affect the degradation because the decomposition is different when different parts of the same plant are attacked microbiologically. Modifications in the physical and chemical characteristics of the habitat can alter either the composition of the microflora or the cellulose-degrading activity of individual organisms.

The application of inorganic nitrogen enhances cellulose breakdown in soil, either ammonium or nitrate salts serving as suitable sources of the element. The rate of decomposition is proportional to the concentration of nitrogen added, but at high application rates, where there is more inorganic nitrogen than needed, cellulose decomposition does not respond to supplemental increments. The point at which additional quantities are no longer beneficial is at a ratio of ca. 1 part of inorganic nitrogen for each 35 parts of cellulose. Manure and organic nitrogen compounds such as urea, amino acids, and casein also increase the conversion rate. The effect of animal manure seems to result from its nitrogen contribution because a similar stimulation is noted when equivalent quantities of ammonium salts are used. The existence of a response to this element suggests that the nitrogen level in soil is limiting. Indeed, it has been proposed that the supply of available nitrogen can be estimated from the quantity of CO_2 evolved when soils are treated with cellulose.

That available nitrogen is a critical factor is apparent from the correlations observed between the cellulose degradation rates and the nitrogen-mineralizing capacity and nitrate content of soil (16). The finding that approximately 1 unit of nitrogen is required for each 35 units of cellulose oxidized suggests that 3 parts of nitrogen are incorporated into microbial protoplasm for 100 parts of cellulose decomposed. Assuming that microbial cells contain 5 to 10 percent

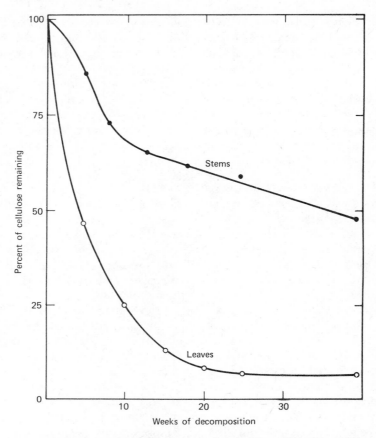

Figure 10.1. Decomposition of the cellulose in corn leaves and stems (1).

nitrogen on a dry weight basis, then 30 to 60 parts of biologically active tissue are synthesized during the aerobic degradation of 100 parts of cellulose. In nature, of course, nutrient elements are continuously recycled as the microorganisms themselves are decomposed; therefore, far more of the polysaccharide is degraded than the available nitrogen supply could account for.

There is no correlation between cellulose mineralization and the level of available phosphorus in many soils nor is there a stimulation if phosphorus is added to various soils low in available phosphate (16). Apparently, the supply of this mineral is usually adequate for microbiological cellulose digestion. On the other hand, a deficiency of phosphorus in some localities may be responsible for low microbial activity in the breakdown of the polysaccharide (11).

The biological utilization of cellulose can proceed from temperatures near freezing to about 65°C. Each of the variety of cellulolytic organisms is affected differently by temperature. Mesophiles dominate at moderate temperatures while a thermophilic microflora adapted to hotter localities can bring about a rapid cellulose dissimilation above 45°C. In addition to temperature-induced changes in the composition of the flora, warming increases the velocity of substrate turnover because of the direct effect of temperature on enzyme action. Season of year has a marked influence on rate of degradation (Figure 10.2), probably a result in large part of the temperature and moisture changes with time.

Aeration also governs the composition of the active flora, aerobes dominating oxygenated environments and anaerobic bacteria being favored by decreasing partial pressures of O_2. Because of the energetics of anaerobic processes, the rate of cellulose metabolism in environments deficient in O_2 is significantly reduced by comparison with aerated habitats. Oxygen disappears at high soil

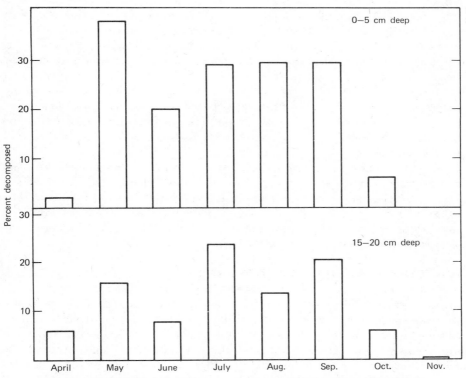

Figure 10.2. Effect of season on rate of cellulose decomposition in soil (18).

moisture levels so that poor drainage is associated with proliferation of the anaerobic cellulolytic bacteria while the numbers of fungi and actinomycetes utilizing cellulose decline. At moderate moisture levels, conditions are conducive to growth of the cellulolytic fungi and aerobic bacteria although certain strains tolerate suboptimal moisture.

In environments of neutral to alkaline pH, many microorganisms are capable of growing and liberating the appropriate enzymes for the hydrolysis of the polysaccharide; under acid conditions, the disappearance of cellulose is mediated largely by filamentous fungi. Although the process is rapid below pH 5.0 and occasionally below 4.0, soils with lower hydrogen ion concentrations degrade cellulose more readily (17). Coincident with man-made changes in pH by liming, there is a shift in the composition of the active flora, one that will be discussed below.

Many microorganisms grow poorly in media containing purified cellulose as the sole source of carbon, yet, on sterile plant material, the same organisms vigorously utilize the polysaccharide. Addition of readily metabolizable organic substances to soil likewise accelerates cellulose decomposition. Since cellulose is only poorly available at best, populations develop slowly; hence, the disappearance of purified cellulose is not rapid. If a population can develop to large size at the expense of some more available carbonaceous nutrient, the flora may adapt to the cellulose once the supply of the second nutrient becomes limiting, the net effect being an increase in cellulose hydrolysis. The phenomenon of greater decay in the presence of other carbohydrates occurs in natural materials, and it should not be confused with the decreased cellulose breakdown noted in pure culture when simple compounds are present. The sparing action in the latter instance is the result of a preferential digestion of the more readily utilized substrate.

Lignin is found in the cell wall in close proximity to cellulose, and this plant constituent apparently slows the rate of cellulose destruction. Lignin is itself not toxic because its addition to cellulolytic cultures results in no inhibition. The influence of lignin in reducing the susceptibility of cellulose to decomposition is probably a physical effect resulting from the close structural interlinkage between cellulose and lignin in the cell wall.

AEROBIC MESOPHILIC MICROFLORA

Cellulolytic microorganisms are common in field and forest soils, in manure, and on decaying plant tissues. The physiological heterogeneity of the responsible microflora permits the transformation to take place in habitats with or without O_2, at acid or at alkaline pH, low or high moisture levels, and from temperatures just above freezing to the thermophilic range. Among the cellulose-utilizing species are aerobic and anaerobic mesophilic bacteria, filamentous fungi, basidiomycetes, thermophilic bacteria, and actinomycetes. Although many of these organisms have been studied only in pure culture, the action in nature is clearly the result of a complex community. At best, it is difficult to compare pure cultures with the many populations active in vivo since, in the latter circumstance, there is an intense microbiological competition for nutrients and sequential changes in the composition of the microflora with time.

A diverse group of fungi utilizes cellulose for its carbon and energy sources (Table 10.1). Following treatment of soil with cellulose, there is a significant increase in the numbers of fungi, particularly if the nitrogen supply is adequate. Plate counts of filamentous fungi in excess of 10^6 per gram of soil during the decomposition of straw plus $NaNO_3$ are not uncommon. Strongly cellulolytic fungi are represented by species of the genera *Aspergillus, Chaetomium, Curvularia, Fusarium, Memnoniella, Phoma, Thielavia,* and *Trichoderma*. It has been proposed that fungi are the main agents of cellulose degradation in humid soils while bacteria are of greater significance in semiarid localities. In the destruction of forest litter, wood, and woody tissues, cellulolytic basidiomycetes are especially prominent, but the basidiomycetes have received scant attention because they thrive poorly on conventional media. Indeed, many fungi seem able to decompose cellulose. This is in great contrast to the bacteria, a group in which possession of the requisite enzymes is a comparative rarity.

The numbers of aerobic, mesophilic bacteria metabolizing cellulose vary enormously from location to location, sometimes being less than 100 and sometimes more than 10 million per gram. The abundance is far greater in

TABLE 10.1

Some Microbial Genera Capable of Utilizing Cellulose

Fungi	Bacteria	Actinomycetes
Alternaria	*Bacillus*	*Micromonospora*
Aspergillus	*Cellulomonas*	*Nocardia*
Chaetomium	*Clostridium*	*Streptomyces*
Coprinus	*Corynebacterium*	*Streptosporangium*
Fomes	*Cytophaga*	
Fusarium	*Polyangium*	
Myrothecium	*Pseudomonas*	
Penicillium	*Sporocytophaga*	
Polyporus	*Vibrio*	
Rhizoctonia		
Rhizopus		
Trametes		
Trichoderma		
Trichothecium		
Verticillium		
Zygorhynchus		

manured fields and sometimes in proximity to plant roots. Bacterial genera that contain representatives digesting cellulose are listed in Table 10.1. A novel aerobic, cellulolytic bacterium which appears as a long, flexuous rod with pointed ends is *Cytophaga* (Figure 10.3). The cytophagas are important in the aerobic decomposition of the polysaccharide and are abundant in soils receiving straw or manure. Members of the genus *Sporocytophaga* also utilize cellulose; these differ from *Cytophaga* species by their capacity to form microcysts. In addition to *Cytophaga* and *Sporocytophaga*, other myxobacteria classified as species of *Angiococcus* and *Polyangium* will develop upon cellulose.

Occasional species of *Pseudomonas*, *Vibrio*, and *Bacillus* utilize cellulose, but this physiological attribute is uncommon to most species of these genera. *Bacillus* contains aerobic, spore-forming, gram-positive rods while the other genera include non-spore-forming, gram-negative aerobes. *Cellulomonas*, on the other hand, is a cellulolytic genus made up of short, gram-negative rods commonly producing yellow, water-insoluble pigments; these organisms are straight or somewhat curved, but occasional pleomorphic forms are found. In addition, some cellulolytic bacteria are long, slender rods that exhibit a slight curving, and others are characterized by spindle- or sickle-shaped cells (20).

Actinomycetes that grow on cellulose have received little attention despite their presence during the decay of cellulosic materials. Many *Streptomyces* isolates develop, frequently with conspicuous pigments, on cellulose agar supplemented with inorganic nutrients. Often, a clear zone appears around the colony, the halo indicating that the responsible enzyme is functioning at a distance from the organism producing it. The halo effect is characteristic of extracellular catalysts. In addition to *Streptomyces*, species of *Micromonospora*, *Streptosporangium*, and *Nocardia* are cellulolytic. That the activity is common to this group is evident from a typical study in which 11 to 65 percent of the actinomycetes were found

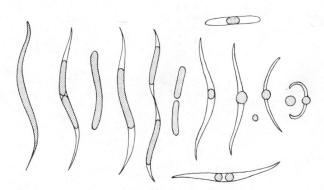

Figure 10.3. Morphological stages in the development of a cytophaga (5).

to have the capacity to attack cellulose (8). Nevertheless, although many actinomycetes have the necessary complement of enzymes, they are much slower in attacking the polysaccharide than most fungi and true bacteria and may not be good competitors for the substrate.

A prominent factor governing the composition of the aerobic, mesophilic flora decomposing cellulose is the pH of the environment. In soils of near neutral reaction, ca. pH 6.5 to 7.0, the active population contains both vibrios and fungi. In soils of slightly greater acidity, about pH 5.7 to 6.2, there are fewer vibrios and more cytophagas. In land more acid than pH 5.5, the flora is dominated by the filamentous fungi. Even when tested in pure culture, the cellulolytic vibrios have their optimum at pH 7.1 to 7.6 and will not grow below pH 6.0 while the cytophagas develop at slightly higher hydrogen ion concentrations (9). Thus, the fungi are active in acid habitats while both the fungi and the bacteria are the causative organisms at reactions greater than pH 6.0. Consequently, the addition of the pure polysaccharide or of cellulosic crop residues to soils of pH greater than ca. 6.0 results in a marked increase in the abundance of fungi and bacteria. In aerated soils of high hydrogen ion concentrations, only the filamentous fungi respond to such treatments (Table 10.2).

Cellulose is on occasion degraded more rapidly in mixed than in pure culture, even when the associated organisms are unable by themselves to attack the polysaccharide. The secondary population probably favors the primary flora

TABLE 10.2

Effect of Cellulose and Nitrogen on the Microbial Community of Soil (19)

| | | No./g of Soil $\times 10^3$ | | |
| | | | | |
Soil	Treatment	Fungi	Bacteria	Actino-mycetes
Unlimed, pH 5.1	None	116	3,900	1,300
	N	116	3,900	1,300
	Cellulose	160	3,600	600
	Cellulose + N	4,800	2,500	400
Limed, pH 6.5	None	25	7,700	2,800
	N	25	7,700	2,800
	Cellulose	47	18,000	2,200
	Cellulose + N	290	47,000	3,200

Cellulose was added at a rate of 1 percent and nitrogen as the nitrate salt (0.1 percent).

Incubation period of 17 days.

by removing the breakdown products and thereby preventing the metabolic wastes from causing inhibitions.

ANAEROBIC MESOPHILIC MICROFLORA

Several microorganisms are capable of decomposing cellulose in the total absence of molecular oxygen, and the polysaccharide disappears under anaerobiosis whether supplied as the purified chemical or in the form of plant materials. The production of large quantities of ethanol and organic acids such as acetic, formic, lactic, and butyric is typical of the anaerobic cleavage of the cellulose molecule. When a soil becomes anaerobic, the decomposition proceeds through the action of bacteria that do not require O_2. Fungi or actinomycetes are not significant in anaerobic environments. Differing from the transformation in air, the anaerobic conversion is not detectably affected by added inorganic nitrogen. Since anaerobic decomposition supplies little energy, the bacteria must degrade large quantities of the substrate in order to assimilate a small amount of carbon. Consequently, there is a proportionally small demand for nitrogen for assimilation into microbial cells, less than the amount usually present in plant residues.

The isolation and maintenance of pure cultures of cellulolytic anaerobes is difficult, and many early investigators undoubtedly never had pure cultures. At present, several types of anaerobic cellulose decomposers are known: spore-forming mesophiles, spore-forming thermophiles, non-spore-forming rods, cocci, and several actinomycetes and fungi that grow anaerobically. At best, cellulolysis without O_2 is slow, regardless of the group concerned. The anaerobic bacteria are rarely numerous in unamended, well-drained soils although peats, marshes, and manure often support a sizable population. Commonly, from 10^2 to 10^3 anaerobic cellulose-fermenting bacteria are found per gram of nonflooded soil. On the other hand, the presence of a fermentable substrate or the exclusion of air stimulates this flora. The low numbers detected in well-drained soils probably represent the spores of the predominant bacteria. Differing from the aerobic bacteria, these organisms are not too sensitive to acidity, and they have been found in soils of pH 4.3.

The most common anaerobic cellulose fermenters in nature appear to be members of the genus *Clostridium*. These bacteria are found in soil, compost, manure, river mud, and sewage. Many *Clostridium* species are cellulolytic, a capacity not too rare in the genus. To isolate such clostridia, a soil suspension is heated at 80°C for 10 minutes, and dilutions are introduced into cellulose media that are then incubated anaerobically. The method takes advantage of the heat resistance of the spores and the anaerobic and cellulolytic nature of the vegetative cell. The technique is, therefore, a relatively specific means of obtaining isolates of the genus.

Non-spore-forming cellulolytic anaerobes can be demonstrated in soil or in sewage sludge, but they are not abundant. Certain fungi such as *Merulius* and

Fomes have the capacity to develop slowly on cellulose in the absence of O_2, and these may play some role in soil. An occasional actinomycete may grow anaerobically, albeit slowly, in cellulosic media.

THERMOPHILIC DECOMPOSITION

Thermophilic cellulolytic bacteria can be readily obtained from soil and manure. For the demonstration of the presence of thermophilic microorganisms, inocula of soil or manure are placed in a medium containing filter paper as a cellulose source, inorganic salts, and $CaCO_3$, and the enrichment is incubated at 65°C. In the decomposition, the filter paper disintegrates and frequently assumes a brownish-yellow color. Since such thermophilic bacteria do occur in soil, cellulose breakdown will take place at elevated temperatures. Despite the widespread distribution of thermophiles, it is likely that their role in cellulose decomposition in nature is minor. An exception is the compost heap, however, in which thermophiles are active agents in the decay.

Both aerobic and anaerobic microorganisms can function in thermophilic transformations. Because of the high temperatures concerned, the cellulolysis is especially vigorous. A common thermophile is *Clostridium thermocellum*, a spore-forming anaerobe that produces acetic acid, ethanol, CO_2, and H_2 and requires low oxidation-reduction potentials for proliferation. The optimum temperature for cellulose degradation by thermophilic clostridia is from 55 to 65°C with little activity below 50°C and no growth above 68°C. The optimal pH is in the vicinity of neutrality (14).

BIOCHEMISTRY OF CELLULOSE DECOMPOSITION

Aerobic bacteria generally convert cellulose to two major products: CO_2 and cell substance. There are no significant accumulations of carbonaceous intermediates, and the concentration of organic acids rarely reaches an appreciable level. The major products of the fungal and actinomycete decomposition of the polysaccharide are CO_2 and cell carbon, but certain groups probably release small amounts of organic acids. The initial hydrolysis of cellulose is the rate-limiting reaction in the microbiological oxidation of the carbohydrate so that intermediates that normally would appear when aerobes are utilizing readily available sugars never accumulate.

The conversion is entirely different with mesophilic and thermophilic anaerobes. These bacteria are incapable of metabolizing even simple substrates to completion, and a number of organic compounds are released as end products. The main substances that accumulate in the absence of O_2 with these genera are CO_2, H_2, ethanol, and acetic, formic, succinic, butyric, and lactic acids (Table 10.3). Early microbiologists reported the finding of a third gas during the bacterial fermentation, CH_4, but modern investigations of cultures whose purity is beyond doubt have established that none of the cellulolytic anaerobes produces CH_4. In enrichment cultures as in soil, however, the anaerobic dissimilation of carbohydrates is accompanied by the evolution of much CH_4. Methane is not produced by the bacteria utilizing the polysaccharides but by a second bacterial group that metabolizes the organic acids liberated by the primary microflora.

The initial step in cellulose destruction is the enzymatic hydrolysis of the polymer. The enzyme system, which actually involves a group of different enzymes, has been given the name of *cellulase*. Cellulase catalyzes the conversion of insoluble cellulose into simple, water-soluble mono- or disaccharides, a reaction characteristic of the entire cellulolytic flora. The steps subsequent to the initial cellulose hydrolysis vary with the individual organisms concerned, the simple sugars being metabolized to CO_2 by the aerobes and to organic acids and alcohols by the anaerobes.

The microbial cell is impermeable to the cellulose molecule so the organism must excrete extracellular enzymes in order to make the carbon source available. The extracellular catalysts act hydrolytically, converting the insoluble material to soluble sugars that penetrate the cell membrane. Once inside the cell, the simple sugars are oxidized and provide energy for biosynthetic reactions. The zone of clearing in agar media that contain the polysaccharide is a result of the extracellular enzyme acting at a distance from the colony that formed it.

The cellulase enzyme system catalyzes the hydrolysis of an entire class of compounds in which the glucose units are connected from carbon number 1 of one sugar unit to carbon number 4 of the adjacent sugar, the linkage being of the β-type. The linkage type is thus known as β-$(1 \rightarrow 4)$.

structure with β-$(1 \rightarrow 4)$ linkages

Included in the group are cellobiose, cellotriose, cellotetraose, and cellulose, the molecules containing two, three, four, and many glucose units, respectively. The molecules with several glucose building blocks are known as oligomers, and they

TABLE 10.3

Products of the Anaerobic Decomposition of Cellulose

Bacterium	Products
	Mesophile
Clostridium cellobioparum	CO_2, H_2, ethanol, acetic, lactic, and formic acids
	Thermophile
Clostridium thermocellum	CO_2, H_2, ethanol, acetic, lactic, formic, and succinic acids

are important because they are intermediates in the conversion of the polymer (cellulose) to the monomer (glucose).

The catalytic system required for a microorganism to convert cellulose to the simple sugars that penetrate the cell typically includes three types of enzymes: (*a*) an as yet poorly characterized enzyme usually termed C_1; (*b*) β-(1 → 4) glucanase or, as it is sometimes called, C_x; and (*c*) β-glucosidase. The total breakdown of the polymer requires the joint action of these catalysts, but the individual enzyme types sometimes have common substrates.

C_1 acts on undegraded (or native) cellulose, the C_1 enzymes of most organisms studied to date having little or no action on the partially degraded polysaccharide or on the oligomers. This enzyme is found in the true cellulolytic species, although many more species can utilize the partially decomposed molecule because they have C_x. It is common to find that a single population will excrete more than one C_1 enzyme; these represent proteins that have slightly different structures, but all presumably function in the same way.

The β-(1 → 4) glucanases known commonly as C_x enzymes do not hydrolyze native cellulose but instead cleave the partially degraded polymers. This type of enzyme is reasonably widespread among fungi, bacteria, and actinomycetes, and most of these glucanases act on molecules containing many glucose units as well as the oligomers like cellotetraose and occasionally cellotriose. On the other hand, the rate of hydrolysis is slower on the oligomers than on the molecules of higher molecular weights. This hydrolysis, which involves the addition of water to the insoluble substrate, ruptures the linkages between the sugar building blocks in the chain. Soluble products—usually cellobiose and the corresponding oligomers—are ultimately formed, but the precise final products appear to depend on the organism. As with C_1, a single species may excrete several proteins of different structure, all of which function in the same way, however. Two modes of cleavage are known among the glucanases: one in which the bonds between glucose units within the long chain are broken more or less at random, and a second in which the enzyme acts on the linkages only near the end of the chain. The former cleavage yields cellobiose, various oligomers, and sometimes glucose as the polymer is broken at random; this type of catalyst is known as an *endo* enzyme. The latter cleavage yields only the fragment that is broken off sequentially from the end of the large substrate, and this fragment, and hence the product, is usually cellobiose; such a catalyst is known as an *exo* enzyme. Exo and endo actions—that is, acting from the end versus internally between repeating units—are characteristic not only of glucanases but also of enzymes acting on other polysaccharides and even other kinds of polymers, and this designation should not be confused with that applied to intracellular versus extracellular sites at which enzymes function relative to the cell that produced them.

The oligomer or even cellobiose that is finally formed from cellulose must be further decomposed before the cell can use the carbon for energy and synthetic purposes. This last phase in transforming cellulose to glucose is catalyzed by β-glucosidase, an enzyme that hydrolyzes cellobiose, cellotriose, and other low-molecular-weight oligomers to glucose. Cellobiase is an old name for β-glucosidase, but the term is inappropriate because cellobiose is not the only substrate. Typical products generated by a fungal cellulase system in the presence and absence of β-glucosidase are shown in Figure 10.4.

Mixtures containing C_1 and β-$(1 \rightarrow 4)$ glucanases often act more readily on native cellulose than C_1 alone. It is still not clear how one enzyme enhances the reaction catalyzed by the second, and indeed it is not quite evident how C_1 actually functions on native cellulose. Furthermore, some organisms may excrete a single enzyme able to release cellobiose as the sole product in the hydrolysis of the undegraded polymer (7), thus combining the functions of C_1 and the glucanase. Many other lines of evidence confirm that the pattern of cellulose breakdown is not the same in all heterotrophs; for example, some bacteria do not simply cleave cellobiose hydrolytically to yield two glucose molecules but convert it to cellobiose phosphate, the latter being transformed to glucose and glucose phosphate (15).

The cellulase complex is inducible in most microorganisms and is synthesized in the presence of cellulose or carbohydrates that are structurally similar to the polysaccharide or its sugar-containing products. Inducers include the polysaccharide, cellobiose, and a few other simple carbohydrates containing glucose in the molecule. The active organisms have a unique way to regulate how much of the enzyme system they produce, a trait of considerable ecological significance since production of too much of the catalysts would yield considerable amounts of sugars, which in turn would stimulate nearby heterotrophs and permit them to compete with the cellulolytic population. This regulatory mechanism is known as *catabolite repression*, a process in which products of the reaction repress or inhibit the synthesis of additional enzyme molecules so that, should the rate of product formation greatly exceed the rate of its utilization or

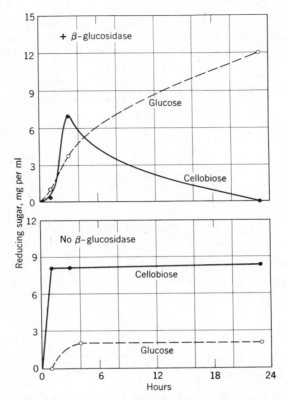

**Figure 10.4. Formation of cellobiose and glucose
by the cellulase system of *Trichoderma viride*
with and without β-glucosidase (12).**

assimilation by the cell, further synthesis of the enzymes is suppressed. Thus, when the concentration of glucose or cellobiose liberated from the polymer increases, the rate of formation of the enzyme declines (2).

Activity of the cellulase system or of individual enzymes in the complex can be measured directly in soil. For this purpose, undegraded or partially degraded cellulose is incubated with soil for a given period, and the quantity of sugar generated is then determined (10). The activity of the enzymes is affected by the presence of roots and the type of plant in the vicinity (3).

An important factor concerned in the biochemistry of cellulose breakdown in soil is the effect of clay minerals. Cellulose and its products can be adsorbed by clay minerals, the polysaccharide derivatives of higher molecular weight generally being adsorbed to a lesser degree than the short-chain compounds. But possibly of greater importance is the partial inactivation of cellulase by certain clays, an effect which has great significance because the enzyme system is extracellular, and therefore it can be altered in its activity by clays. The enzyme inactivation and the substrate adsorption phenomena may account, at least in part, for the protective action of montmorillonite clay on cellulosic materials subjected to microbial decomposition (13). Cellulase is also inhibited by melanin, a constituent of the walls of a number of microorganisms (4), and probably by other complex organic substances or colloids in soil.

REFERENCES

Reviews

Emert, G. H., E. K. Gum, Jr., J. A. Lang, T. H. Liu, and R. D. Brown, Jr. 1974. Cellulases. In J. R. Whitaker, ed., *Food related enzymes*. American Chemical Society, Washington, pp. 79–100.

Hajny, G. J. and E. T. Reese, eds. 1969. *Cellulases and their applications*. American Chemical Society, Washington.

Imshenetsky, A. A. 1967. Decomposition of cellulose in the soil. In T. R. G. Gray and D. Parkinson, eds., *The ecology of soil bacteria*. Liverpool Univ. Press, Liverpool, pp. 256–269.

Whitaker, D. R. 1971. Cellulases. In P. D. Boyer, ed., *The enzymes*. Academic Press, New York, vol. 5, pp. 273–290.

Literature Cited

1. Amberger, A. 1971. *Proc. Intl. Symp. Soil Fertility Evaluation*, 1:773–780.
2. Berg, B., B. Van Hofsten, and G. Pettersson. 1972. *J. Appl. Bacteriol.*, 35:201–214.
3. Bernhard, K. 1965. *Zent. Bakteriol.*, II, 119:566–569.
4. Bull, A. T. 1970. *Enzymologia*, 39:333–347.
5. Gray, P. H. H. 1957. *Can. J. Microbiol.*, 3:897–903.
6. Gupta, U. C. and F. J. Sowden. 1964. *Soil Sci.*, 97:328–333.
7. Halliwell, G., and M. Griffin. 1973. *Biochem. J.*, 135:587–594.
8. Ishizawa, S., M. Araragi, and T. Suzuki. 1969. *Soil Sci. Plant Nutr.*, 15:104–112.
9. Jensen, H. L. 1931. *J. Agr. Sci.*, 21:81–100.
10. Kong, K. T., J. Balandreau, and Y. Dommergues. 1971. *Soil Biol.*, No. 13, pp. 26–27.
11. Kong, K. T. and Y. Dommergues. 1970. *Rev. D'Ecol. Biol. Sol*, 7:441–456.
12. Levinson, H. S., G. R. Mandels, and E. T. Reese. 1951. *Arch. Biochem. Biophys.*, 31:351–365.
13. Lynch, D. L. and L. J. Cotnoir. 1956. *Soil Sci. Soc. Amer. Proc.*, 20:367–370.
14. McBee, R. H. 1950. *Bacteriol. Rev.*, 14:51–63.
15. Palmer, R. E. and R. L. Anderson. 1972. *J. Biol. Chem.*, 247:3415–3419.
16. Ruschmeyer, O. R. and E. L. Schmidt. 1958. *Appl. Microbiol.*, 6:115–120.
17. Schmidt, E. L. and O. R. Ruschmeyer. 1958. *Appl. Microbiol.*, 6:108–114.
18. Schnetter, M.-L. 1971. *Z. Pflanzenernaehr. Bodenk.*, 130:1–8.
19. Waksman, S. A. and O. Heukelekian. 1924. *Soil Sci.*, 17:275–291.
20. Winogradsky, S. 1929. *Ann. Inst. Pasteur*, 43:549–633.

11
Microbiology of the Hemicelluloses

Polysaccharides known as hemicelluloses are one of the major plant constituents added to soil, second only in quantity to cellulose, and they consequently represent a significant source of energy and nutrients to the microflora. Furthermore, since the hemicelluloses make up a large portion of plant tissue, the rate of decomposition of structurally associated organic materials will be greatly affected by the disappearance of the hemicelluloses. Because of their abundance and susceptibility to microbiological degradation, they are thus important in the dry weight loss of crop residues.

Most of the hemicelluloses are found in close physical proximity to cellulose in the primary and secondary cell walls of higher plants, but although these polysaccharides are thus associated structurally with cellulose in the plant, the term *hemicellulose* is an unfortunate choice because these molecules bear no structural relationship to cellulose. As a class, the polysaccharides present in the cell walls of higher plants, except for cellulose and pectin, are designated as hemicelluloses. In these polymers, the simple sugars (or monosaccharides)—or uronic acids, which are derivatives of the simple sugars—are bound together into the large molecule. The complete chemical hydrolysis of the polymer gives the monosaccharides or uronic acids as products.

CHEMISTRY

Hemicelluloses have an *an* at the end of their names, and all are considered to be *glycans*, the prefix *glyc* referring to an unspecified simple sugar. This class of polymers is divided into two categories.

(*a*) Homoglycans. These contain only a single monosaccharide type, but they usually are not the major hemicelluloses in a plant. Typical homoglycans are xylan, mannan, or galactan, which are polymers containing xylose, mannose, or galactose units.

(*b*) Heteroglycans. These frequently abundant polysaccharides contain more than one kind of monosaccharide or uronic acid, and 2 to 4 or occasionally 5 or 6 different sugars coexist in the molecule.

The heteroglycans are named on the basis of the sugars or uronic acids in the polymer, with the most abundant sugar being the last one in the name; thus, plants contain glucomannans, arabinoxylans, arabinogalactans, or arabinoglucuronoxylans in their cell walls.

The structure of such polymers is usually complex. Some may have 50 to 200 sugar units made up from the several sugars characteristic of these wall constituents. A polysaccharide may exist as a linear chain as with cellulose, but the hemicelluloses are usually branched. Some branched structures are depicted in Figure 11.1. The arrangement of different units in the heteroglycan is not chaotic, even in the branched molecules, and distinct and ordered arrangements are typical. Furthermore, only a few sugars and uronic acids are common. These are:

(*a*) Pentoses (five-carbon sugars): xylose and arabinose.
(*b*) Hexoses (six-carbon sugars): mannose, glucose, and galactose.
(*c*) Uronic acids: glucuronic and galacturonic acids.

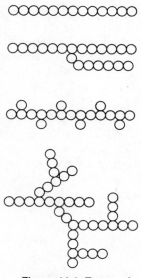

Figure 11.1. Types of branched and linear structures that occur in polysaccharides (5). Each circle represents a simple sugar or uronic acid unit.

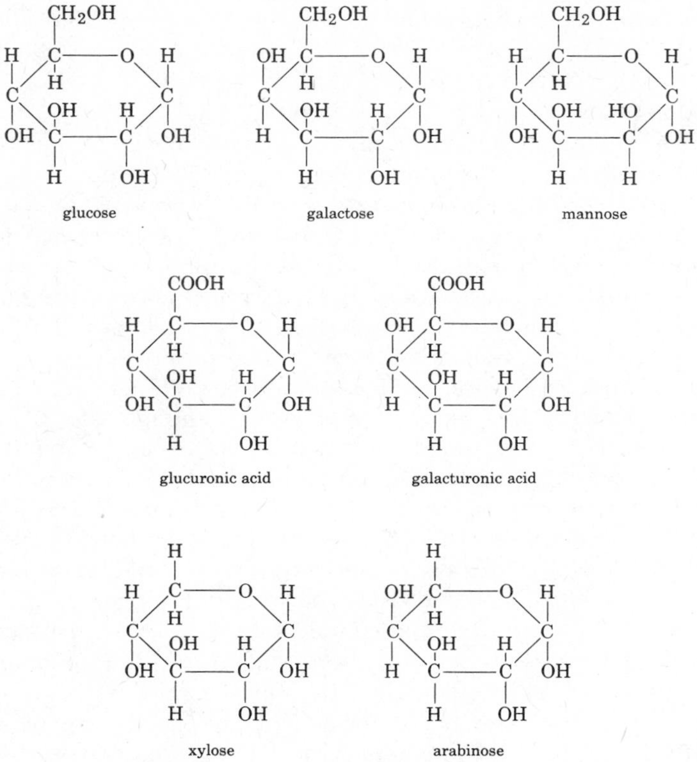

Figure 11.2. Structure of constituents of hemicelluloses.

These monomers are shown in Figure 11.2. The proportion of uronic acids in such molecules is generally small.

The structure of only some hemicelluloses has been determined adequately. Much attention has been given to the xylans, particularly because they may account for 7 to 30 percent of the plant weight. Polysaccharides with only xylose are rare, and most xylans have side chains containing arabinose, glucuronic acid, or other sugars, although the abundance of these monomers is generally small. In the xylans, the xylose units constitute the backbone of the linear chain of the molecule, and the sugars are linked from C-1 of one unit to C-4 of the next pentose. Some plants contain galactans with essentially only galactose units in the structure, and mannans in which mannose is essentially the only sugar are also present in some species. Most of the hemicellulose fraction of certain woody tissues, by contrast, is a glucomannan, with the mannose:glucose ratio being in the vicinity of 2:1. Also present in certain tissues are arabans, galactomannans, arabinoxylans, and arabinogalactans, to mention only a few of the hemicelluloses with less than four different monomer types.

DECOMPOSITION

When a plant residue is added to soil, its hemicellulose fraction disappears initially at a rapid rate, but the subsequent degradation appears to be more slow. The change in decomposition rate is probably a result of the chemical heterogeneity of the hemicellulose fraction, some portions decaying slowly, others rapidly. The effect may also be attributed in part to the presence within the microorganisms of polysaccharides that are formed in soil during the period of decay. These polysaccharides, when formed in soil, can not be readily distinguished from the remaining hemicelluloses derived from the plant remains. Such a microbiological synthesis would be reflected in an apparently slow disappearance of the total hemicelluloses although the plant-derived carbohydrates may be transformed quickly. Polysaccharides are abundant in the cells and excretions of the microbial community, and thus a significant but as yet unknown portion of the material in soil that might otherwise be designated as hemicellulosic material is in fact derived from the synthetic activities of the microflora. Methods for the extraction, purification, and determination of the structure of these soil polysaccharides have been developed (10).

As hemicelluloses are decomposed, the carbon is converted to CO_2 and microbial cells. To overcome some of the problems in differentiating between the disappearance of plant hemicelluloses and the appearance of microbial polysaccharides, analyses can be performed of the metabolism of individual sugars in the plant polymer. For example, the xylose and arabinose components of the rye straw hemicellulose are readily destroyed in soil, and only 5 percent of the xylose constituent remains after 56 days. At the same time, microbial polysaccharides containing glucose and mannose are synthesized as the micro-flora assimilates some of the substrate carbon and makes new cells (2). The microbial polysaccharides themselves are also subject to attack, and they are converted to other cell constituents as well as to complex aromatic components of humus (18).

The metabolism of hemicelluloses is governed by the physical and chemical characteristics of the habitat, and pH and temperature affect this process in a manner similar to their effects on the decomposition of plant residues. The enhancement of degradation by increasing temperature is illustrated in Figure 11.3. Persistence is greater in the absence than in the presence of O_2, and the disappearance is most marked when the environment is aerobic, less in waterlogged habitats, and least under conditions of complete anaerobiosis. Important also is the availability of inorganic nutrients, especially nitrogen, for which there is a great demand by the microflora. Similarly, age of plants has a distinct bearing on the decay, and the hemicelluloses of more mature plants are degraded more slowly than those in younger tissue. In part, the effect of crop

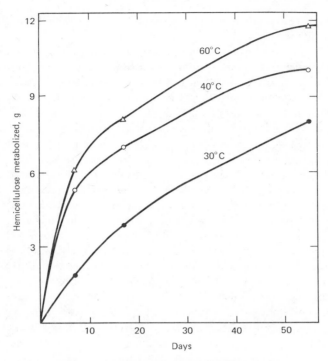

Figure 11.3. Influence of temperature on the decomposition of the hemicellulose fraction of rice straw (1).

maturity may result from an alteration in the structure of the polysaccharide, but it may at the same time reflect a change with age in the physical or chemical relationships among the various carbonaceous constituents.

MICROORGANISMS
Many aerobic and anaerobic microorganisms of the soil utilize hemicelluloses for growth and cell synthesis, and more microbial species are active in destroying hemicelluloses than cellulose. The responsible flora contains a broad cross section of taxonomic, morphological, and physiological groups, and the organisms do not exhibit the substrate specificity that is associated with a portion of the cellulolytic microflora. Thus, in addition to polysaccharides, these microorganisms use organic acids and many simple sugars. The populations concerned in the utilization and destruction of the hemicelluloses of different plants may be expected to differ from one another because of the great chemical dissimilarities of the various hemicelluloses and because the associated

components of the plant tissues undoubtedly have a modifying effect on the composition of the community.

Various fungi, bacteria, and actinomycetes in pure culture can decompose hemicelluloses, frequently using them as the sole sources of carbon and energy. The capacity to degrade these carbohydrates is present in the major fungal groups—Hyphomycetes, Zygomycetes, Pyrenomycetes, and Hymenomycetes. Many fungal genera have the capacity to use one or several of these polysaccharides, and the rate of utilization varies from slow to extremely rapid. Although not as extensively investigated, species of the bacterial genera *Bacillus, Cytophaga, Erwinia, Pseudomonas, Sporocytophaga, Xanthomonas*, and undoubtedly others are also able to utilize hemicelluloses as carbon sources. Species of *Streptomyces* and undoubtedly other actinomycete genera have similar physiological capabilities. Representative genera containing species attacking one or more of the purified hemicelluloses are listed in Table 11.1. In addition, utilization of such polysaccharides can be shown when sterilized plant tissues are inoculated with a variety of heterotrophs (23). The list of organisms able to grow on hemicelluloses is far from complete, but it is clear even now that this activity is not a rare attribute. The activity of some organisms is initially low in culture and the organisms develop slowly in media containing hemicelluloses, but rapid growth can sometimes be achieved by growing the organisms in media containing a sugar and the polysaccharide until acclimation to rapid degradation occurs (6).

TABLE 11.1

Microorganisms Utilizing Hemicelluloses

Organism	Substrate	Reference
Bacteria		
Bacillus	Mannan, galactomannan, xylan	4, 13, 19
Cytophaga	Galactan	21
Erwinia	Xylan	17
Pseudomonas	Xylan	17
Streptomyces	Mannan, xylan	14, 15
Fungi		
Alternaria	Arabinoxylan, xylan	11, 14
Aspergillus	Araban, arabinoxylan, mannan	11, 12, 19
Chaetomium	Arabinoxylan	11
Fusarium	Araban, arabinoxylan	11, 12
Glomerella	Xylan	14
Penicillium	Araban, mannan	12, 19
Trichoderma	Araban, arabinoxylan	11, 12

SPECIFIC HEMICELLULOSES

One of the hemicellulosic-type compounds that has received considerable attention is xylan. The breakdown of xylose-containing carbohydrates is important in the disappearance of many plant residues from the soil because these polysaccharides make up a large proportion of the total carbohydrate content of grasses and woody plants. Soil contains a large number of fungi, bacteria, and actinomycetes capable of xylan degradation, and pure culture representatives of this flora have been shown to utilize the purified carbohydrate as well as the xylan in natural products. Dominant during the preliminary stages of the decomposition in acid soils are the filamentous fungi whereas the initial population in neutral to alkaline environments and in the rotting of straw and manure appears to consist of strains of *Bacillus*. At temperatures of 60 to 65°C, the rapid xylan metabolism results from the activities of thermophilic, aerobic, spore-forming bacilli.

A method for estimating xylan-hydrolyzing activity entails the incubation of a soil-buffer mixture with purified xylan. Toluene is added to prevent appreciable microbial proliferation, and the reaction is studied by measuring the formation of reducing sugars. Determined in this manner, the xylanase content of soil varies with the crop and the supply of organic materials (Figure 11.4); the enzyme is especially concentrated in land treated with barnyard manure. The presence of a xylan-containing crop such as wheat results in an enhanced xylanase activity in soil when compared with locations supporting a crop containing little or no xylan (20).

Mannan, the polysaccharide composed of mannose in polymeric combination, is also rapidly metabolized. This fact can be confirmed experimentally by inoculating a sample of soil into an inorganic salts medium containing mannan as sole carbon source and observing the rate of microbiological growth proceeding at the expense of the carbohydrate. Among the organisms shown to utilize mannan in plant tissue or purified mannan polysaccharides are bacteria of the genera *Bacillus* and *Vibrio*, a number of actinomycetes, and fungi placed in the genera *Aspergillus*, *Chaetomium*, *Penicillium*, *Rhizopus*, *Trichoderma*, and *Zygorhynchus*.

Galactans are suitable carbon sources for a number of basidiomycetes, aerobic and anaerobic bacteria, actinomycetes, and the fungi *Aspergillus*, *Cunninghamella*, *Humicola*, *Penicillium*, *Rhizopus*, *Trichoderma*, and *Zygorhynchus*. Both fungi and actinomycetes use purified galactan in media where it is the sole organic nutrient, and the same organisms similarly cause the breakdown of galactan in Irish moss when the plant material is added to inorganic nutrient solutions. Whether the organism be a fungus, actinomycete, or bacterium, galactans frequently tend to be less readily dissimilated than mannans and xylans (Table 11.2), but such observations on relative rates are based largely on pure culture studies. It has been demonstrated, however, that the release of

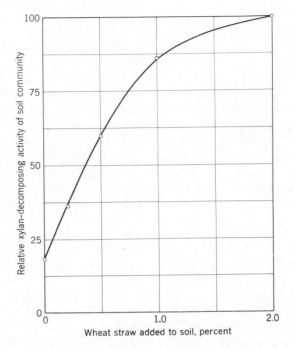

Figure 11.4. Effect of the addition of straw on the
xylanase activity of soil (20).

TABLE 11.2

Microbiological Decomposition of Polysaccharides in Liquid Media (24)

	% of Substrate Decomposed		
Microorganism	Mannan	Xylan	Galactan
Aspergillus sp.	94	55	37
Penicillium sp.	98	55	34
Rhizopus sp.	96	19	34
Streptomyces 26	47	38	29
Streptomyces 40	93	35	29
Streptomyces 50	49	18	6.8
Trichoderma sp.	29	56	37

Incubation period of 42 days.

CO_2 from carbohydrate-amended soils is most rapid with starch and slowest with galactan while mannan and xylan occupy intermediary positions (7).

BIOCHEMISTRY OF HEMICELLULOSE DECOMPOSITION

Because the hemicelluloses have high molecular weights and fail to pass through the microbial cell membrane, they must be converted into simpler compounds prior to utilization as carbon sources. The active flora must therefore first hydrolyze the polysaccharide by means of extracellular enzymes to shorter carbohydrate fragments that the cell can assimilate. The sole peculiarity of hemicellulolytic populations is thus their ability to catalyze the initial hydrolysis. Following the preliminary breakdown, however, secondary populations develop upon the metabolic products. The latter group of organisms responds to the compounds liberated from the long polymers by the hemicellulolytic flora.

Many different enzymes are concerned in hemicellulose breakdown, and the various polysaccharides included in this class of plant constituents require special groups of enzymes for their utilization. As a rule, three types of catalysts may be involved: (*a*) endo enzymes, which randomly cleave the bonds between building blocks in the polymer; (*b*) exo enzymes, which cleave either a single dimer or monomer from the end of the polysaccharide chain; and (*c*) enzymes known collectively as *glycosidases*. The glycosidase hydrolyzes the oligomers or the disaccharides produced from the hemicellulose by the polysaccharide-cleaving enzymes and produces the simple sugar or uronic acid; the latter are then metabolized within the cell to yield energy. The original polysaccharide is a large molecule so that a given quantity of hemicellulose has few ends accessible for attack; hence, endo enzymes are chiefly responsible for the initial breakdown. However, once the ends become numerous through the action of the endo enzymes, which often do not act on di- or trisaccharides or act on them slowly, then exo enzymes gain importance. Since the complex polymers are often branched, the degradation is frequently hindered owing to the inability of many of the hemicellulases to cleave the branch so that the low molecular weight products still are somewhat complex; in this event, the branch in the product is generally removed by another enzyme. Alternatively, the branch may be removed enzymatically before the endo or exo enzymes begin to function. As in the case of cellulose, the microbial formation of such catalysts is controlled by catabolite repression so that if the rate of extracellular hydrolysis of the polymer exceeds the rate that the products are assimilated by the cell, the simple sugars that accumulate will repress the synthesis of additional enzyme.

Xylanases of many heterotrophs have been characterized. The products of their action depend on the particular type of xylanase, but characteristically the endo enzymes give rise to xylobiose (the disaccharide) and higher oligosaccharides (13), and the exo enzymes usually yield xylose. With time the oligosaccharides are converted to xylose, which the organism then metabolizes intracellularly. The small quantities of sugars other than xylose in xylans are ultimately released and assimilated (17).

Mannanases are produced extracellularly by fungi and bacteria. The enzyme in some organisms is constitutive, but in other species it is inducible. Mannans of several different categories are known, and a mannanase hydrolyzing one mannan is often inactive on another (22). On the other hand, the enzyme is not restricted to polysaccharides containing solely mannose units, for galactomannan and glucomannan—in which the major building block still is mannose—are also attacked (8). In the absence of a glycosidase, often the dimer and oligomers accumulate so that the complete reaction requires the depolymerizing and the oligosaccharide-cleaving catalysts.

Galactans may have β-(1 → 4) or β-(1 → 3) linkages between the galactose units, and rupture of these linkages requires different enzymes, both of which are termed galactanases. The enzymes produce galactose, galactobiose, and/or galactotriose, the product depending on how the substrate is cleaved (9). Arabanases of *Aspergillus, Sclerotium*, and *Clostridium* have also been investigated, and these enzyme preparations ultimately give rise to arabinose.

As with the polysaccharide-cleaving enzymes, the glycosidases are specific for their substrates and are named for them. Thus, xylosidase acts on xylose oligomers to produce xylose, and the mannosidase generates mannose from the oligosaccharides containing this sugar. Following this, the last step in making the large molecule into water-soluble sugars, the cell incorporates the monomer and uses it for biosynthetic purposes.

Although the enzymes are specific for their substrates, a single organism may nevertheless make use of many hemicelluloses. This is accomplished by the excretion of an array of enzymes. For example, *Fusarium oxysporum* growing on components of tomato tissues synthesizes arabanase, xylanase, galactanase, and glycosidases (3).

Since polysaccharide-splitting enzymes are extracellular, their activity in soil is affected by the diverse number of inanimate materials present. Such external influences are in addition to factors affecting growth of the responsible organisms and factors governing availability of the substrate. Therefore, it is interesting to note that a hemicellulose-cleaving enzyme has a reduced activity in the presence of montmorillonite clay (16). This may result from adsorption of the enzyme, the substrate, or both. Such inhibitory effects are of profound significance in altering the transformations catalyzed by the many extracellular enzymes required for the depolymerization of polysaccharides found in natural materials.

REFERENCES

Reviews

Aspinall, G. O. 1973. Carbohydrate polymers of plant cell walls. In F. Loewus, ed., *Biogenesis of plant cell wall polysaccharides*. Academic Press, New York, pp. 95–115.

Reese, E. T. 1968. Microbial transformation of soil polysaccharides. In *Matiere organique et fertilite du sol*. Scripta Varia, Vatican City, pp. 535–582.

Towle, G. A. and R. L. Whistler. 1973. Hemicelluloses and gums. In L. P. Miller, ed., *Phytochemistry*. Van Nostrand Reinhold, New York, vol. 1, pp. 198–248.

Whistler, R. L. and E. L. Richards. 1970. Hemicelluloses. In W. Pigman and D. Horton, eds., *The carbohydrates*. Academic Press, New York, vol. 2A, pp. 447–469.

Literature Cited

1. Abd-el-Malek, Y. and M. Monib. 1969. *J. Soil Sci., Un. Arab Repub.,* 9:13–24.
2. Cheshire, M. V., C. M. Mundie, and H. Shepherd. 1974. *J. Soil Sci.,* 25:90–98.
3. Cooper, R. M. and R. K. S. Wood. 1975. *Physiol. Plant Pathol.,* 5:135–156.
4. Courtois, J. E. and P. LeDizet. 1970. *Bull. Soc. Chim. Biol.,* 52:15–22.
5. Danishevsky, I., R. L. Whistler, and F. A. Bettelheim. 1970. In W. Pigman and D. Horton, eds., *The carbohydrates*. Academic Press, New York, vol. 2A, pp. 375–412.
6. Dekker, R. F. H. and G. N. Richards. 1974. *Carbohydr. Res.,* 38:257–265.
7. Diehm, R. A. 1930. *Proc. Comm. III, 2nd Intl. Cong. Soil Sci.,* Leningrad, pp. 151–157.
8. Emi, S., J. Fukumoto, and T. Yamamoto. 1972. *Agr. Biol. Chem.,* 36:991–1001.
9. Emi, S. and T. Yamamoto. 1972. *Agr. Biol. Chem.,* 36:1945–1954.
10. Finch, P., M. H. B. Hayes, and M. Stacey. 1971. In A. D. McLaren and J. Skujins, eds., *Soil biochemistry*. Marcel Dekker, New York, vol. 2, pp. 257–319.
11. Flannigan, B. 1970. *Trans. Brit. Mycol. Soc.,* 55:277–281.
12. Fuchs, A., J. A. Jobsen, and W. M. Wouts. 1965. *Nature,* 206:714–715.
13. Horikoshi, K. and Y. Atsukawa. 1973. *Agr. Biol. Chem.,* 37:2097–2103.
14. Iizuka, H. and T. Kawaminami. 1969. *Agr. Biol. Chem.,* 33:1257–1263.
15. Lemos-Pastrana, A. and J. Ortigoza-Ferado. 1971. *Rev. Latinoamer. Microbiol.,* 13:291–295.
16. Lynch, D. L. and L. J. Cotnoir. 1956. *Soil Sci. Soc. Amer. Proc.,* 20:367–370.
17. Maino, A. L., M. N. Schroth, and N. J. Palleroni. 1974. *Phytopathology,* 64:881–885.
18. Martin, J. P., K. Haider, W. J. Farmer, and E. Fustec-Mathon. 1974. *Soil Biol. Biochem.,* 6:221–230.
19. Reese, E. T. and Y. Shibata. 1965. *Can J. Microbiol.,* 11:167–183.
20. Sorensen, H. 1955. *Nature,* 176:74.
21. Turvey, J. R. and J. Christison. 1967. *Biochem. J.,* 105:311–316.
22. Vojtkova-Lepsikova, A., D. Sikl, L. Masler, and S. Bauer. 1970. *Folia Microbiol.,* 15:437–441.
23. Waksman, S. A. 1931. *Arch. Mikrobiol.,* 2:136–154.
24. Waksman, S. A. and R. A. Diehm. 1931. *Soil Sci.,* 32:97–117.

12
Lignin Decomposition

The third most abundant constituent of plant tissues is commonly lignin. It is superseded in relative quantity only by cellulose and the hemicelluloses. Certain plants, the woody species in particular, contribute large amounts of lignin to the material to be degraded through the activities of the soil microflora. In forests alone, vast quantities of lignin are continually deposited on the soil as wood waste, and these must be destroyed either by burning or by biological means.

The large amount of lignin annually entering the soil in plant residues does not accumulate, but rather it slowly and perceptibly disappears. Nevertheless, little is known of the microbiology and decomposition of lignin or of the environmental variables governing its loss. Three factors account for the inadequate state of knowledge: difficulties arising from the chemical complexity of the lignin molecule; difficulties in assaying for this substance; and problems related to the isolation of a purified lignin fraction suitable for use as a microbiological substrate. Much of the early research is of doubtful value because the conclusions were based on unreliable methods for quantitatively estimating lignin in plant residues. In addition, a considerable portion of the early work on lignin breakdown is open to question since the drastic conditions used to isolate and purify the lignin fraction resulted in preparations decidedly different from those in the original tissue. These difficulties have now been largely overcome.

Within the plant, lignin is found in the secondary layers of the cell wall and also to some extent in the middle lamella. The lignin content of young plants is relatively low, but the quantity increases as the plant matures. Lignin probably never occurs free; usually it is combined with the polysaccharides. It is especially plentiful in woody plants, whereas its concentration is quite low in succulent tissue. Young, immature grasses and legumes commonly contain from 3 to 6 percent lignin on a dry weight basis while chemical analyses of wood samples from a variety of trees give figures ranging from about 15 to 35 percent. Lignins or ligninlike molecules are found not only in higher plants but also in certain fungi (5) and algae (14).

CHEMISTRY OF LIGNIN

Lignins found in the many species, genera, and families of the plant kingdom are clearly different chemically from one another, and hence they cannot be considered as substances of uniform structure. Even in a single plant, the chemical composition may change to some extent with the stage of maturity. Prominent among the chemical properties of lignin is its strong resistance to acid hydrolysis, concentrated mineral acids having little effect on the molecule. It is also insoluble in hot water and neutral organic solvents but is solubilized by alkali. Solutions containing lignin give characteristic bands in the ultraviolet region of the light spectrum, with an absorption maximum in the vicinity of 280 nm. The results of ultraviolet absorption studies show that lignin is a modified benzene derivative.

The lignin molecule contains only three elements—carbon, hydrogen, and oxygen—but the structure is aromatic rather than being of the carbohydrate type as typified by cellulose and the hemicelluloses. The molecule is a polymer of aromatic nuclei with either a single repeating unit or several similar substances as the basic building blocks. The basic unit in lignin seems to be a phenyl-propane (C_6-C_3) type of structure, which may exist in three types,

in which (*a*) R and R′ are H, (*b*) R is H and R′ is a methoxyl (-OCH_3), and (*c*) both R and R′ are methoxyl groups. The relative abundance of the three major building blocks varies with the plant species; for example, conifers contain a high percentage of the second kinds, dicotyledonous angiosperms are rich in the second and third kind, whereas grasses show large amounts of the first and second types. These repeating units may be linked together by strong ether bonds (—C—O—C—) or by C—C bonds, and the linkages may be between two benzene rings, two propane (C_3) side chains, or between a ring and an adjacent side chain. The ether linkage involves the O of the OH on the ring and the C of a propane moiety. The C—C and ether bonds between units appear to be distributed at random, and indeed adjacent phenylpropane monomers may be bound together by more than a single linkage. Thus, by contrast with the polysaccharides, the same bonds do not recur at regular intervals in the polymer, and this random structure coupled with the strong linkages between the monomers make lignin notably resistant to both microbial and chemical degradation.

The actual polymer, moreover, is highly branched and particularly complex in structure. Molecular weight determinations often fail to give single values, but figures of from about 10,000 to many fold higher have been noted so that the actual polymer contains a large number of the appropriate monomers.

Several models have been proposed to account for the information available on the structure of the lignin molecule (Figure 12.1). Although the

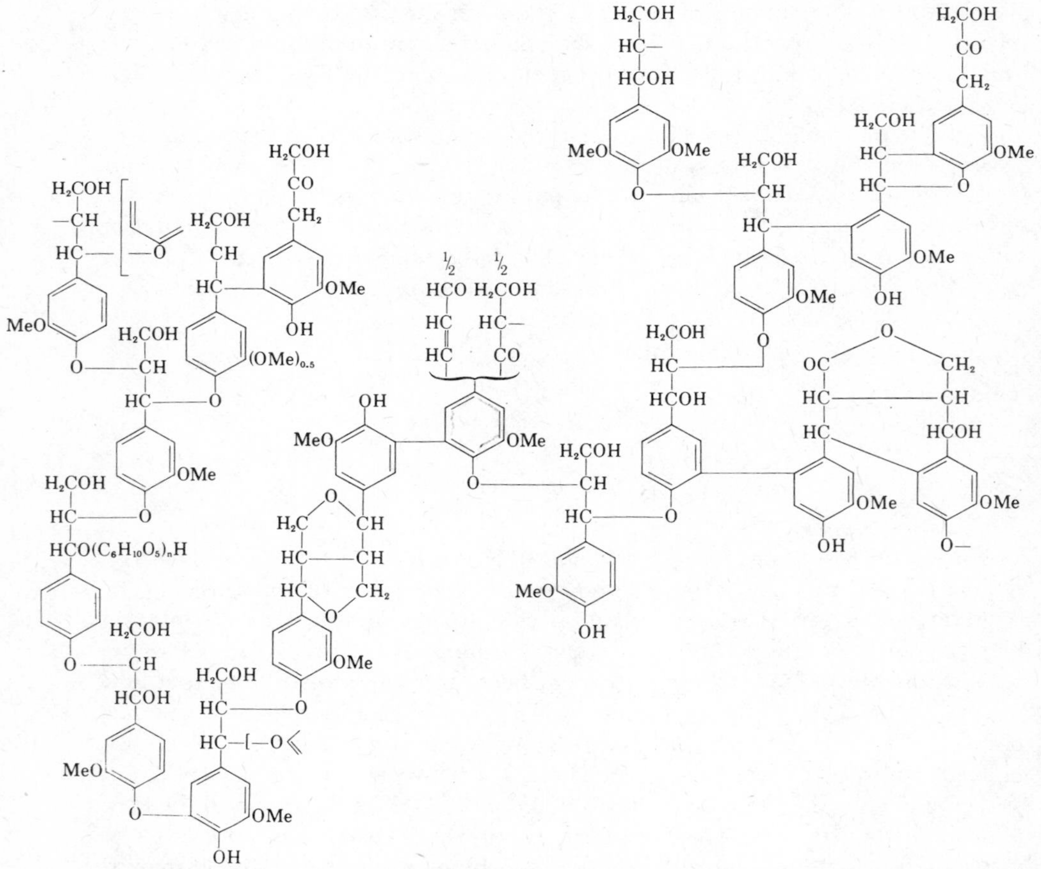

Figure 12.1. Schematic formula showing units linked in a molecule of spruce lignin (10). MeO- refers to CH₃O-. Copyright 1965 by the American Association for the Advancement of Science.

structure shown in Figure 12.1 is probably still only a working model, it does illustrate the complexity of the compound and the basic C_6—C_3 phenyl-propane building block, the presence of an aromatic nucleus, methoxyl groups, and an oxygen bridge.

DECOMPOSITION

The outstanding microbiological characteristic of lignin is its resistance to enzymatic degradation. The decomposition of lignin proceeds either in the presence or in the absence of O_2, but the rate of loss in both circumstances is characteristically less than that observed for cellulose, hemicelluloses, and other carbohydrates. In short-term experiments, little loss is observed, but slowly, in a period of months, the lignin does disappear (Figure 12.2). Despite its resistance, this material quite clearly must be metabolized; otherwise it would have accumulated in vast quantities wherever plant remains are subjected to decay.

During decomposition, the microflora consumes the individual organic components of natural materials at different rates. Lignin is the last to show

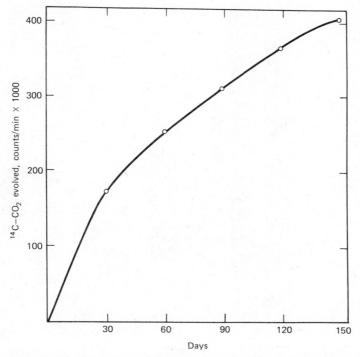

Figure 12.2. Decomposition of ^{14}C-lignin in soil. After 150 days, 21.7% of the lignin-C has been converted to CO_2 (23).

appreciable oxidation. Thus, as the complex organic substrates are metabolized and the water-soluble constituents, cellulose, and hemicelluloses disappear, the lignin content of the decaying residue rises. As a consequence, well-decomposed materials have a high percentage of lignin (37). For example, during the aerobic decomposition of corn stalks, two-thirds of the total dry matter is lost in 6 months but only about one-third of the lignin. Consequently, a residue initially containing 14.8 percent lignin would contain approximately twice that amount one-half year later (Figure 12.3). Such lignin, however, is modified to a certain extent. Among the chemical changes that have occurred in the molecule are a removal of some of the side chains of the aromatic nucleus and a decrease in the number of methoxyl groups. At the same time, the number of phenolic

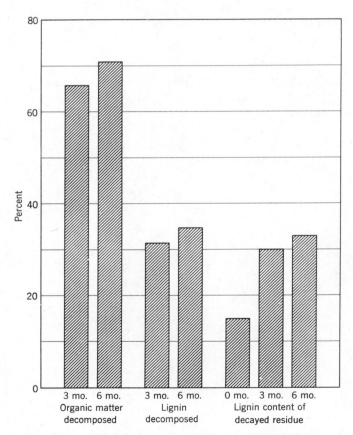

Figure 12.3. Changes in the lignin fraction of corn stalks during aerobic decomposition (3).

hydroxyls and carboxyls increases. Direct examination of tissues undergoing attack confirms that lignified walls and plant constituents are slowly destroyed (2).

The rate and extent of lignin decomposition are affected by temperature, availability of nitrogen, anaerobiosis, and by constituents of the plant residue undergoing decay. Under optimum conditions in the laboratory and when the temperature is maintained in the vicinity of 30°C, it is common to find not more than one-third of the lignin metabolized in a period of six months and only about half to have been lost at the end of a year. Differences do exist in the microbial degradation of lignins found in different plants. For example, the lignins of alfalfa, oats, and corn stalks are more readily decomposed than the same constituents in wheat and rye straw or in the leaves and needles of trees (31, 35). Some of the differences in rate may arise from structural dissimilarities between the lignins, but the association of the aromatic constituent with polysaccharides may alter its susceptibility in a manner analogous to the stimulation of cellulose decomposition by readily oxidizable carbohydrates. Here too, the presence of an accessible source of energy can provide nutrients to support a larger microflora that could bring about greater losses. Age of plant is another important variable. The lignin of young tissue disappears more rapidly than that in mature plants, but the explanation for the differences remains obscure. Part of the influence of maturity on lignin turnover may result from the supply of elements other than carbon in the tissue that is undergoing decay, but physical and chemical changes in the carbonaceous nutrients themselves cannot be discounted.

There has been some controversy in the past regarding losses of lignin under anaerobiosis, but it is now clear that the anaerobic conversion does occur. Nevertheless, the organisms concerned have not yet been identified. Bacteria are probably the responsible agents but, until the microorganisms are isolated and characterized, no unequivocal statement can be made. As with the aerobic transformation, the anaerobic decomposition is slower for lignin than for cellulose and the hemicelluloses. Anaerobic lignoclastic activity is invariably slow, and half of the substrate is still recovered after a year and a half (1, 4).

Temperature has a profound influence on the rate and extent of breakdown. Little loss occurs at 7°C, and progressively higher temperatures favor the active microflora. At 37°C, the oxidation is extensive, and appreciable quantities disappear. Above the mesophilic temperature range, thermophiles participate in lignin destruction, and enrichment cultures of thermophilic bacteria degrade the lignin of finely ground wood in a relatively short period. Unfortunately, the thermophilic organisms involved in the oxidation of lignin also have never been isolated in pure culture.

Several chemical alterations occur during the decomposition of lignin (Figure 12.4). Since methoxyl groups are prominent in the molecule, it is not unexpected that they are metabolized as the lignin is degraded. Because of their

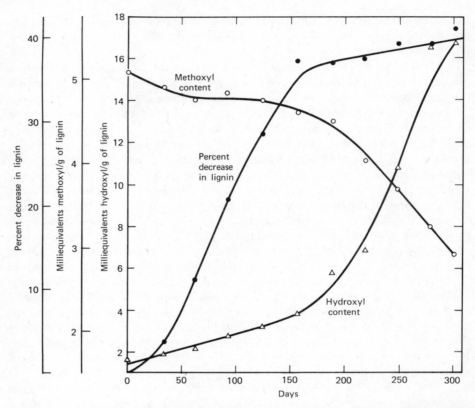

Figure 12.4. Changes in methoxyl and hydroxyl content during the decomposition of lignin (36).

exposed position, moreover, methoxyl groups are particularly prone to enzymatic cleavage, and they are oxidized more readily than the rest of the molecule. The rate of methoxyl decrease, when expressed as a percentage of the lignin remaining, is greatest under anaerobiosis, less in waterlogged circumstances, and least in aerobic environments. This suggests an enhancement of the preferential metabolism of methoxyls in the absence of O_2, that is, a magnification of the differential breakdown between various portions of the molecule (27). In investigations of the rotting of natural organic materials rather than fractions isolated therefrom, it is necessary to differentiate between methoxyl groups derived from lignin and those associated with carbohydrate fractions. A large part of the methoxyl loss may originate in the breakdown of the latter substrates. Paralleling the methoxyl cleavage during decomposition, there is an increase in the hydroxyl group content of the lignin.

The lignin in most natural organic materials protects associated carbonaceous substances from destruction. The protective influence does not appear to arise from any toxicity of the lignin since lignin preparations do not retard microbiological decomposition. To study this phenomenon directly, delignified fractions of plant materials containing varying percentages of lignin and cellulose can be prepared. These may range from preparations composed almost entirely of cellulose to those with none of the lignin removed. If these substances are placed in culture media into which are inoculated pure cultures of various bacteria, the total amount of organic material degraded and the percentage of cellulose oxidized are inversely proportional to the lignin concentration, and the lowest carbon and cellulose losses occur in the preparations with the highest percentage of lignin. In addition, the hemicelluloses in plants having a high lignin content are less susceptible to microbial digestion (1, 11). As a rule, natural products very rich in lignin are resistant to biological decay, woody materials with less lignin are somewhat more susceptible, while tissues low in lignin are most available to microorganisms. The lignin retardation is not physiological, at least not in the sense of a toxicity. Its effect in retarding the microbiological degradation of organic constituents of plant remains probably results from a physical or physicochemical barrier set up by the close interlinkage between lignins and the hemicelluloses and cellulose of the plant cell wall, possibly by means of a lignin encrustation which mechanically separates the microorganism from the carbohydrate.

MICROBIOLOGY

The genera active in breakdown of this polyaromatic molecule in soil are still not well defined. The lack of complete understanding results in part from the inadequate methodology and in part from concern almost solely with groups of higher fungi that are active on lignin of woody plants.

Some studies of microbiological lignin metabolism have used as a criterion of utilization the disappearance of methoxyl groups or of sodium lignosulfonate from culture media. However, the decrease in the concentration of substrate in the culture filtrate often can be accounted for by the adsorption of the organic molecule onto the mycelium or cells of the microorganism. There is also evidence that a number of filamentous fungi common in soil can synthesize ligninlike substances (5). The presence in microbial cells of such compounds, whether structurally similar to plant lignins or merely related to them in only certain properties, presents a problem to the microbiologist concerned with the oxidations carried out by a mixed flora.

Many of the basidiomycetes are capable of degrading lignin, but the reaction is invariably slow. Because of the slow growth of the organisms, little is yet known of the cultural or environmental factors affecting the oxidation. The limited information presently available indicates that the microflora functioning in aerobic lignin decomposition at moderate temperatures includes a variety of the higher fungi. In addition to the genera cited in Table 12.1, *Agaricus,*

TABLE 12.1

Decomposition of Lignin from Sweetgum Wood by Several Fungi (19)

Organism	Days of Incubation	Lignin Decomposed, %
Collybia velutipes	88	22.3
Fomes officinalis	79	21.1
Pleurotus ostreatus	57	11.4
Polyporus fumosus	50	13.5
Poria taxicola	37	9.4
Trametes heteromorpha	88	21.7

Armillaria, Clavaria, Clitocybe, Coprinus, Cortinellus, Ganoderma, Lenzites, Marasmius, Mycena, Panus, Pholiota, Polystictus, Schizophyllum, Stereum, and *Ustulina* contain active representatives.

Two methods have been in general use to demonstrate lignoclastic capacities. One entails a determination of the disappearance of lignin when portions of sterile plants are inoculated with pure cultures of suspected organisms. The other technique involves the use of isolated and purified lignin preparations. For example, species of *Coriolus* and *Pleurotus* are able to use purified lignin in culture (32). With certain of these fungi, however, the organisms must be adapted to utilize the aromatic compound by prolonged cultivation in glucose-lignin media. The adaptation is accomplished by making each serial transfer into culture solutions with progressively lower concentrations of glucose (17).

Most fungi that attack lignin also utilize cellulose. Commonly, the latter material is the more available of the two, and growth is sparse in lignin-containing media. In one study, for example, 44 of a total of 46 soil-inhabiting basidiomycetes could degrade both lignin and cellulose. This population included species of *Clitocybe, Collybia, Mycena,* and *Marasmius*. Even in pure cultures of these fungi, periods of six to seven months were required for half the lignin to be metabolized. On the other hand, some higher fungi are somewhat specialized for the consumption of lignin, for they decompose up to twice as much lignin as cellulose. Furthermore, occasional fungi degrade the lignin of vegetable matter but are without effect on the cellulose (22).

Filamentous fungi are characteristically found intimately associated with the lignin in plant tissues undergoing decay, but it is rarely clear whether they are growing at the expense of the lignin or of products liberated by the basidiomycetes. The latter organisms are unquestionably important in attacking the polymer in wood, but their role in soil has yet to be well characterized. Nevertheless, pure cultures of the lower fungi have rarely been shown to metabolize the polyaromatic, although it has been reported that species of *Aspergillus, Fusarium,* and *Penicillium* can use a lignin derived from wheat straw (13).

Aerobic bacteria also are able to bring about some degradation of lignin, but little attention has been given to them. For example, bacteria multiply in soil to which purified lignin is added, and their populations can bring about a slow degradation of the molecule. Among the active organisms are strains of *Arthrobacter, Flavobacterium, Micrococcus, Pseudomonas,* and *Xanthomonas* (6, 18, 29, 30). Their relative importance to the process in nature and whether they can effect extensive depolymerization, however, remain unsure.

BIOCHEMISTRY OF LIGNIN DECOMPOSITION

The biochemistry of lignin degradation, the mechanism of the process, and the enzymes involved have still to be clearly established. From the rate of substrate turnover, it is clear that the products of depolymerization must be formed slowly, and these probably are oxidized almost as rapidly as they appear. Because the molecule is a polymer of aromatic nuclei, moreover, the likeliest intermediates are low molecular weight aromatic substances.

In order to simplify the difficulties inherent in studies of the enzymatic breakdown of the lignin molecule, analogous but far simpler structures have sometimes been used as microbial substrates. It is assumed that the larger molecule is degraded to give products similar to or identical with the simpler structures. In typical studies with these model compounds, *Agrobacterium radiobacter* was found to form vanillin and coniferyl alcohol (34), and a strain of *Pseudomonas* was observed to produce vanillic acid (8). The finding that these metabolites are formed from the simpler substrates takes on added significance in light of the discovery that the same products are generated from the polymer itself.

During the attack on purified lignins supplied to microbial cultures, a number of simple aromatic compounds appear. The precise substances and the amounts that accumulate depend on the organism in question, but the following products—presumably intermediates in the degradation—are frequently recovered from culture filtrates: vanillic, *p*-hydroxybenzoic, *p*-hydroxycinnamic, ferulic, syringic, and 4-hydroxy-3-methoxyphenylpyruvic acids, vanillin, dehydrodivanillin, coniferaldehyde, *p*-hydroxycinnamylaldehyde, syringylaldehyde, guaiacylglycerol, and guaiacylglycerol-β-coniferyl ether (15, 28). All these compounds contain benzene rings, and some have methoxyl (—OCH$_3$), carboxyl (—COOH), aldehyde (—CHO), or hydroxyl (—OH) groups. The existence of such products suggests that, coinciding with the depolymerization, there is a cleavage of the ether bonds and the appearance of hydroxyl groups. Methoxyls may be retained in the monomers, but these may be cleaved either from the monomer or the polymer, probably with the conversion of —OCH$_3$ to —OH. At the same time, the complex, insoluble polymer is converted, presumably by extracellular enzymes, to the water-soluble simple products containing the benzene ring.

On the basis of the products identified to date, a scheme has been postulated for lignin decomposition. The pathway is illustrative and in its outlines probably applies to several lignins, but steps in the decomposition of dissimilar lignins probably differ somewhat. The pathway assumes that attack on the polymer gives rise to guaiacylglycerol-β-coniferyl ether, the ether bond in this molecule is then cleaved to yield guaiacylglycerol and coniferyl alcohol, and then two of the three carbons of the side chain are removed with the generation of vanillin (Figure 12.5). The vanillin is then oxidized to vanillic acid, the ring ultimately being cleaved to yield the simple organic molecules the active organisms assimilate and use as sources of energy.

Considerable controversy still exists on the identities of the enzymes involved, but in no instance is the enzymology well characterized. A single lignin-degrading enzyme, a so-called "lignase," has yet to be described. However, enzymes known as phenol oxidases, laccases, and peroxidases have been postulated frequently as being the responsible catalysts, particularly because they are common among the lignin-utilizing basidiomycetes (33). The phenol oxidases oxidize aromatic compounds containing one (phenols) or two hydroxyls, while the laccases oxidize only those aromatic compounds with more than one hydroxyl group. Peroxidases can oxidize aromatic molecules also, but they do so in the presence of hydrogen peroxide. These enzymes are said to be able to cleave ether linkages, remove side chains, and cause the loss of methoxyl groups. Nevertheless, no direct experimental evidence currently exists for their involvement in lignin degradation, although they may be implicated in the destruction of the simple aromatic products that are toxic and would, if they accumulated, inhibit the active organisms (12). It is also conceivable that heterotrophs may modify slightly the lignin molecule without causing extensive destruction.

The simple aromatic compounds produced from lignin sustain the growth of many fungi and bacteria in culture media. Vanillic acid, vanillin, ferulic acid, and related compounds thus can be used as carbon sources, and these compounds are converted to protocatechuic acid (7, 16, 25). The latter then is the actual compound subject to ring cleavage (Figure 12.5). In soil, too, vanillin is rapidly oxidized, and it is initially converted to vanillic acid and then protocatechuic acid (20). Under certain circumstances, some of these breakdown products (e.g., vanillin and vanillic, ferulic, and protocatechuic acids) may be detectable in soil (21), but their concentration never is great inasmuch as they are rapidly destroyed. Moreover, such products are not necessarily derived from lignin, because these or similar intermediates may be formed in humus breakdown.

Lignin or lignin-derived molecules have long been considered to be of significance in the formation of humus. This view became popular when early

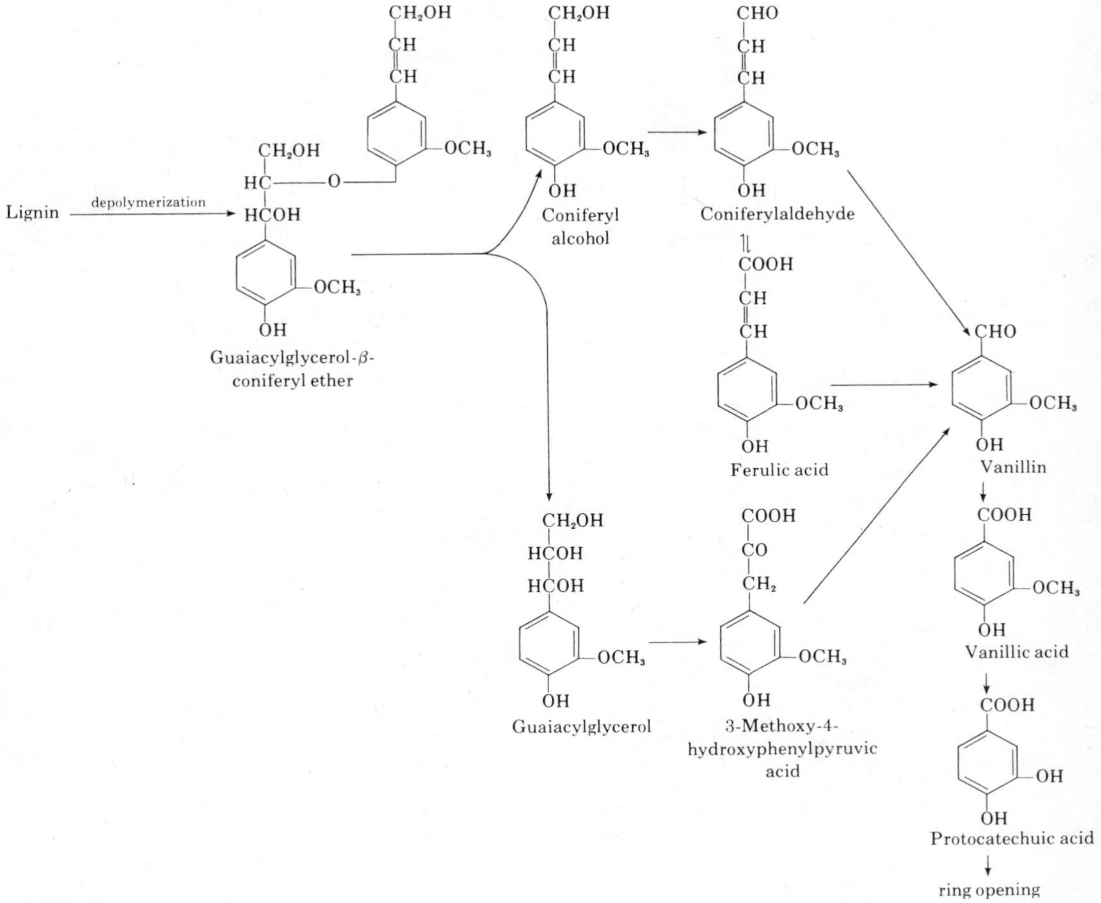

Figure 12.5. Proposed pathway for the microbial metabolism of lignin (26).

chemists found that the relative content of lignin in decomposing plant remains increased as the decomposition proceeded and humus began to appear. The hypothesis is strengthened by analytical data showing that monomers of portions of humus are similar to constituents of lignin (24). On the other hand, appreciable chemical differences also exist. Although the relation of lignin to the biosynthesis of humus components remains uncertain, it is possible either that simple aromatics released in the microbial attack on lignin polymerize to yield constituents of the soil organic fraction or that partially altered lignin itself gives rise to humus constituents (9, 29).

REFERENCES

Reviews

Freudenberg, K. and A. C. Neish. 1968. *Constitution and biosynthesis of lignin.* Springer-Verlag, New York.

Higuchi, T. 1971. Formation and biological degradation of lignins. *Advan. Enzymol.,* 34:207–283.

Kirk, T. K. 1971. Effects of microorganisms on lignin. *Annu. Rev. Phytopathol.,* 9:185–210.

Sarkanen, K. V. and C. H. Ludwig, eds. 1971. *Lignins: Occurrence, formation, structure, and reactions.* Wiley-Interscience, New York.

Schubert, W. J. 1965. *Lignin biochemistry.* Academic Press, New York.

Literature Cited

1. Acharya, C. N. 1935. *Biochem. J.,* 29:1459–1467.
2. Babel, U. 1964. In A. Jonerius, ed., *Soil micromorphology.* Elsevier Publishing Co., Amsterdam, pp. 15–22.
3. Bartlett, J. B. and A. G. Norman. 1938. *Soil Sci. Soc. Amer. Proc.,* 3:210–216.
4. Boruff, C. S. and A. M. Buswell. 1934. *J. Amer. Chem. Soc.,* 56:886–888.
5. Bu'Lock, J. D. and H. G. Smith. 1961. *Experientia,* 17:553–554.
6. Cartwright, N. J. and K. S. Holdom. 1973. *Microbios,* 8:7–14.
7. Cartwright, N. J., A. R. W. Smith, and J. A. Buswell. 1970. *Microbios,* 2:113–143.
8. Crawford, R. L., T. K. Kirk, J. M. Harkin, and E. McCoy. 1973. *Appl. Microbiol.,* 25:322–324.
9. Flaig, W. and K. Haider. 1968. *Trans. 9th Intl. Cong. Soil Sci.,* 3:175–182.
10. Freudenberg, K. 1965. *Science,* 148:595–600.
11. Fuller, W. H. and A. G. Norman. 1943. *J. Bacteriol.,* 46:291–297.
12. Gierer, J. and A. E. Opara. 1973. *Acta Chem. Scand.,* 27:2909–2922.
13. Gulyas, F. 1967. *Agrokem. Talajtan,* 16:137–150.
14. Gunnison, D. and M. Alexander. 1975. *Appl. Microbiol.,* 29:729–738.
15. Hata, K., W. J. Schubert, and F. F. Nord. 1966. *Arch. Biochem. Biophys.,* 113:250–251.
16. Henderson, M. E. K. 1965. *Plant Soil,* 23:339–350.
17. Ishikawa, H., W. J. Schubert, and F. F. Nord. 1963. *Arch. Biochem. Biophys.,* 100:131–139.
18. Jaschhof, H. 1964. *Geochim. Cosmochim. Acta,* 28:1623.
19. Kirk, T. K. and A. Kelman. 1965. *Phytopathology,* 55:739–745.
20. Kunc, F. 1971. *Folia Microbiol.,* 16:41–50.
21. Kunze, C. 1969. *Plant Soil,* 31:389–390.
22. Lindeberg, G. 1947. *Proc. 4th Intl. Cong. Microbiol.,* Copenhagen, pp. 401–403.
23. Mayaudon, J. and L. Batistic. 1970. *Ann. Inst. Pasteur,* 118:191–198.
24. Ogner, G. 1973. *Soil Sci.,* 116:93–99.
25. Ribbons, D. W. 1970. *FEBS Lett.,* 8:101–104.
26. Schubert, W. J. 1968. In M. Florkin and E. H. Stutz, eds., *Comprehensive biochemistry.* Elsevier Publishing Co., Amsterdam, vol. 20, pp. 193–230.
27. Sircar, S. S. G., S. C. De, and H. D. Bhowmick. 1940. *Indian J. Agr. Sci.,* 10:152–157.
28. Sopko, R. 1965. *Mater. Organismen, Beih.,* pp. 187–196.
29. Sorensen, H. 1962. *J. Gen. Microbiol.,* 27:21–34.
30. Sundman, V., T. Kuusi, S. Kuhanen, and G. Carlberg. 1964. *Acta Agr. Scand.,* 14:229–248.
31. Tenney, F. G. and S. A. Waksman. 1929. *Soil Sci.,* 28:55–84.
32. Trojanowski, J. and A. Leonowicz. 1969. *Microbios,* 1:247–251.
33. Trojanowski, J., A. Leonowicz, and B. Hampel. 1966. *Acta Microbiol. Polon.,* 15:17–22.

34. Trojanowski, J., M. Wojtas-Wasilewska, and B. Junosza-Wolska. 1970. *Acta Microbiol. Pol., Ser. B.,* 19:13–22.
35. Waksman, S. A. and I. J. Hutchings. 1936. *Soil Sci.,* 42:119–130.
36. Wojtas-Wasilewska, M., J. Trojanowski, and Z. Stepniewska. 1973. *Acta Microbiol. Polon., Ser. A,* 5:37–48.
37. Yamane, I. 1974. *Sci. Rep. Res. Inst., Tohoku Univ., Ser. D.,* 25:25–30.

13
Microbiology of Other Polysaccharides

The decomposition of two plant polysaccharide types, cellulose and the hemicelluloses, has already been considered. Together with lignin, these two carbohydrates make up the bulk of the plant remains that undergo decay. Several other polysaccharides, however, are found in plant tissues or in microorganisms, and their decomposition occupies a degree of prominence in soil. The metabolism of four additional polysaccharide types will be discussed in detail: starch, the pectic substances, inulin, and chitin.

STARCH

Starch is second only to cellulose as the most common hexose polymer in the plant realm. Starch serves the plant as a storage product, and as such it is the major reserve carbohydrate. This material occurs in large amounts in leaves carrying out photosynthesis, but the polysaccharide is distributed in the xylem, phloem, cortex, and pith of the stems of many species as well as in tubers, bulbs, corms, underground rhizomes, fruits, and seeds. Typically, the starch of higher plants accumulates in definite granules that may vary from several to 150 μm in diameter, the size depending on the species. Microorganisms may also accumulate the polysaccharide.

Plant starches contain two components, amylose and amylopectin. The former has an essentially linear structure built up of several hundred or more glucose units linked together by an α-(1 $\rightarrow$ 4)-glucosidic bonding.

In amylopectin, the individual glucose units are likewise bound together by α-(1 $\rightarrow$ 4) linkages, but the molecule is branched and has side chains attached through α-(1 $\rightarrow$ 6)-glucosidic linkages. The two components of starch differ both physically and chemically, and they can be separated by a number of simple procedures. Starches commonly contain 70 to 90 percent amylopectin and 10 to 30 percent amylose, but exceptions are not uncommon; for example, the starch within the seeds of waxy corn contains no amylose, whereas that in wrinkled peas is essentially free of amylopectin.

Starch disappears rapidly when subjected to the activity of the soil community, and its decomposition proceeds at a greater rate than the microbiologically induced losses of cellulose, hemicelluloses, and a variety of other polysaccharides. Under conditions of limiting O_2, a fermentation occurs with the formation of appreciable lactic, acetic, and butyric acids. The process of degradation goes on at a good pace even under total anaerobiosis, and considerable methane may be evolved.

Bacteria, fungi, and actinomycetes have the capacity to hydrolyze starch, and the physiological heterogeneity of the active flora suggests that the decomposition can take place in diverse environments. From 3 to 90 percent of the bacteria and actinomycetes appearing on dilution plates can utilize the polysaccharide; therefore, soils frequently contain 10^5 to 10^7 or more starch hydrolyzers per gram (8). The organisms are sometimes particularly numerous in proximity to the plant root system.

Some of the more ubiquitous genera implicated in starch utilization are listed in Table 13.1. Among the bacteria are gram-positive and gram-negative genera, spore formers and nonspore formers, aerobes and obligate anaerobes, as well as representatives of many physiologically different groups (26). Starch is generally an excellent carbon source for most actinomycetes, sometimes 25 to 96 percent of the isolates using the polysaccharide (10), and strains of *Streptomyces, Nocardia,* and *Micromonospora* make use of the carbohydrate. Many filamentous fungi are also capable of excreting the appropriate hydrolytic enzymes; yeasts, on the other hand, rarely attack the polysaccharide. Organisms

TABLE 13.1

Some Microbial Genera Utilizing Starch

Bacteria		Actinomycetes	Fungi
Bacillus	Flavobacterium	Micromonospora	Aspergillus
Chromobacterium	Micrococcus	Nocardia	Fomes
Clostridium	Pseudomonas	Streptomyces	Fusarium
Cytophaga			Polyporus
			Rhizopus

using starch as a carbon and energy source exhibit no particular substrate specificities, many attacking simple sugars, organic acids, pentosans, and sometimes cellulose. When pure starch is added to soil in the presence of sufficient inorganic nitrogen, the diversity of microbial types becomes evident. After approximately 16 hours, the community is dominated by large rods, fungi, and actinomycetes, and these increase in abundance with time. Water-saturated samples develop an anaerobic community made up largely of species of *Clostridium*. Starch-hydrolyzing thermophiles and psychrophiles are also found readily and sometimes are notably abundant.

Starch-hydrolyzing enzymes, the amylases, are usually extracellular and remain in the culture fluid after removal of the microorganism. Three amylases may be concerned in the breakdown of starch, α-amylase, β-amylase, and glucoamylase. β-Amylase acts on both amylose and amylopectin, cleaving every second glucose-glucose bond from the end of the molecule. Hence, maltose fragments are liberated from the straight chain of amylose, and significant quantities of other intermediates do not accumulate. The type of action of β-amylase indicates that it is an exo enzyme. β-Amylase is incapable of catalyzing the hydrolysis of branch points of amylopectin, however, and a residual dextrin fraction in addition to the disaccharide, maltose, remains. In contrast, α-amylase is an endo enzyme that acts randomly on the $1 \rightarrow 4$-linkages throughout the amylose and amylopectin molecules. With amylose as substrate, α-amylase hydrolyzes the polysaccharide and the fragments produced from it with the ultimate formation of large amounts of maltose and often small quantities of glucose and the trisaccharide known as maltotriose. This enzyme also acts on amylopectin and generates the same products from the linear portions of the molecule as it does from amylose; however, because it cannot attack the branch points, high-molecular-weight dextrins containing the branches accumulate. Some organisms possess a different exo enzyme, glucoamylase, which hydrolyzes glucose units from the ends of the molecule; some glucoamylases may be incapable of cleaving the branch point so that the dextrins likewise are generated, but others slowly hydrolyze this linkage. The dextrins, or oligosaccharides left as the amylopectin is degraded, do not remain for long because several other microbial enzymes hydrolyze the branch point positions. Diagrammatically, the action of α- and β-amylases may be visualized as shown in Figure 13.1. So far as presently known, β-amylases are not common in microorganisms. The maltose, maltotriose, and low-molecular-weight linear oligosaccharides that are formed are converted to glucose by mediation of the enzyme α-glucosidase so that starch is transformed ultimately to glucose. The simple sugars are water-soluble and penetrate the cell, there to be used as energy sources for growth and protoplasmic synthesis.

Starch-hydrolyzing enzymes are usually inducible, but the ability of microorganisms to form amylolytic enzymes depends on the type of starch. Many

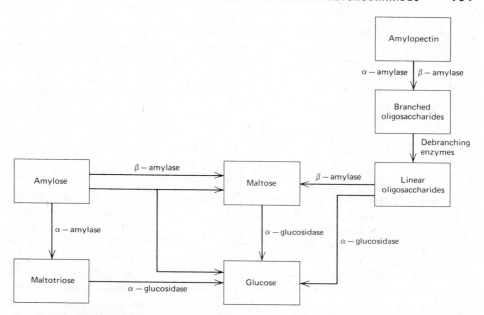

Figure 13.1. Enzymatic conversion of starch to glucose initiated by α- and β-amylases.

amylolytic isolates are capable of growing on the polysaccharide obtained from one plant but not from another, and some bacteria are highly specific for one or a few related starches (32).

Measurements of starch-decomposing activity have been made by determining the accumulation of sugars when toluene-treated soils are incubated with starch. Such assays show that the activity often varies with kind of vegetation, moisture, and soil type (25, 27). The responsible enzymes are usually extracellular and can be adsorbed to clays (1), but the decomposition of starch by the community is not always appreciably affected by the presence of various clays (16).

PECTIC SUBSTANCES

Pectic substances never make up a large portion of the dry matter of plants, commonly less than 1 percent. The importance of these polysaccharides rests on their relationship to the physical structure of the plant and to certain diseases produced by soil-borne and other pathogens. The pectic substances are found abundantly in the middle lamella, a tissue constituent located between individual cells, and the pectin that thus lies between the cells is important in binding them together. The primary and the secondary cell walls also contain polysaccharides of this type.

The pectic carbohydrates are complex polysaccharides composed of galacturonic acid units bound to one another in a long chain. They are thus polygalacturonic acids, although sometimes traces of sugars may be present also.

The carboxyl of the galacturonic acid building block may be partially or completely esterified with methyl groups and may be partially or entirely neutralized by various cations. There are four types of pectic substances: (*a*) protopectin, a water-insoluble constituent of the cell wall; (*b*) pectin, a water-soluble polymer of galacturonic acid that contains many methyl ester linkages; (*c*) pectinic acids, colloidal pectic substances that are also galacturonic acid polymers, but these molecules contain few methyl ester linkages; and (*d*) pectic acids, the water-soluble galacturonic acid polymers that are essentially devoid of methyl ester linkages. Each of these four classes represents a group of closely related substances. The parent pectic substance, protopectin, can be converted by simple and mild treatment to pectin or pectinic acids. Pectic acids can be produced from pectins by treatment with dilute alkali, a conversion of $(RCOOCH_3)_n$ to $(RCOOH)_n$ where RCOOH designates the free galacturonic acid and $RCOOCH_3$ designates the methyl ester of a single galacturonic acid unit. The various pectic substances also differ from one another in the quantities of components other than galacturonic acid associated with the polymer and in molecular weights.

Bacteria, fungi, and actinomycetes are capable of hydrolyzing the pectic substances, using the polysaccharides as carbon and energy sources to support proliferation. As a rule, pectic substances are readily decomposed by microorganisms in soil and in culture media. The diversity of microbial types and the ability of active genera to colonize a variety of soils and substrates can be taken as an indication that pectolytic microfloras are large. Soils generally contain 10^5 to 10^6 pectolytic microorganisms per gram, many of which are actinomycetes, although lower counts are occasionally observed. In a typical study, from less than 1 to 11 percent of the bacterial colonies appearing on dilution plates prepared from various soils were found to use pectin (21). Microorganisms utilizing pectic substances are common not only in soil but in the root zone, and counts in excess of 10^7 pectolytic bacteria per gram have been reported in soil immediately adjacent to plant roots. Among the bacteria, species of *Arthrobacter, Bacillus, Clostridium, Flavobacterium, Micrococcus,* and *Pseudomonas* are often particularly abundant in this activity, but other genera contain representatives utilizing these polymers as well. The addition of the polysaccharide selectively stimulates the particular groups that are especially suited to grow in that soil (9). The capacity to utilize the polysaccharide is also common to many fungi (17) and also to *Streptomyces, Micromonospora, Actinoplanes, Microbispora,* and *Streptosporangium* among the actinomycetes (11). Acidity affects the composition of the flora; thus, the population of pectolytic streptomycetes is favored by high pH while the fungi become more significant with a fall in pH.

The enzymes involved in the breakdown of pectic substances can be divided into three major categories: (*a*) pectinesterases, (*b*) enzymes that carry out a hydrolytic cleavage of the polymer, and (*c*) enzymes catalyzing a *trans*-eliminative cleavage. The pectinesterases cause only a modest change in the molecule inasmuch as their function is to remove the methyl groups from the methyl esters, thereby converting pectin or pectinic acids to pectic acid.

$$(RCOOCH_3)_n + n\ H_2O \rightarrow (RCOOH)_n + n\ CH_3OH \qquad (I)$$

This reaction does not destroy the polymer but rather alters its availability to the enzyme concerned in the subsequent step in decomposition. The enzymes acting by hydrolytically cleaving the molecule are of two types. (*a*) If pectin is attacked more readily than pectic acid, the catalyst is called polymethylgalacturonase. (*b*) If pectic acid is hydrolyzed at a faster rate than pectin, the enzyme is designated polygalacturonase. The process is a simple depolymerization that, in the case of pectic acid, may ultimately yield galacturonic acid.

On the other hand, the catalysts causing *trans*-eliminative cleavage, although also bringing about a depolymerization, yield a modified form of galacturonic acid if the reaction proceeds to give rise to the monomer form.

Galacturonic acid

4-Deoxy-L-threo-5-hexoseulose uronic acid
(product of pectate lyase)

These enzymes also are of two sorts. (*a*) If pectin is attacked more readily than pectic acid, the catalyst is called pectin lyase. (*b*) If pectic acid is attacked at a faster rate than pectin, the enzyme is designated pectate lyase. Each depolymer-izing enzyme can theoretically act at random between building blocks of the polysaccharide (an endo enzyme) or remove portions solely from the end of the molecule (an exo enzyme). The existence of a unique catalyst acting on protopectin has yet to be verified, and attack on the parent substance probably is effected by one of the other enzymes. The final products of the reaction are commonly the monomer, dimer, or trimer, but sometimes oligomers accumu-late; the actual products depend on the particular organism. It is noteworthy that pectins, although containing methyl ($-CH_3$) groups, can be depolymerized without the intervention of pectinesterases.

A listing of these enzymes and some organisms possessing them is given in Table 13.2. The catalysts are usually extracellular and inducible, but intracellu-lar and also constitutive ones are also known; however, only the extracellular enzymes are probably important in the breakdown of pectic substances in plant

TABLE 13.2
Some Microbial Genera Producing Pectic Enzymes

Enzyme	Fungi		Bacteria
Polygalacturonase[a]	*Aspergillus* *Fusarium* *Monilia*	*Penicillium* *Rhizoctonia* *Rhizopus*	*Bacillus* *Erwinia* *Pseudomonas* *Xanthomonas*
Pectate lyase[a]	*Fusarium* *Geotrichum* *Rhizoctonia*		*Arthrobacter* *Bacillus* *Clostridium* *Corynebacterium* *Flavobacterium* *Pseudomonas*
Polymethylgalacturonase[b]	*Aspergillus* *Botrytis*	*Fusarium* *Rhizoctonia*	
Pectin lyase[b]	*Aspergillus* *Fusarium*	*Penicillium* *Rhizoctonia*	*Arthrobacter* *Clostridium* *Corynebacterium* *Flavobacterium* *Micrococcus* *Xanthomonas*
Pectinesterase	*Alternaria* *Fusarium*		*Clostridium* *Pseudomonas* *Xanthomonas*

[a] Endo or exo enzyme.
[b] Only endo enzyme.
(Note: Lyases are sometimes called transeliminases. The words pectic acid, polygalacturonic acid, or polygacturonate are sometimes used in place of pectate.)

residues reaching the soil. A single organism can and often does make several of the extracellular pectic enzymes, moreover. Of the enzymes listed, the exo pectate lyase forms only the dimer and the exo polygalacturonase forms usually only the monomer (i.e., galacturonic acid) or only the dimer from pectic acid; this is expected in view of the repeated cleavage of the same fragment from the end of the polysaccharide. In contrast, the endo enzymes act to rupture bonds between building blocks situated at a distance from the polymer's end, and the products may include the monomer, dimer, and up to the pentamer, the yield depending on the organism and enzyme in question. By use of other enzymes, however, the oligomers are further degraded so that the galacturonic acid or related structural unit penetrates the cell, where it is used as a source of carbon and energy.

Little is known about the significance or behavior of these enzymes in nature, although some of them do not seem to be affected by the presence of clays, which frequently lower the activity of extracellular enzymes (16). Assay methods have been devised to determine the rate of pectin breakdown in soil (12), but the process has yet to be investigated in detail.

INULIN

Inulin is a polysaccharide composed of fructose units, that is, a fructosan. The inulin molecule has approximately 25 to 28 fructose residues in the carbohydrate chain bound in $(2 \rightarrow 1)$-linkage. The substance occurs in a number of plants as the storage carbohydrate, replacing starch in this regard, and it has been reported in tubers, roots, stems, and leaves.

There is also a polysaccharide made up of fructose units in which the sugar units are bound together through the number 2 carbon of one fructose and the number 6 carbon of the next sugar in the chain. This is a $(2 \rightarrow 6)$-fructosan.

Many microorganisms utilize inulin. Bacteria of the genera *Pseudomonas, Flavobacterium, Micrococcus, Cytophaga, Arthrobacter*, and *Clostridium*, many streptomycetes, and a heterogeneous group of fungi use the fructose polysaccharide as carbon source for growth. The enzyme converting inulin to smaller fructose units is called inulinase. Little attention has been given to the transformation despite the large population active on the carbohydrate, but it has been shown that inulinase, an extracellular catalyst, is highly active in several fungi and bacteria. As a rule, inulinases are exo enzymes and remove single fructose units from the end of the molecule, but the product of exo inulinase of some organisms may be a disaccharide containing two fructose units (34). On the basis of the rapid turnover exhibited in vitro, it is likely that fructosans are readily transformed in soil.

Certain microorganisms possess two fructosanases—inulinase, which by the structure of its substrate is a $(2 \rightarrow 1)$-fructosanase, and a $(2 \rightarrow 6)$-fructosanase. These enzymes are both extracellular. Some fungi produce both the $(2 \rightarrow 1)$- and the $(2 \rightarrow 6)$-fructosanases while others produce only one or the other of the polysaccharide-splitting enzymes. The enzymes are generally not formed when the organisms are grown on simple sugars but are excreted in appreciable amounts in the presence of the fructosan. The mechanism of action, however, is somewhat different for the two catalysts. The $(2 \rightarrow 6)$-fructosanase of a species of *Streptomyces* leads to the accumulation of levanbiose, a sugar containing two fructose units. The hydrolysis of fructosan by the $(2 \rightarrow 1)$-fructosanase (inulinase) of *Aspergillus fumigatus*, on the other hand, results in the accumulation of slightly longer chains of fructose units. In contrast, the fructosanases of *Fusarium monoliforme* and *P. funiculosum* transform the fructosan entirely to fructose with no accumulation of significant quantities of intermediary compounds (15). Apparently, there are several end products of fructosanase action—short fructose polymers, the disaccharide, or free fructose. Nevertheless, the ultimate fate of the large molecule is an enzymatic degradation to units small enough to enter the cell.

CHITIN

Chitin is the most common polysaccharide in nature whose basic unit is an amino sugar. This polysaccharide is a structural constituent giving mechanical strength to organisms containing it. Chitin is insoluble in water, organic solvents, concentrated alkali, or dilute mineral acids, but it can be solubilized and degraded either enzymatically or by treatment with concentrated mineral acids. In structure, chitin consists of a long chain of N-acetylglucosamine units in linear arrangement. The empirical formula is $(C_6H_9O_4 \cdot NHCOCH_3)_n$, and the pure compound contains 6.9 percent nitrogen. The polymer is similar to cellulose, with one of the hydroxyl groups of each glucose in cellulose being replaced by an acetylamino ($-NHCOCH_3$) unit.

Some chitins have varying amounts of chitosan associated with them. Chitosan is similar to chitin but it has lost the acetyl (—$COCH_3$) groups so that it is a long chain of glucosamine units ($C_6H_9O_4 \cdot NH_2$) linked together as in the original chitin.

Chitin is undoubtedly an important substance in the carbon cycle in soil not only because of its availability to microbiological attack but also as a result of its continuous microbial biosynthesis. Soil chitin originates in the remains of insects that spend part or all of their life underground, but it also arises during growth of fungi and possibly other organisms. Consequently, even in the absence of added chitin, the aminopolysaccharide is formed in soil by the biosynthetic action of the community. Here, then, is a native component of the soil's organic fraction, one whose supply is replenished as a result of microbial cell synthesis. The supply is thus independent of insect remains.

Chitin is produced by members of both the plant and animal kingdoms. In the cell walls of many filamentous fungi, considerable amounts of the polysaccharide are deposited, and the polymer is important to maintenance of the structural integrity and hence viability of these organisms. The carbohydrate is also part of the cellular structure of many basidiomycetes, some yeasts, and possibly some protozoa and algae, but it is not found in the cell walls of bacteria and actinomycetes. The chitin component of the fungus hyphal wall can make up a large proportion of the total cellular material, but growth conditions, age, temperature, and pH alter the concentration. Chitin is found in the skeletons of a number of invertebrate animals, and a substantial quantity of the polysaccharide becomes incorporated into soil in the form of insect remains. The chitin in insects is linked intimately with proteins in a complex that stabilizes the chitin against enzymatic degradation.

The application of chitin to soil stimulates the microflora, and this chemically stable compound is mineralized in relatively short periods. The rates of decomposition of chitin and cellulose are compared in Figure 13.2. Chitosan is also degraded at reasonable rates, and 53 percent of the chitosan carbon added to one soil was converted to CO_2 in a period of eight weeks (4). Pure chitin has a narrow C:N ratio so that the nitrogen supply exceeds the microbiological requirement. Consequently, inorganic nitrogen will accumulate provided that the organisms are active. In periods of less than two months, 30 to 60 percent of the chitin nitrogen is mineralized in aerobic environments, but the chitin is decomposed more slowly than proteins or nucleic acids (5, 35). Many of the tests of decomposition that have been conducted may be misleading, however, because they used purified chitin preparations. Thus, although chitin-rich insect wings are notably refractory to destruction in soil, removal of the protein from the wings yields a largely chitinous residue that is much more readily destroyed microbiologically (23).

Particulate substrates like chitin pose a special problem in the soil environment. In contrast with liquid media or aquatic environments, soils contain particulates such as sand, silt, and clay that may not only retain certain

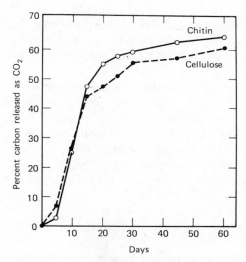

Figure 13.2. The decomposition of chitin and cellulose in soil as measured by CO_2 formation (22).

compounds but may physically impede the movement of organisms to their nutrients. Because of the mechanical barriers, individuals or populations often exist in microenvironments that are adjacent to sites containing nutrient sources but from which they cannot gain ready access to utilizable substrates. Such a physical hindrance to microbial metabolism of chitin has been established using particles the size of silt and a strain of *Pseudomonas* that grows at the expense of chitin (24).

Arable soils contain large numbers of chitinoclastic organisms. Up to 10^6 microorganisms per gram of soil utilize the polysaccharide. In contrast to the relatively small proportion of actinomycetes in the total microflora, approximately 90 to 99 percent of the chitinoclastic isolates from certain soils may be actinomycetes. Only a fraction of the chitin digesters in such soils are bacteria, and less than 1 percent are fungi. The actinomycetes thus are sometimes the dominant segment of the aerobic chitinoclastic flora of untreated areas. Similarly, when such soils are amended with chitin, more than 90 percent of the chitin-decomposing organisms are also actinomycetes provided that aeration is ample. Counts of chitin digesters in these samples may be as high as 700 million per gram. *Streptomyces* and lesser numbers of *Nocardia* constitute the bulk of the community (25). The capacity of so many *Streptomyces, Nocardia, Micromonospora, Actinoplanes,* and *Streptosporangium* isolates to make use of chitin has led to the proposal that media containing the polymer be used for the selective isolation of actinomycetes (13, 14).

On the other hand, fungi and bacteria are prominent in other soils. Fungi of the genera *Mortierella, Trichoderma, Verticillium, Paecilomyces,* and *Gliomastix,* as well as *Bacillus* and *Pseudomonas* among the bacteria are often prominent (7). Chitin utilization is an attribute of numerous other genera of fungi, many of which are prominent and probably important in the process in acid regions, and of bacterial species of the genera *Chromobacterium, Cytophaga, Flavobacterium,* and *Micrococcus.* Bacteria are dominant in poorly drained habitats, and *Clostridium* is known to metabolize the molecule in the absence of O_2.

Growth of chitinoclastic microorganisms in chitin-containing agar media is often characterized by a zone of clearing around the colonies. The halo results from the destruction of the insoluble substrate. Many soil isolates utilize chitin as a carbon source, others as a nitrogen source, while not a few satisfy both their carbon and nitrogen needs from the compound.

Because of the key role of chitin in the cell wall of many fungi, species that bring about the digestion of hyphal walls characteristically excrete chitin-depolymerizing enzymes. The walls contain a second major polysaccharide, however, usually a glucose polymer, so that digestion of the wall to the extent that its structural integrity is lost and the cell is lysed requires both chitin- and glucan-cleaving enzymes (3, 29).

The presence of chitin in the walls of many plant pathogenic fungi and the stimulation by the polysaccharide of actinomycetes, many of which excrete antifungal products, prompted attempts to bring about biological control of disease agents. The assumption in these investigations was that chitin amendment would selectively enhance the development of populations which could either strip part of the cell walls of the pathogens or produce toxins harmful to them. The results were completely gratifying, and the severity of diseases

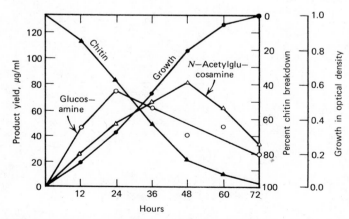

Figure 13.3. Growth of *Cytophaga johnsonae* in media containing chitin and appearance of products (33).

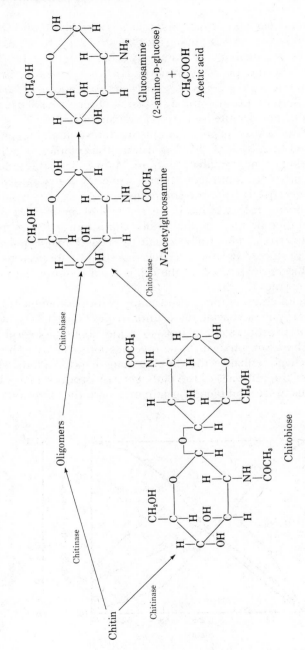

Figure 13.4. Pathway of chitin breakdown.

caused by soil-borne *Fusarium* was markedly reduced (6, 18). Other fungal pathogens or the diseases they cause have also been similarly affected, but not all fungi can be controlled in this manner, possibly because they contain no chitin in their walls or are insensitive to the toxins. It is still not clear, moreover, how this type of biological control actually operates, and enzymes that cause lysis, complex antifungal inhibitors, and simple nitrogenous compounds that suppress the fungi have all been implicated (18, 28, 31).

The breakdown of chitin involves the conversion of an insoluble, crystalline molecule to water-soluble products that penetrate the cells and serve to provide energy and carbon or sometimes nitrogen for growth. The responsible enzymes are extracellular and usually inducible, and common products that appear in culture during the degradation are *N*-acetylglucosamine and glucosamine (Figure 13.3). The transformation usually involves the following enzymes: (*a*) chitinase, which catalyzes a depolymerization of the chain to yield oligomers having several of the *N*-acetylglucosamine units as well as the dimer, the latter having the trivial name of chitobiose, and (*b*) a second enzyme—called chitobiase or more appropriately *N*-acetylglucosaminidase—that hydrolyzes the oligomers and chitobiose to yield *N*-acetylglucosamine. The chitinases receiving most attention randomly cleave the bonds between structural units of the polymer and so are endo enzymes, but exo enzymes may also exist. Chitobiose is sometimes the dominant product. In addition to these two types of enzymes, a third may be required for the initial attack on crystalline chitin, making the native molecule a suitable substrate for chitinase (20). This mode of attack does not take into account the fact that insect chitin is not free but is strongly bound to proteins, and it is still totally unclear how this chitin is made available to microorganisms in nature. The *N*-acetylglucosamine ultimately generated is then converted to acetic acid and glucosamine, and the ammonia is liberated from the latter compound or one of its subsequent derivatives (Figure 13.4). The glucosamine or a compound formed from it is readily attacked and serves within the cell to supply carbon and energy.

Recent research has suggested a second possible mode of depolymerization. This involves an enzyme that removes the acetyl groups from the *N*-acetylglucosamine units in the chitin polymer to yield chitosan (2). The chitosan is then hydrolyzed to simple products by chitosanase (19). Whether this process is important in soil or not remains to be resolved.

As an extracellular enzyme, chitinase is subject to adsorption or complexing with clay or humus constituents. The adsorption can lead to a diminution in activity, both the retention and the activity of the protein being affected by pH (30).

REFERENCES

Reviews

Doesburg, J. J. 1973. The pectic substances. In L. P. Miller, ed., *Phytochemistry*. Van Nostrand Reinhold Co, New York, vol. 1, pp. 270–296.

Greenwood, C. T. and E. A. Milne. 1968. Starch degrading and synthesizing enzymes. *Advan. Carbohydr. Chem.,* 23:281–366.

Manners, D. J. 1974. The structure and metabolism of starch. In P. N. Campbell and F. Dickens, eds., *Essays in biochemistry.* Academic Press, London, vol. 10, pp. 37–71.

Rombouts, F. M. and W. Pilnik. 1972. Research on pectin depolymerases in the sixties—A literature review. *CRC Crit. Rev. Food Technol.,* 3:1–26.

Literature Cited

1. Aomine, S. and Y. Kobayashi. 1966. *Soil Sci. Plant Nutr.,* 12:7–12.
2. Araki, Y. and E. Ito. 1975. *Eur. J. Biochem.,* 55:71–78.
3. Ballesta, J.-P. G. and M. Alexander. 1972. *Trans. Brit. Mycol. Soc.,* 58:481–487.
4. Bondietti, E., J. P. Martin, and K. Haider. 1972. *Soil Sci. Soc. Amer. Proc.,* 36:597–602.
5. Bremmer, J. M. and K. Shaw. 1954. *J. Agr. Sci.,* 44:152–159.
6. Buxton, E. W., O. Khalifa, and V. Ward. 1965. *Ann. Appl. Biol.,* 55:83–88.
7. Gray, T. R. G. and P. Baxby. 1968. *Trans. Brit. Mycol. Soc.,* 51:293–309.
8. Hankin, L., D. C. Sands, and D. E. Hill. 1974. *Soil Sci.,* 118:38–44.
9. Hirte, W. F. 1972. *Symp. Biol. Hungar.,* 11:221–227.
10. Ishizawa, S., M. Araragi, and T. Suzuki. 1969. *Soil Sci. Plant Nutr.,* 15:104–112.
11. Kaiser, P. 1971. *Compt. Rend. Acad. Sci., Ser. D,* 272:501–504.
12. Kaiser, P. and M. S. Monzon de Asconegui. 1971. *Soil Biol.,* No. 14, pp. 16–19.
13. Kuznetsov, V. D. and I. V. Yangulova. 1970. *Mikrobiologiya,* 39:902–906.
14. Lingappa, Y. and J. L. Lockwood. 1962. *Phytopathology,* 52:317–323.
15. Loewenberg, J. R. and E. T. Reese. 1957. *Can. J. Microbiol.,* 3:643–650.
16. Lynch, D. L. and L. J. Cotnoir. 1956. *Soil Sci. Soc. Amer. Proc.,* 20:367–370.
17. Malan, C. E. and L. Leone. 1962. *Allionia,* 8:195–208.
18. Mitchell, R. and M. Alexander. 1962. *Soil Sci. Soc. Amer. Proc.,* 26:556–568.
19. Monaghan, R. L., D. E. Eveleigh, R. P. Tewari, and E. T. Reese. 1973. *Nature New Biol.,* 245:78–80.
20. Monreal, J. and E. T. Reese. 1969. *Can. J. Microbiol.,* 15:689–696.
21. Naguib, A. I., S. H. Elwan, and M. R. Rabie. 1971. *UAR J. Bot.,* 14:173–187.
22. Okafor, N. 1966. *Soil Sci.,* 102:140–142.
23. Okafor, N. 1966. *Plant Soil,* 25:211–237.
24. Ou, L.-T. and M. Alexander. 1974. *Soil Sci.,* 118:164–167.
25. Pancholy, S. K. and E. L. Rice. 1973. *Soil Sci. Soc. Amer. Proc.,* 37:47–50.
26. Remacle, J. 1966. *Rev. D'Ecol. Biol. Sol,* 3:563–570.
27. Ross, D. J. 1966. *J. Soil Sci.,* 17:1–15.
28. Schippers, B. and L. C. Palm. 1973. *Neth. J. Plant Pathol.,* 79:279–281.
29. Skujins, J. J., H. J. Potgieter, and M. Alexander. 1965. *Arch. Biochem. Biophys.,* 111:358–364.
30. Skujins, J., A. Pukite, and A. D. McLaren. 1974. *Soil Biol. Biochem.,* 6:179–182.
31. Sneh, B. and Y. Henis. 1972. *Phytopathology,* 62:595–600.
32. Stark, E. and P. A. Tetrault. 1951. *Can. J. Bot.,* 29:104–112.
33. Sundarraj, N. and J. V. Bhat. 1972. *Arch. Mikrobiol.,* 85:159–167.
34. Uchiyama, T. 1975. *Biochim. Biophys. Acta,* 397:153–163.
35. Veldkamp, H. 1955. *Meded. Landbouw. Wageningen,* 55:127–174.

14
Transformation of Hydrocarbons

Among the innumerable reactions of the carbon cycle resulting in a mineralization of organic carbon is found a series of transformations that lead to the degradation of hydrocarbons. Numerous hydrocarbons or their derivatives are added to or synthesized within the soil, and their mineralization and formation are of significance in the general cycle of carbon. Many herbicides, insecticides, and fungicides are modified hydrocarbons, and the decomposition of these synthetic compounds in soil has a bearing on the duration of pesticidal action. Therefore, the microbiological utilization of hydrocarbons and related compounds is both agronomically and ecologically important.

FORMATION OF METHANE

In the aerobic decomposition of organic matter, the main gas evolved is CO_2. At low O_2 tensions, appreciable CH_4 is released in the degradation of cellulose, hemicelluloses, proteins, organic acids, and alcohols. When CH_4 production is most vigorous, its volume may be a reasonable percentage of the quantity of CO_2 liberated. Yet, the capacity to form CH_4 is uncommon in the biological realm, but the few active species are widely distributed in environments containing little O_2.

In waterlogged soils, considerable CH_4 is formed during the anaerobic decomposition of carbonaceous substrates. Although H_2 is also a common end product of anaerobic metabolism, little is lost to the atmosphere from flooded fields since it is probably used by the CH_4-producing bacteria as a source of energy for growth. Flooded soils planted to rice frequently evolve less CH_4 than corresponding uncropped sites, but the incorporation of readily available organic materials greatly increases CH_4 production (34). The formation of CH_4 and CO_2 in a soil kept under anaerobic conditions is depicted in Figure 14.1. Not only do simple organic materials and plant remains serve to promote

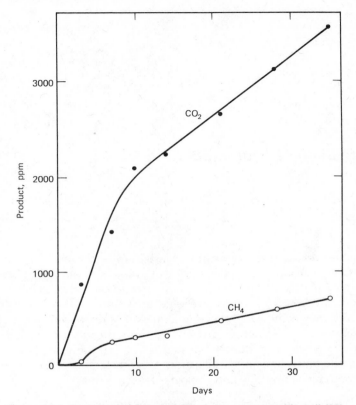

Figure 14.1. Evolution of CH$_4$ and CO$_2$ from an anaerobic soil (23).

release of the gas, but CH$_4$ may also be generated under feedlots, where decomposition of the animal manure provides microbial nutrients and results in O$_2$ depletion (11). The CH$_4$-evolving activity is widespread and quite pronounced, so great in fact that probably more than 10^{11} kg of the gas is discharged each year from paddy fields and swamps (10).

The biosynthesis of this, the simplest hydrocarbon is markedly affected by environmental conditions. The presence of O$_2$ at the sites where the organisms are functioning abolishes CH$_4$ release, but some would be evolved in poorly drained land exposed to air because the O$_2$ fails to penetrate to many of the sites of microbial activity. The transformation is also suppressed by nitrogen compounds bacteria can use as electron acceptors; thus, nitrate, nitrite, N$_2$O, but not ammonium bring about an inhibition (9); such a suppression would not be long-lived, however, because the nitrate, nitrite, and N$_2$O are usually reduced readily when O$_2$ is absent. Methane formation is also influenced markedly by acidity. The maximum rate occurs in environments of near neutral pH, and there is little activity below pH 6.0. Analogous effects of acidity have been reported in studies with pure cultures of the responsible microorganisms.

The biosynthesis of CH_4 is limited to a rather specialized physiological group of bacteria living in waterlogged soils, marshes, swamps, manure piles, the intestines of higher animals, and in marine and fresh water sediments. All of these bacteria are strict anaerobes, and they will not proliferate in the presence of O_2. It is difficult to obtain pure cultures of the responsible bacteria, and many early investigations were carried out with impure or enrichment cultures. Mixed populations contain many organisms that are themselves unable to produce the gas.

The CH_4-forming bacteria have many physiological properties in common, but they are heterogeneous in cellular morphology. Some are rods, some cocci, while others occur in clusters of cocci known as sarcinae. Thermophiles are also represented. All of the CH_4-forming bacteria, despite their morphological dissimilarities, are related biochemically and are grouped into a single family, Methanobacteriaceae. The family is divided into genera on the basis of cytological differences.

I. Rod-shaped bacteria. *Methanobacterium*
II. Spherical cells
   A. Sarcinae. *Methanosarcina*
   B. Not in sarcinal groups. *Methanococcus*

Spiral-shaped CH_4-forming bacteria (*Methanospirillum*) have been isolated from sewage, but such organisms have yet to be found in soil.

Anaerobes that evolve CH_4 are unable to utilize the conventional carbohydrates and amino acids available to most heterotrophs. Glucose and other sugars are not fermented by pure cultures, and polysaccharides are also resistant to attack. The only organic substrates metabolized by the bacteria are short-chain fatty acids such as formic, acetic, propionic, *n*-butyric, and *n*-valeric acids and simple alcohols such as methanol, ethanol, *n*- and isopropanol, *n*- and isobutanol, and *n*-pentanol. In nature, however, a mixed flora is the rule, and many other compounds serve in the CH_4 fermentations of natural communities. However, the conversion of sugars, proteins, cellulose, and hemicelluloses to CH_4 requires two or more microbial groups, one acting on the complex nutrients and the methane bacteria fermenting the organic acids and alcohols produced by the primary microflora.

In flooded soils, for example, simple organic acids accumulate as the O_2 is consumed by the resident aerobes, and the quantity of acetic, *n*-butyric acid, and often formic and propionic acids typically increases prior to or concomitant with CH_4 biogenesis. Furthermore, the organic acids may accumulate to a greater extent if CH_4 producers are not too active, probably because they are not metabolizing these substrates (31).

A variety of reactions can lead to the evolution of CH_4 in O_2-depleted environments. The processes have been characterized in mixtures of populations and frequently in pure cultures. Typical are the following:

$$CO_2 + 4H_2 \rightarrow CH_4 + 2H_2O \tag{I}$$
$$4HCOOH \rightarrow CH_4 + 3CO_2 + 2H_2O \tag{II}$$
$$CH_3COOH \rightarrow CH_4 + CO_2 \tag{III}$$
$$2CH_3CH_2OH \rightarrow 3CH_4 + CO_2 \tag{IV}$$

It should be noted that the major products of the degradation of simple molecules are CO_2 and CH_4, but the ratio of the two gases varies with the substrate in question.

A common mechanism for CH_4 formation is the reduction of CO_2; that is, the simple organic molecule is fermented and CO_2 is utilized as the electron (or hydrogen) acceptor. If CO_2 is not supplied to the microbial fermentation, it is produced in the decomposition of the organic substrate and subsequently reduced. In the case of acetic acid, this occurs as follows:

$$CH_3COOH + 2H_2O \rightarrow 2CO_2 + \quad 8(H)$$

$$8(H) + CO_2 \rightarrow CH_4 + 2H_2O$$

$$Net:\ CH_3COOH \rightarrow CH_4 + \quad CO_2$$

The CO_2-reduction pathway has been established for the fermentation of a variety of carbonaceous compounds, and a generalized equation for the process can be formulated.

$$4H_2R + CO_2 \rightarrow 4R + CH_4 + 2H_2O \tag{V}$$

The CH_4 here arises by reduction of CO_2 at the expense of the hydrogen donor, H_2R, which is thereby converted to product R. Should the CO_2 supplied be radioactive, then the CH_4 recovered will contain the isotope. On the other hand, should the carbon of the organic molecule be tagged, then the CH_4 will not bear the isotopic label.

The substance represented as H_2R is frequently organic, but it need not be. As early as 1920, Harrison proposed that H_2 evolution is inconsequential in paddy fields solely because the gas reacts with CO_2 to yield CH_4. A number of bacteria have subsequently been isolated that, in pure culture, are capable of using molecular H_2 for the reduction of CO_2 in a manner identical with the type reaction given by equation V.

$$4H_2 + CO_2 \rightarrow CH_4 + 2H_2O \tag{VI}$$

A limited number of substrates are fermented by a second mechanism. A species of *Methanosarcina*, for example, decomposes methanol and acetate, the

CH_4 arising from the methyl groups of the alcohol or the acetate. This has been confirmed by the use of the radioisotope ^{14}C.

$$^{14}CH_3COOH \rightarrow {}^{14}CH_4 + CO_2 \tag{VII}$$

This pathway for acetate fermentation is essentially a simple decarboxylation (28). Not only does the methyl carbon of acetate go into CH_4, but the hydrogens attached to the carbon atom are transferred with it. Little attention has been given to ascertaining which of the two mechanisms—reduction of CO_2 or conversion of methyl groups to CH_4—applies to soil. In a study of one flooded soil, however, much of the CH_4 came from the methyl of acetic acid added to the sample, as in equation VII, but some was also derived by a reduction of CO_2, as in equation V (30).

The intermediates formed as CO_2 is reduced to CH_4 are essentially unknown. Enzymes derived from cells of the methanogenic anaerobes can catalyze the reduction of formic acid (HCOOH), formaldehyde (HCHO), and methanol (CH_3OH) to CH_4, but these or other products have not been shown to be synthesized as CO_2 is reduced by the bacteria or their enzymes. Nevertheless, because they or compounds related to them are likely intermediates, a pathway may be postulated in which some unknown compound (RH) combines with CO_2, and the resulting product is reduced in a series of steps involving the formation of molecules structurally related to the one-carbon compounds listed.

$$CO_2 + RH \rightarrow RCOOH \rightarrow RCHO \rightarrow RCH_2OH \rightarrow RCH_3 \rightarrow CH_4 + RH \tag{VIII}$$

A second simple and volatile hydrocarbon is often evolved from soil, ethylene ($H_2C{=}CH_2$). It has been known for some time that a few fungi evolve the gas in culture, but recent assays have disclosed that the capacity for its synthesis is not rare. Thus, species of *Agaricus, Alternaria, Aspergillus, Cephalosporium, Chaetomium, Fusarium, Mucor, Penicillium,* and *Scopulariopsis* among the fungi, *Candida* and *Trichosporon* among the yeasts, *Pseudomonas,* spore-forming bacteria, and actinomycetes have the ability, in vitro at least. Ethylene generation in soil is of particular interest because it affects root elongation and the development of lateral roots, stimulates the sprouting of bulbs, and enhances seed germination. Analyses of soil in the laboratory and field have shown that ethylene is indeed generated in quantities sufficient to influence higher plants. The process is biological inasmuch as only a trace of ethylene is formed in sterilized soil, and the production is most pronounced when little O_2 is present (22, 27). The addition of organic nutrients and excess soil moisture levels also favor ethylene evolution. The agents responsible for the biosynthesis of this hydrocarbon in nature are not fully known, but prime interest has centered on certain spore-forming bacteria and fungi. The ethylene that is synthesized may not only affect growth of underground portions of plants, moreover, inasmuch as it can move upward and enter the shoots, where a variety of responses may take place (17). In addition, the compound even at low concentrations is toxic to some fungi so its production may lead to changes in composition of the community.

Soils may also give rise to other volatile hydrocarbons. Ethane, propane, propylene, and n-butane have been detected in the atmosphere over soils maintained under anaerobiosis (27). Enrichment cultures developing anaerobically produce ethane and sometimes acetylene, propylene, and propane, and certain fungi in pure culture synthesize traces of propylene, ethane, propane, and acetylene. Undoubtedly, other organisms can do likewise. With the exception of CH_4 and ethylene, however, gaseous hydrocarbons have no known significance in soil.

OXIDATION OF ALIPHATIC HYDROCARBONS

The microflora responds to the addition to soil of paraffin, petroleum, petroleum products, and other aliphatic hydrocarbons, and the resultant community causes the added substrate to disappear. These transformations are of great significance in the terrestrial cycle of carbon because waxes and other constituents of plant tissue contain aliphatic hydrocarbons. It has been estimated that approximately 0.02 percent of plant tissues may be considered as hydrocarbon or hydrocarbonlike in structure. Another source of supply is the soil microflora itself that can synthesize a variety of hydrocarbons or hydrocarbonlike molecules; for example, some species of bacteria and algae and the spores of fungi contain either aliphatic hydrocarbons or materials structurally similar to hydrocarbons. Hydrocarbon oxidizers also probably metabolize the oils used as carriers for pesticide sprays, which, even when applied to the foliage, ultimately reach the soil. In addition, the soil under asphalt-paved highways possesses a large bacterial flora capable of utilizing the asphalt.

The short persistence of hydrocarbons of many types is indicative of vigorous populations, and counts in excess of 10^5 per gram have been recorded when paraffin is used as the growth substrate. Among the substances used by the flora are paraffin, kerosene, gasoline, mineral and lubricating oils, asphalts, tars, and natural and synthetic rubbers. Methane, ethane, propane, butane, pentane, hexane, and many other aliphatic hydrocarbons of the type structure C_nH_{2n+2} are decomposed as well (Figure 14.2). Of the hydrocarbons with a type structure C_nH_{2n+2}, those with one to four carbon atoms are gases at room temperature and the somewhat larger molecules are liquid at temperatures below 30°C, and these are all biodegradable. Even solids with this type structure, such molecules having about 20 or more carbon atoms, are metabolized, and chemicals like dotriacontane ($C_{32}H_{66}$) are substrates for components of the community (6). However, the rate of decomposition is markedly affected by the length of the hydrocarbon chain.

Concern with environmental pollution has led to many inquiries on the

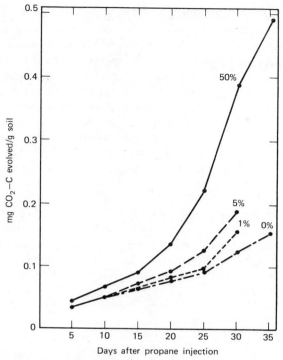

Figure 14.2. Production of CO$_2$ in soil incubated in an atmosphere with different propane concentrations (18). Reproduced by permission of the American Society of Agronomy.

microbial utilization of aliphatic hydrocarbons. Ethylene is one of the major hydrocarbon pollutants of the atmosphere, and millions of metric tons are discharged to the air, chiefly from the combustion of gasoline in automobiles. Despite the vast quantities generated, however, the levels in the atmosphere in nonurban areas are not increasing and typically are below 0.005 ppm. The lack of any rise in the ambient concentrations while the rate of emission of the pollutant has been constantly increasing prompted a search to define the mechanism of ethylene destruction. The outcome of this quest was the finding that soils have a remarkable capacity to remove ethylene from the overlying gas phase, a removal that can be prevented by heating the soil or treating it with germicides to destroy the community. The metabolic process requires O$_2$, for ethylene metabolism ceases under anaerobiosis (1).

Leaks of natural gas from underground pipes have become a more frequent, albeit local pollution problem as the use of this fuel becomes more widespread in urban and suburban areas. These gas leaks may sometimes be responsible for killing more trees in cities and towns than all natural causes combined. Natural gas commonly contains a high percentage of methane, small amounts of ethane, plus traces of various other volatile substances. Bacteria able to oxidize the volatile hydrocarbons proliferate in the vicinity of the leak (Figure 14.3), and in the process they consume O_2, create local O_2-poor regions, and synthesize products some of which are probably phytotoxic. The trees that succumb are likely affected not by the natural gas itself but by the O_2 deficiency, the high CO_2 levels, and perhaps the microbiologically formed toxicants (2).

Soils may receive hydrocarbons as crude oil at sites adjacent to oil wells or from leaking pipes. In such grossly contaminated localities, plants are drastically affected, and commercial agriculture may be impossible. Soils may also be used for the disposal of oily wastes, such a means of disposal being an effective way of preventing gross environmental damage. In these land areas, the hydrocarbon users proliferate to reach levels of millions per gram. Manipulating the treated site to favor microbial growth promotes the degradation and hence alleviates the actual or potential pollution. On the other hand, the low temperature in arctic regions from which crude oil is sometimes pumped is an apparently insurmountable obstacle to high rates of microbial metabolism, and hence long periods will elapse before soil pollution from accidental oil spills in the tundra will be eliminated.

Environmental conditions govern the rate of transformation of aliphatic hydrocarbons. The oxidation is sensitive to temperature, but the decomposition occurs from as low as 0°C to about 55°C. Where thermophiles are present, the rate of reaction at elevated temperatures is quite rapid. The oxidation at about

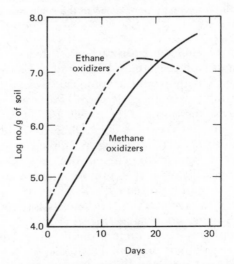

Figure 14.3. Increase in microbial abundance in soil exposed to natural gas (2).

0°C is, of course, extremely slow. Many of the strains are sensitive to acidity and frequently show little growth below pH 5. The availability of other carbonaceous substrates affects hydrocarbon destruction, a depression in rate being common in the presence of a readily metabolized substrate. The depression probably results from a preferential utilization of the second carbon source. Hydrocarbon decomposition will proceed in the absence of free O_2 provided that an alternate electron acceptor like sulfate is supplied. However, as anaerobiosis favors the reduction of sulfate, little hydrocarbon loss can be expected in O_2-depleted environments. Although sulfate-reducing bacteria degrade straight-chain hydrocarbons in the absence of O_2, few denitrifying bacteria are capable of utilizing such substrates for the reduction of nitrate.

Many bacteria metabolize aliphatic hydrocarbons of long or short chain length, but there are characteristically one or more of the simpler gaseous compounds, particularly CH_4, that cannot be oxidized by these microorganisms. For example, many of the bacteria isolated from soils receiving oil are able to utilize the gaseous hydrocarbons ethane (C_2H_6), propane (C_3H_8), and butane (C_4H_{10}) but not CH_4 (21). The metabolism of ethane can be accomplished by species of *Mycobacterium, Nocardia, Streptomyces, Pseudomonas, Flavobacterium*, cocci, and several filamentous fungi. Especially common in the utilization of the simple compounds are the mycobacteria. High-molecular-weight hydrocarbons are consumed by a variety of microorganisms including *Mycobacterium, Nocardia, Pseudomonas, Streptomyces, Corynebacterium, Acinetobacter, Bacillus*, the yeasts *Candida* and *Rhodotorula*, and many fungi. A particular organism is quite restricted in the range of molecules on which it can grow, some using only the low-molecular-weight gases, others the compounds with chains of 10 to 16 carbon atoms, and still others utilizing molecules having 18 to 32 carbons. It was long assumed that the linear aliphatic hydrocarbons with more than 32 carbon atoms could not be used, but it now appears that even $CH_3(CH_2)_{42}CH_3$ is a good substrate (14). The abundance of organisms in unamended soil that can multiply at the expense of such chemicals varies with the chain length and the soil type, but counts from less than 10^3 to greater than 10^5 per gram have been recorded. That the activity is not a rare attribute is evident in one report that from 3 to 17 percent of the organisms in surface horizons are able to use compounds ranging from $CH_3(CH_2)_6CH_3$ to $CH_3(CH_2)_{14}CH_3$ (19).

Many microorganisms metabolize aliphatic hydrocarbons that they cannot use as carbon sources for growth. This is a reflection of the phenomenon known as *cometabolism*, the metabolism by a microorganism of a compound that the cell is unable to use as a source of energy or an essential nutrient. Because the cometabolism of hydrocarbons commonly involves an oxidation, sometimes the term *cooxidation* is applied. In demonstrating the cometabolism of these molecules, usually a carbon source supporting growth is provided to the organisms together with a second substrate; the latter is then oxidized concurrently with the former. For example, a bacterium grown on CH_4 is able to effect the following reactions.

$$CH_3CH_3 \rightarrow CH_3COOH \qquad\qquad\qquad (IX)$$
$$CH_3CH_2CH_3 \rightarrow CH_3CH_2COOH \qquad\qquad\qquad (X)$$
$$CH_3CH_2CH_2CH_3 \rightarrow CH_3CH_2CH_2COOH \qquad\qquad (XI)$$

Because the products cannot be assimilated and further degraded, they accumulate in the liquid. The ecological significance of hydrocarbon cometabolism remains unknown.

A common mechanism for the microbial degradation involves the oxidation of the hydrocarbon in a series of steps that yield the corresponding alcohol, aldehyde, and fatty acid.

$$CH_3(CH_2)_nCH_2CH_2CH_3 \rightarrow CH_3(CH_2)_nCH_2CH_2CH_2OH \rightarrow$$
$$CH_3(CH_2)_nCH_2CH_2CHO \rightarrow CH_3(CH_2)_nCH_2CH_2COOH \quad (XII)$$

In the sequence, only the terminal carbon is initially oxidized. The first reaction in equation XII requires O_2, and it is because of this requirement that such molecules are usually degraded only when O_2 is available. The fatty acid then is commonly decomposed by a reaction sequence termed β-oxidation in which two carbons are removed from the end of the chain as acetic acid.

$$CH_3(CH_2)_nCH_2CH_2COOH \rightarrow$$
$$CH_3(CH_2)_nCOCH_2COOH \rightarrow CH_3(CH_2)_nCOOH + CH_3COOH \quad (XIII)$$

The acetic acid is then degraded in the cell to yield carbon and energy for assimilatory purposes. The remaining fatty acid, now with two fewer carbons, is attacked again and again in a sequence identical to that in equation XIII, yielding acetic acid fragments and an increasingly small chain. Thus, the molecule is cleaved into small fragments that the cell can use for energy and biosynthetic processes. The initial intermediates in this oxidation, as shown in equation XII, rarely reach high concentrations, even in culture, except for organisms bringing about cometabolism. On occasion, however, the alcohol (RCH_2OH) produced in equation XII and one or more of the fatty acids ($R'COOH$) formed in equations XII or XIII combine to give esters (R'-CO-OCH_2R)

$$R'COOH + RCH_2OH \rightarrow R'CO\text{-}OCH_2R + H_2O \qquad\qquad (XIV)$$

Ester formation is known to occur during active heterotrophic proliferation in soil (3).

On occasion with large molecules, the fatty acid generated enzymatically in equation XII is acted on in a different way. The organisms oxidize the other end of the molecule before cleaving it. The steps are essentially the same as in equation XII, but the substrate is $CH_3(CH_2)_nCH_2CH_2COOH$ and the product is $HOOC(CH_2)_nCH_2CH_2COOH$. The latter is then degraded by removal of two carbon fragments (as acetic acid) from the ends of the chain, so again the organism is ultimately provided with small products that it uses within the cell.

Sometimes heterotrophs metabolize the hydrocarbon in a different fashion. Instead of acting on the terminal carbon, they oxidize the adjacent carbon.

$$CH_3(CH_2)_nCH_2CH_2CH_3 \rightarrow$$
$$CH_3(CH_2)_nCH_2CHOHCH_3 \rightarrow CH_3(CH_2)_nCH_2COCH_3 \quad (XV)$$

Compounds having the structure of the final product of this reaction, known as methyl ketones, are also generated in soil (3).

OXIDATION OF METHANE

Methane is biologically unique among the gaseous hydrocarbons in two ways. First, it is the only one produced in large amounts through microbial action. Second, it is metabolized by microorganisms that are often inactive on the larger hydrocarbon molecules. Moreover, the status of CH_4 as a carbonaceous substrate not having the carbon-carbon linkage typical of biologically synthesized organic molecules provides an interesting problem in the delineation of heterotrophy and chemoautotrophy.

Paddy soils frequently contain a surface film that brings about the oxidation of CH_4. To isolate the biological agents of the oxidation from the film, advantage is taken of their capacity to develop in inorganic media incubated in an atmosphere of CH_4 and O_2. Methane oxidizers also occur in well-drained soils, particularly in surroundings containing both CH_4 and O_2. The organisms probably subsist on the CH_4 released from the lower regions of the soil in anaerobic decomposition, and they seem to be more abundant in lower layers than near the surface.

In the process of CH_4 oxidation, O_2 is consumed and CO_2 produced. For each mole of CH_4 that disappears, two moles of O_2 theoretically are required.

$$CH_4 + 2O_2 \rightarrow CO_2 + 2H_2O \quad\quad (XVI)$$

Since CH_4 is also a carbonaceous nutrient for the organism, some of the gas will go into the formation of cell substance. The experimental values therefore do not necessarily agree with the theoretical equation because of carbon assimilation. The more efficient the bacterium is in assimilation, that is, efficient in utilizing the energy released by oxidation for cell synthesis, the greater will be the disparity from the theoretical ratio of $1CH_4 : 2O_2$. However, when CH_4 oxidation in nature is proceeding for some time and the population density is large and probably reasonably constant, a ratio of nearly $1 : 2$ is achieved (15).

Two schools of thought exist on the organisms responsible for CH_4 oxidation. One school maintains that CH_4 utilization is restricted to specialized bacteria that use CH_4 and methanol but no other aliphatic hydrocarbon as carbon and energy sources for multiplication. These bacteria, sometimes termed methylotrophs, are strict aerobes of various sizes and shapes, and they are classified in the genera *Methylomonas, Methylococcus, Methylosinus, Methylobacter,* and *Methylocystis.* A few physiologically related but morphologically different CH_4-oxidizing bacteria are also known. Although unable to grow on other simple compounds, these organisms may cometabolize them.

A second view holds that other heterotrophs can oxidize CH_4. These organisms, many of which are species of *Mycobacterium,* use CH_4 as well as other aliphatic hydrocarbons as carbon sources. In addition, bacteria of a few additional genera as well as fungi like strains of *Cephalosporium* and *Penicillium* are reported to use CH_4, and many of these are able to grow in media containing sugars or organic acids as carbon sources.

Cell suspensions of CH_4 oxidizers metabolize methanol, formaldehyde, and formic acid in addition to the gaseous hydrocarbon. Methanol, formaldehyde, and formic acid have been found to accumulate if the conditions of incubation are varied or if inhibitors are used. The results of such studies on the oxidation and accumulation of these one-carbon compounds indicate that the pathway of CH_4 oxidation proceeds as shown in equation XVII.

$$CH_4 \rightarrow CH_3OH \rightarrow HCHO \rightarrow HCOOH \rightarrow CO_2 \qquad \text{(XVII)}$$

DECOMPOSITION OF AROMATIC COMPOUNDS

Aromatic compounds, although rarely dominating the organic substrates reaching the soil, represent an important group of substances subjected to attack by the microflora. The lignin decomposed in vast amounts has a structure based on aromatic building blocks as do humus constituents. Plant tissues contain simple, monocyclic compounds with single benzene rings and also more complex molecules such as flavonoids, alkaloids, terpenes, and tannins. Fungi and actinomycetes often produce melanins, and these are polymers of aromatic units. Several of the amino acids in proteins and many of the synthetic chemicals used for pest control are modified aromatic hydrocarbons. These various substances present the community with a broad array of substrates for utilization and decomposition. Moreover, as many of the aromatic substances are phytotoxic, their destruction is essential in order to prevent the creation of unfavorable conditions for plant growth. Indeed, under certain conditions where active aerobic metabolism is hindered, such compounds may accumulate at sites in soil to levels that do injury to higher plants (32).

The aromatic compounds are thus constantly being made available to the microflora as humus, plant tissues, and microbial cells decompose and as synthetic chemicals are applied in connection with farm operations. For example, p-hydroxybenzoic, vanillic, p-coumaric, ferulic, and syringic acids are released as rice straw decays (20), and soils may contain the same compounds, sometimes in concentrations of 2 to 8 μmol/g of soil (8). Benzene, toluene, xylene, ethylbenzene, and naphthalene are also present in various soils (26). These molecules usually exist in a dynamic state, being both synthesized and destroyed, but the absence of toxicity demonstrates that the heterotrophs continue to serve as effective and potent agents of detoxication.

Laboratory tests have shown that a diverse group of simple aromatic compounds is decomposed when added to soil. Typical data are given in Table 14.1. The extent and rate of degradation are commonly assessed by measurements of CO_2 evolution or O_2 consumption or by assay of the disappearance of the test substance. In the latter instance, gas chromatography or other chromatographic techniques are frequently employed. These trials show that the type, number, and position of substituents on the benzene ring markedly affect the rate of mineralization. The substituents of interest have been the OH, CH_3, COOH, CH_2OH, NH_2, and SO_3H groups.

Many members of the soil microflora destroy aromatic hydrocarbons and their derivatives. Specific microorganisms decompose molecules such as phenol, naphthalene, and anthracene containing one, two, and three benzene rings, respectively. Apparently, bacteria are the dominant microbial group concerned in the mineralization of compounds of this type, largely species of *Pseudomonas, Mycobacterium, Acinetobacter, Arthrobacter,* and *Bacillus,* but *Nocardia* frequently appears prominently. Under some conditions, the fungi and streptomycetes may participate in the breakdown of aromatic hydrocarbons. The filamentous microorganisms may also be important in the decomposition of certain aromatic humus constituents. An individual isolate may even be capable of growing at the expense of 5, 10, or often more of such substrates, in culture at any rate.

Bacteria concerned in the aerobic decomposition are widespread, and almost every soil contains organisms growing on a variety of these compounds. Not only are the microorganisms widely distributed but the diversity of organic substances utilized by them is immense. Every naturally occurring aromatic hydrocarbon and many of those created by the chemist in the laboratory are metabolized. The species dominating the transformations are not substrate-

TABLE 14.1

Decomposition of Aromatic Compounds in Greenfield Sandy Loam (13)

	Quantity Decomposed, %		
Aromatic Acid	7 days	14 days	28 days
Benzoic	68	71	74
Caffeic	38	55	59
Protocatechuic	32	62	65
Vanillic	52	61	65
Veratric	59	65	69

specific for they utilize simple sugars or organic acids as well as hydrocarbons. Counts of bacteria potentially capable of using hydrocarbons such as benzene, toluene, vanillin, and phenanthrene depend on the location and the particular chemical, but values range from about 10^2 to 10^6 per gram of soil.

A consideration of the metabolism of aromatics is made difficult because of the baffling array of natural products and synthetic chemicals that are attacked. Moreover, the latter category includes molecules with nitro, chloro, and other substituents that are rare in tissue components; these substituents are typical of many pesticides and industrial pollutants. However, although initial phases of the pathways of degradation differ, the reactions are funneled in a direction that only a few common and key intermediates are produced, and these several are then metabolized by only a few and essentially similar processes. The most common of these focal intermediates are catechol, protocatechuic acid, and to a lesser degree, gentisic acid.

Catechol   Protocatechuic   Gentisic
                    acid   acid

The three molecules have in common the presence of two hydroxyls.

The first phase of aromatic metabolism is frequently the modification or removal of substituents on the benzene ring and the introduction of hydroxyl groups. Several generalizations apply to this first phase, generalizations that are necessary in light of the many and diverse pollutants and natural aromatics. (*a*) Methyl groups are often converted to carboxyl groups before ring cleavage, a reaction that proceeds stepwise (See A in Figure 14.4).

$$RCH_3 \rightarrow RCH_2OH \rightarrow RCHO \rightarrow RCOOH \qquad \text{(XVIII)}$$

Sometimes, however, the methyl is not removed before the ring is opened. (*b*) The carboxyl is often but not always removed before ring cleavage. (*c*) The methoxyl is replaced by a hydroxyl and gives rise to formaldehyde.

$$ROCH_3 + \tfrac{1}{2}O_2 \rightarrow ROH + HCHO \qquad \text{(XIX)}$$

See B in Figure 14.4. (*d*) Long aliphatic chains are usually shortened to give a residue with one or two carbon atoms left. This occurs stepwise, often by β-oxidation, as discussed above.

$$R(CH_2)_9COOH \rightarrow RCH_2COOH \qquad \text{(XX)}$$

(*e*) The chlorines present in many herbicides either are replaced by hydroxyls or hydrogens or stay on the ring until after it is opened, and then they are removed. (*f*) Nitro groups (—NO$_2$) characteristic of some pesticides and industrial pollutants can be replaced by hydroxyls (See C in Figure 14.4) and then they appear as nitrite, or the nitro substituent may be reduced to an amino group (—NH$_2$). Ultimately, therefore, the substituents are modified or removed and hydroxyls are added, giving rise to the focal intermediates. Representative reactions are depicted in Figure 14.4.

To obtain energy and carbon from the transformation, the active organism

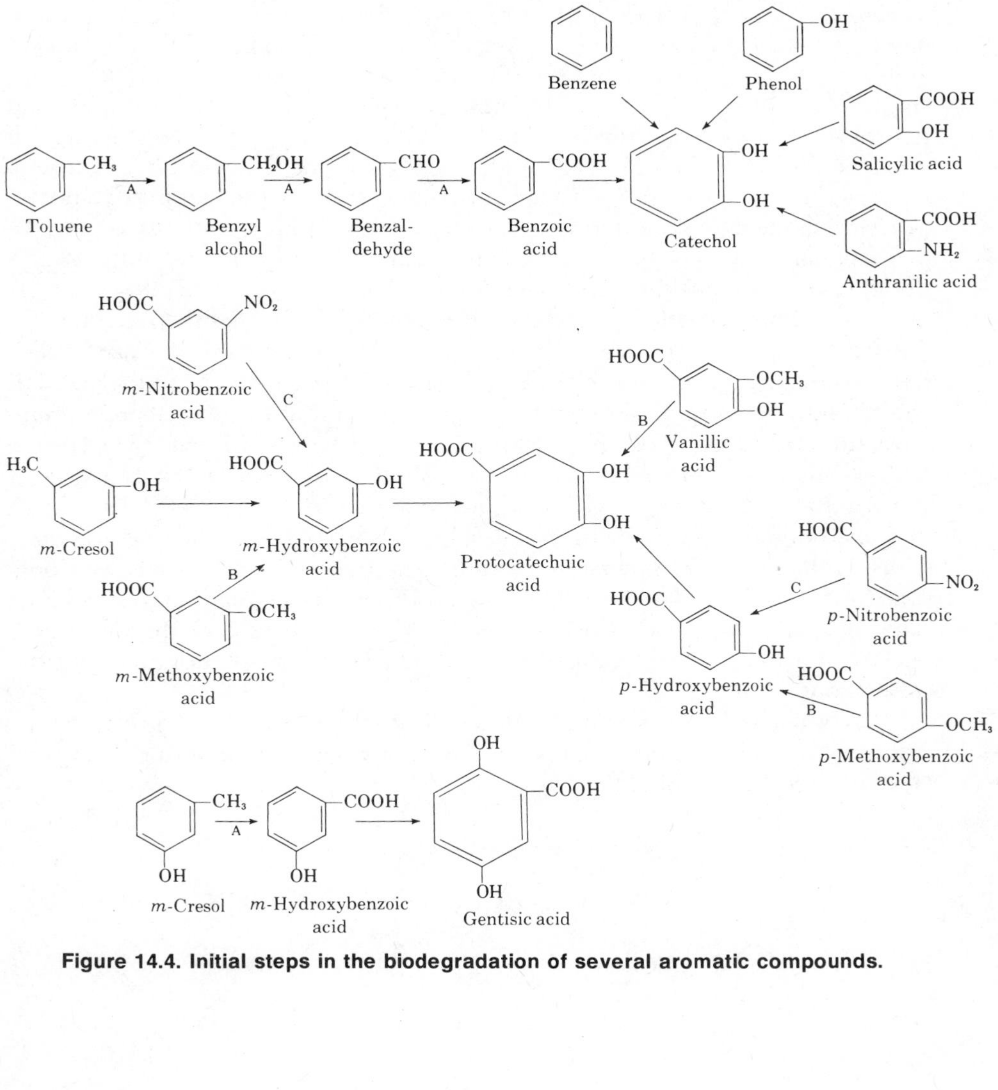

Figure 14.4. Initial steps in the biodegradation of several aromatic compounds.

must break the ring and convert the cleavage products to substrates that enter the metabolic pathways concerned with energy production and biosynthesis. The ring opening almost always requires the addition of oxygen that comes from O_2, a fact confirmed by finding that some of the $^{18}O_2$ provided in experimental trials appears in the product. This need for O_2 dictates that the organisms involved be aerobes. The three focal intermediates are degraded by five essentially similar pathways as shown in Figure 14.5. The five products of this phase—succinic, fumaric, pyruvic, and acetic acids and acetaldehyde—are easily and readily used by the cell.

Many microorganisms, when provided with aromatic substrates, bring about cometabolism but cannot proliferate by virtue of the reactions they catalyze because the process does not go sufficiently far to yield carbon and energy for growth. The process, nevertheless, may give high yields of products, and indeed sometimes nearly all of the substrate cometabolized is recovered as a single compound (16). Many organisms, especially among the fungi, add hydroxyls without being able to open the ring, but ring opening, cleavage of ether bonds, and removal of nitro groups also can be brought about through cometabolic conversions. Such transformations have particular importance in pesticide biodegradation, and toxic molecules arising apparently by cometabolism appear and may persist in soil for years or, on occasion, more than a decade. In view of the enormous practical importance and the body of information on pesticide microbiology, the topic will be considered separately.

Microorganisms are provided not only with monocyclic substrates, which have a single ring, but also bicyclic and polycyclic nutrients. Naphthalene and naphthol are structures with two rings, whereas anthracene and phenanthrene have three, and each of these can sustain the growth of many species of heterotrophs. Pesticides with two rings are often readily attacked in nature as in culture. From the pathway of breakdown outlined in Figure 14.6, it is evident that microorganisms act on these molecules to remove one cyclic unit at a time. Thus, anthracene and phenanthrene are converted to intermediates with only two rings, and the latter are further oxidized to benzene derivatives like catechol and gentisic acid, which are the focal intermediates already considered. Species of *Pseudomonas, Mycobacterium*, and other bacterial genera carry out the conversions.

Chlorinated biphenyls and diphenylalkanes have assumed great importance because the former include the PCB class widely used in industry and the latter include DDT and related pesticides.

Biphenyl

DDT

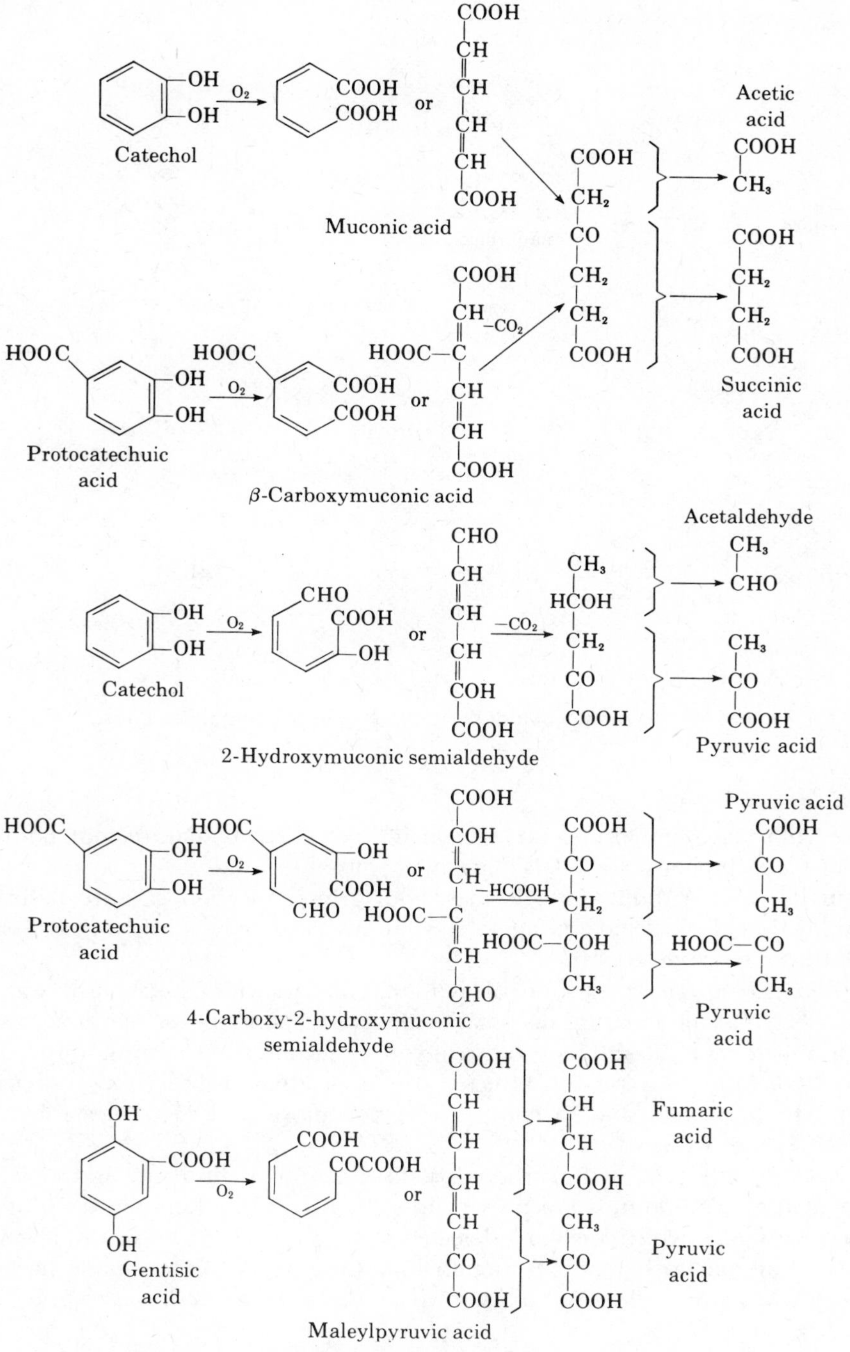

Figure 14.5. Pathways of cleavage of benzene rings.

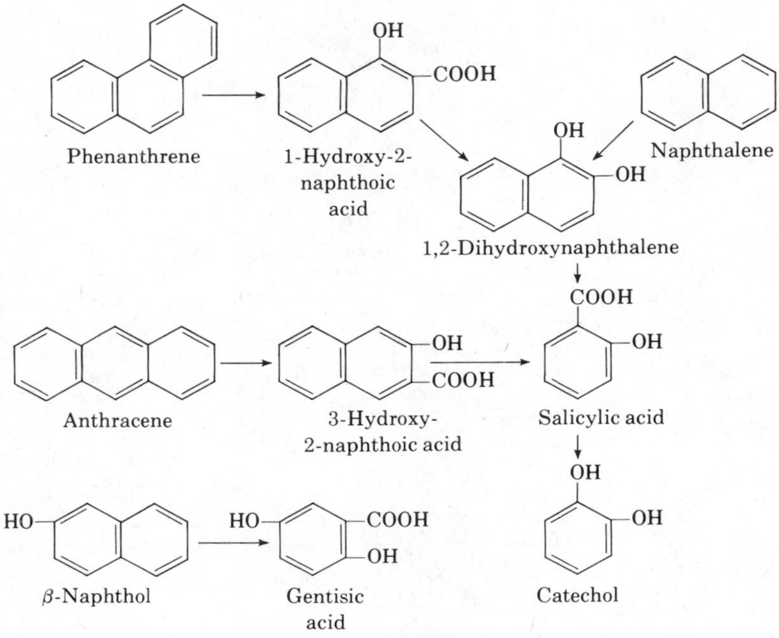

Figure 14.6. Decomposition of compounds with two and three rings.

The chlorinated molecules are reasonably persistent in nature, but both the chlorinated biphenyls and DDT are metabolized to some extent by microorganisms (4, 25). Without the chlorines, both groups of chemicals are destroyed quickly, but the chlorinated analogues are attacked only slowly and largely, if not solely, by cometabolism.

An unexpected outcome of research on pesticide metabolism was the discovery that microorganisms occasionally are responsible for the dimerization of aromatic molecules; that is, they convert molecules with one ring to products with two. A microbiologically effected process leading to dimerization occurs in soil with herbicides and in culture with nonchlorinated molecules related to DDT (Figure 14.7).

Some polycyclic hydrocarbons produce cancer in humans, and at least a few of these compounds are present in soils. Particular attention has centered on benzpyrene, which has been detected in many areas of the world, always in trace quantities to be sure. Although how these reach or are made in soil is unknown, microorganisms are known to both form and metabolize them (5, 12).

Field evidence and tests of soil samples in the laboratory reveal that the

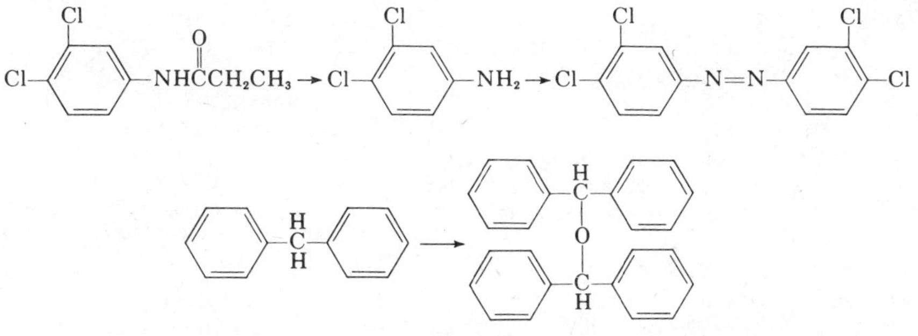

Figure 14.7. Microbial conversion of two compounds to dimers (7, 29).

aromatic ring is quite stable in the absence of O_2. These findings coincide with the need for O_2 in the well-characterized enzymatic reactions opening the ring and the paucity of isolates able to bring about the cleavage under anaerobiosis. Some evidence exists, nevertheless, that a few heterotrophic bacteria may be able to grow on benzoic or protocatechuic acids in the absence of O_2 provided that nitrate is present (24, 33). It is not clear whether such observations have much ecological relevance because nitrate is reduced rapidly in soils or horizons having no O_2 so that the necessary alternate electron acceptor is not available to the species possessing the novel physiological attribute.

REFERENCES

Reviews

Chapman, P. J. 1972. An outline of reaction sequences used for the bacterial degradation of phenolic compounds. In *Degradation of synthetic organic molecules in the biosphere.* National Academy of Sciences, Washington, pp. 17–55.

Klug, M. J. and A. J. Markovetz. 1971. Utilization of aliphatic hydrocarbons by microorganisms. *Advan. Microbial Physiol.,* 5:1–43.

Whittenbury, R. 1971. Hydrocarbons as carbon substrates. In P. Hepple, ed., *Microbiology— 1971.* Institute of Petroleum, London, pp. 13–24.

Wolfe, R. S. 1971. Microbial formation of methane. *Advan. Microbial Physiol.,* 6:107–146.

Literature Cited

1. Abeles, F. B., L. E. Craker, L. E. Forrence, and G. R. Leather. 1971. *Science,* 173:914–916.
2. Adamse, A. D., J. Hoeks, J. A. M. de Bont, and J. F. van Kessel. 1972. *Arch. Mikrobiol.,* 83:32–51.
3. Adamson, J. A., A. J. Francis, J. M. Duxbury, and M. Alexander. 1975. *Soil Biol. Biochem.,* 7:45–50.
4. Ahmed, M. and D. D. Focht. 1973. *Can. J. Microbiol.,* 19:47–52.
5. Andelman, J. B. and M. J. Suess. 1970. *Bull. W.H.O.,* 43:479–508.

6. Antoniewski, J. and R. Schaefer. 1972. *Ann. Inst. Pasteur,* 123:805–819.
7. Bartha, R. and D. Pramer. 1970. *Advan. Appl. Microbiol.,* 13:317–341.
8. Batistic, L. 1974. *Plant Soil,* 41:73–80.
9. Bollag, J.-M. and S. T. Czlonkowski. 1973. *Soil Biol. Biochem.,* 5:673–678.
10. Ehhalt, D. H. 1973. In G. M. Woodwell and E. V. Pecan, eds., *Carbon in the biosphere.* U. S. Atomic Energy Commission, Washington, pp. 144–157.
11. Elliott, L. F. and T. M. McCalla. 1972. *Soil Sci. Soc. Amer. Proc.,* 36:68–70.
12. Gibson, D. T., V. Mahadevan, D. M. Jerina, H. Yagi, and H. J. C. Yeh. 1975. *Science,* 189:295–297.
13. Haider, K., and J. P. Martin. 1975. *Soil Sci. Soc. Amer. Proc.,* 39:657–662.
14. Haines, J. R. and M. Alexander. 1974. *Appl. Microbiol.,* 28:1084–1085.
15. Hoeks, J. 1972. *Soil Sci.,* 113:46–54.
16. Horvath, R. S. and M. Alexander. 1970. *Can. J. Microbiol.,* 16:1131–1132.
17. Jackson, M. B. and D. J. Campbell. 1975. *New Phytol.,* 74:397–406.
18. Johnson, D. R. and L. R. Frederick. 1971. *Agron. J.,* 63:573–575.
19. Jones, J. G. and M. A. Edington. 1968. *J. Gen. Microbiol.,* 52:381–390.
20. Kuwatsuka, S. and H. Shindo. 1973. *Soil Sci. Plant Nutr.,* 19:219–227.
21. Kvasnikov, E. I., O. A. Nesterenko, V. A. Romanovskaya, and S. A. Kasumova. 1971. *Mikrobiologiya,* 40:280–288.
22. Lynch, J. M. and S. H. T. Harper. 1974. *J. Gen. Microbiol.,* 80:187–195.
23. Moraghan, J. T. and K. A. Ayotade. 1968. *Trans. 9th Intl. Cong. Soil Sci.,* 4:699–707.
24. Oshima, T. 1965. *Z. Allg. Mikrobiol.,* 5:386–394.
25. Pfaender, F. K. and M. Alexander. 1972. *J. Agr. Food Chem.,* 20:842–846.
26. Simonart, P. and L. Batistic. 1966. *Nature,* 212:1461–1462.
27. Smith, K. A. and S. W. F. Restall. 1971. *J. Soil Sci.,* 22:430–443.
28. Stadtman, T. C. and H. A. Barker. 1951. *J. Bacteriol.,* 61:81–86.
29. Subba-Rao, R. V. and M. Alexander. Unpublished data.
30. Takai, Y. 1970. *Soil Sci. Plant Nutr.,* 16:238–244.
31. Takai, Y., J. Macura, and F. Kunc. 1969. *Folia Microbiol.,* 14:327–333.
32. Tousson, T. A., A. R. Weinhold. R. G. Linderman, and Z. A. Patrick. 1968. *Phytopathology,* 58:41–45.
33. Williams, R. J. and W. C. Evans. 1975. *Biochem. J.,* 148:1–10.
34. Yamane, I., and K. Sato. 1963. *Soil Sci. Plant Nutr.,* 9:32–36.

THE NITROGEN CYCLE

The biological availability of nitrogen, phosphorus, and potassium is of considerable economic importance because they are the major plant nutrients derived from the soil. Of the three, nitrogen stands out as the most susceptible to microbial transformations. This element is a key building block of the protein molecule upon which all life is based, and it is thus an indispensable component of the protoplasm of plants, animals, and microorganisms. Because of the critical position of the nitrogen supply in crop production and soil fertility, a deficiency markedly reduces yield as well as quality of crops; and because this is one of the few soil nutrients that is lost by volatilization as well as by leaching, it requires continual conservation and maintenance.

Nitrogen undergoes a number of transformations that involve organic, inorganic, and volatile compounds. These transformations occur simultaneously, but individual steps often accomplish opposite goals. The reactions may be viewed in terms of a cycle in which the element is shuttled back and forth at the discretion of the microflora. A small part of the large reservoir of N_2 in the atmosphere is converted to organic compounds by certain free-living microorganisms or by a microbial-plant association that makes the element directly available to the plant. The nitrogen present in the proteins or nucleic acids of plant tissues is used by animals. In the animal body, the nitrogen is converted to other simple and complex compounds. When the animals and plants are subjected to microbiological decay, the organic nitrogen is released as ammonium, which in turn is utilized by the vegetation or is oxidized to nitrate. The latter ion may be lost by leaching, it may serve as a plant nutrient or, alternatively, it may be reduced to ammonium or to gaseous N_2, which escapes to the atmosphere, thereby completing the cycle. The present discussion is concerned with the nitrogen cycle of terrestrial habitats, but the same general sequence occurs in aquatic environments.

The portions of the nitrogen cycle governed by microbial metabolism are composed of several individual transformations. In *nitrogen mineralization*, part of the large reservoir of organic complexes in the soil is decomposed and converted to the inorganic ions used by plants, ammonium and nitrate.

Microbial mineralization results in the degradation of proteins, polypeptides, amino acids, nucleic acids, and other organic compounds. In contrast to the conversion of the complex to the simple substances is *nitrogen immobilization* or assimilation. Microbiological immobilization leads to the biosynthesis of the complex molecules of microbial protoplasm from ammonium and nitrate. The mineralization of organic nitrogen and the microfloral assimilation of inorganic ions proceed simultaneously.

Nitrogen, once in the nitrate form, may be lost from the soil in several ways. Because of its solubility in the soil solution, nitrate readily moves downward out of the zone of root penetration. Nitrate and ammonium will also be removed to satisfy the nutrient demand of the plant cover. The greatest biological leak in the otherwise closed cycle in soil is through *denitrification*, whereby nitrogen is removed entirely from the realm of ready accessibility because the end product of denitrification, N_2, is unavailable to most macro- and microorganisms.

Any leaks in the cycle deplete the soil's nitrogen reserve and eventually could have drastic effects on man's agricultural economy. But since the leakage to the atmosphere is omnipresent, there must exist a reverse process to maintain the balance; otherwise the world's nitrogen reservoir would be diminishing continuously. Although inert as far as plants, animals, and most microorganisms are concerned, N_2 is acted on by certain microorganisms, sometimes in symbiosis with a higher plant, which can use it as a nitrogen source for growth. This process, *nitrogen fixation*, results in the accumulation of new organic compounds in the cells of the responsible organisms. The N_2 thus fixed reenters general circulation when the newly formed cells are in turn mineralized.

By means of these reactions, the subterranean microflora regulates the supply and governs the availability and chemical nature of the nitrogen in soil.

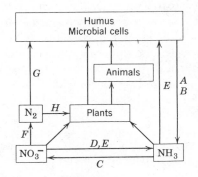

A. Ammonification   E. Immobilization
B. Mineralization   F. Denitrification
C. Nitrification   G. N_2 fixation. Nonsymbiotic
D. Nitrate reduction   H. N_2 fixation. Symbiotic

15
Mineralization and Immobilization of Nitrogen

The soil nutrient that plants require in greatest quantity is nitrogen. Yet, despite its critical role in plant nutrition, nitrogen is assimilated almost entirely in the inorganic state, as nitrate or ammonium. On the other hand, the bulk of the nitrogenous materials found in soil or added in the form of plant residues is organic and, hence, largely unavailable. The release of the bound element and the mobilization of the vast reservoir of organically combined nitrogen are essential to the recycling of the nutrient and therefore to soil fertility.

The conversion of organic nitrogen to the more mobile, inorganic state is known as *nitrogen mineralization*, a process analogous to the liberation of CO_2 from carbonaceous materials in that both transformations result in the release of the elements in inorganic forms. The two processes are also similar in that they are the sole means of regenerating the nutrient in a form usable for the development of green plants. As a consequence of mineralization, ammonium and nitrate are generated and organic nitrogen disappears. These products delineate two distinct microbiological processes, *ammonification*, in which ammonium is formed from organic compounds, and *nitrification*, a term that usually is taken to mean the oxidation of ammonium to nitrate. Ammonium is typically associated with a waste-product overflow in microbial metabolism, the accumulated ammonium representing the quantity of substrate nitrogen in excess of the microbial demand. Nitrification, however, is usually associated with the energy-yielding reactions in the metabolism of autotrophic bacteria.

Mineralization of the nitrogen in humus, proteins, nucleic acids, or related materials is determined by measurements of the formation of inorganic (mineral) nitrogen. Early microbiologists chose to limit their analyses to ammonium, the first inorganic product, but the results were soon found to be misleading. Nitrate production has also been used as a criterion of mineralization. The latter approach is not as objectionable as the former, but it is not too difficult to select localities where ammonium production is rapid and nitrate is not formed at all. Clearly, the most suitable procedure is the assay of all inorganic products—ammonium, nitrite, and nitrate. More sophisticated techniques requiring the stable isotope, ^{15}N, have been developed. Isotopic procedures for the measurement of mineralization rates are indispensable in environments in which the conversion of inorganic nitrogen to microbial proteins is proceeding rapidly.

NITROGEN MINERALIZATION

Almost all of the nitrogen found in surface soil horizons is in organic combination. Nevertheless, the chemical composition of this portion of the soil organic fraction is not completely understood. Detected in extracts or hydrolyzates are essentially all of the known amino acids in combined form, minute amounts of free amino acids, amino sugars such as glucosamine and galactosamine, and several of the purine and pyrimidine bases derived from nucleic acids. Bound amino acids generally constitute from 20 to 50 percent of the total humus nitrogen while from 5 to 10 percent is amino-sugar nitrogen. The bound amino acids are in large molecules, but the characteristics of these molecules are uncertain. The amino acids may be bound in proteins, and although a part probably is, much could exist in large polymers with phenolic compounds or in mucopeptides. At least 20 free amino acids have been detected in soil, but the total quantity is quite low, usually less than 2 ppm. Amides are probably present too. Some of the organic nitrogen, usually less than 1.0 percent, is in the form of purine and pyrimidine bases, and possibly a part—either small or large—of these bases exists in soil as nucleic acids. The chemical identity of the remainder is unknown.

The nitrogenous compounds present in the soil organic fraction persist for long periods in nature, the resistance to attack being so appreciable that only a small proportion of the nitrogen reservoir of the soil is mineralized in each growing season. The anomalous resistance to biological destruction has aroused considerable interest, and several hypotheses have been advanced to account for the slow mineralization. One hypothesis states that nitrogenous compounds form complexes or polymers with phenols or polyphenols, rendering the nitrogen-containing substances less susceptible to digestion. In support of this view are experiments demonstrating that proteins complexed with polyphenols and amino acids linked with phenols into polymers are resistant to microbial attack, by contrast with the free proteins and amino acids (6, 33). A second hypothesis proposes that clay minerals spare nitrogenous substrates by entrapping them within the lattice of the clay crystal. Extracellular proteolytic enzymes, themselves proteins, are adsorbed by clays, and they may thereby be rendered less active. The demonstration of a protective action in vitro, however, does not necessarily signify that the same holds true in nature.

The net change in the amount of inorganic nitrogen, N_i, is expressed by the relationship

$$\Delta N_i = \text{organic nitrogen mineralized} - (N_a + N_p + N_l + N_d) \qquad \text{(I)}$$

where N_a, N_p, N_l, and N_d represent the nitrogen assimilated by the microflora, removed by the plant, lost by leaching, and volatilized by denitrification, respectively. The rate at which organic nitrogen is converted to ammonium and nitrate is termed the mineralization rate, and the velocity of such release in environments receiving nitrogen-rich crop residues ranges from less than 1.0 to 20 ppm nitrogen per day. The rate represents the gross liberation of N_i and is not the amount typically seen in practice. The net quantity of inorganic nitrogen produced for a constant quantity of metabolized substrate is governed by leaching, denitrification, and particularly microbial assimilation. Consequently, though the production of N_i may be great, no increase or sometimes a decrease may be observed in measurements of the ammonium or nitrate content of soil.

The mineralization of organic nitrogen and organic carbon are related to one another. In unamended soil, the two elements are mineralized at parallel rates, and the ratio of CO_2-C to N_i produced is essentially constant at approximately 7 to 15 : 1. Soils active in the one transformation are generally active in the second. The ratio of C mineralized : N mineralized decreases somewhat as the rate of N_i production increases so that a microflora vigorously forming nitrate tends to release carbon and nitrogen in a ratio of ca. 7 : 1 while those least active exhibit ratios near 15 : 1 (36). The equilibrium may be upset by the introduction of external substrates; for example, protein-poor residues favor CO_2 evolution whereas protein-rich materials favor N_i release.

Little is known of the immediate source of the ammonium formed by decomposition of the many nitrogenous complexes found within the soil. With plant materials, on the other hand, the water-soluble nitrogen is often the fraction most readily converted to ammonium, and the rapidity of ammonium release from crop residues is therefore often closely related to the quantity of water-soluble nitrogen.

In light of the essentiality of nitrogen for plant nutrition, considerable effort has been devoted to devising a technique for predicting the quantity of soil nitrogen that will be made available to a crop during a given growing season. If this quantity were known, then the amount of fertilizer that would have to be applied could be predicted, and excessive rates of fertilization could be avoided. The problem of establishing a suitable technique is essentially one of devising a means of predicting the microbial activity in the field by some suitable operation in the laboratory. Two general approaches have been employed: (a) determining the quantity of nitrogen mineralized in a reasonably short incubation period in the laboratory and (b) establishing what fraction of humus nitrogen is correlated with nitrogen mineralization rate. The latter approach is made difficult because the chemical identity of the soil constituents that are mineralized in a growing season is unknown, and hence arbitrary fractionation procedures have been introduced in the hope that one or more would give substances whose yield would correlate with the extent of mineralization in a designated period. Whatever the procedure, its utility must be validated by comparing its results with those obtained when plant nitrogen uptake is assessed, either in the greenhouse or in the field.

The vast effort to devise such a test for nitrogen availability has led to two sorts of results. For some soils and in certain regions or with particular crops, significant correlations have been obtained between nitrogen uptake by plants and either mineralization rate or the yield of nitrogen by some extraction technique. A suitable laboratory test would thus seem to have been found. On the other hand, with other soils, in other regions, or with different crops, no correlation was evident, and hence an acceptable laboratory procedure would appear to be still lacking. Some investigators have proposed the use of a fungus, often an *Aspergillus* strain, to measure the quantity of nitrogen made available by the community, but also in this instance the results have not been entirely satisfying. For the present, therefore, it appears that incubation or extraction procedures reliably predict the nitrogen made available to higher plants in some situations, but for reasons not as yet clear, the very same methods have little predictive value under other circumstances.

Mineralization has also been the focus of considerable recent interest owing to the use of land for the disposal of large quantities of organic materials, either as digested sewage sludge from urban regions or as animal manure from feedlots containing thousands of cattle. If mineralized, the organic nitrogen would be released as ammonium, which would then be oxidized to nitrate. Nitrate entering the groundwater in high concentrations is undesirable from the standpoint of public health. In addition, when large quantities of the wastes are spread in the field, free ammonia may be released in large amounts to the atmosphere, and on occasion, presumably when manure is present in volumes such that aeration is inadequate to maintain aerobic populations, various volatile amines may be evolved as well (18, 23). Indeed, the public concern with environmental quality has led to assessments of the formation of many volatile compounds, and these evaluations have led to estimates that mineralization gives rise to almost 5×10^{12} kg of atmospheric ammonia-N per year, a figure about eight times greater than that associated with the production of all the nitrogen oxides (28). Even over grazed pastures, a considerable quantity of ammonia and possibly volatile amines may be discharged to the atmosphere (7).

MICROBIOLOGY

Innumerable investigations have dealt with the ability of microorganisms to utilize nitrogen-containing organic molecules. With few exceptions, ammonium is a major product of the reaction. Population estimates made on diverse soils reveal that approximately 10^5 to 10^7 organisms per gram are ammonifiers, but such determinations have little importance by themselves because estimates of the population size are governed by the nitrogen compound provided as substrate in the culture medium. A simple amino acid has a vast group of microorganisms acting on it while the nitrogen of chitin is freed by only certain select genera. Proteins serve as carbon or nitrogen sources for a large segment of the microflora, and populations of gelatin, casein, and albumin decomposers from 10^5 to 10^7 have been observed in the surface horizon. The data in Table 15.1 demonstrate the capacity of some of the predominant bacteria to utilize gelatin. It is apparently not uncommon for one-sixth to one-third of the bacteria detected on dilution plates to liquefy the protein, gelatin.

In nature, the breakdown of proteins and other nitrogenous substances is the result of the metabolism of a multitude of microbial strains each of which has some function—large or small—in the pathway of degradation. A diverse flora liberates ammonium from organic nitrogen compounds; indeed, almost all bacteria, fungi, and actinomycetes attack some complex form of the element, but the rate of decomposition and the compounds thus utilized vary with the species and genus. Because the ultimate liberation of ammonium from organic matter is an action associated with many physiologically dissimilar microorganisms, nitrogen is mineralized in the most extreme conditions. The amount of ammonium that accumulates, however, varies with the organism, the substrate, the soil type, and environmental conditions.

Microorganisms synthesize extracellular, proteolytic enzymes for the decomposition of proteins. The initial stages of the transformation are mediated by catalysts functioning outside of the cell, where they cleave the large protein

TABLE 15.1
Utilization of Gelatin by the Bacterial Flora of Canadian Soils (31)

	% of Group That	
Bacterial Group	**Grows on Gelatin**	**Liquefies Gelatin**
Short rods		
Gram positive	57	25
Gram negative	57	14
Gram variable	34	0
Arthrobacter spp.	100	100
Coccoid rods, gram positive	39	17
Cocci	42	33
Long rods, nonsporulating	86	14
Bacillus spp.	67	44

molecule to simpler units. Among the proteolytic organisms are aerobic bacteria, fungi, and actinomycetes as well as certain facultative and strict anaerobes. Several intermediary products are formed in aerobic proteolysis; these quickly disappear so that the major end products of protein breakdown are CO_2, ammonium, sulfate, and water. Anaerobically, foul-smelling compounds are released in the degradation of protein-rich materials, a conversion known as *putrefaction*. The final products of the anaerobic transformation are ammonium, amines, CO_2, organic acids, mercaptans, and hydrogen sulfide.

Pure proteins are decomposed readily by species of *Pseudomonas, Bacillus, Clostridium, Serratia,* and *Micrococcus.* The first two genera are particularly numerous in soil receiving purified proteins. Proteolysis is a common criterion in bacteriological classification, and the digestion of casein and the liquefaction of gelatin are important taxonomic characters. In spite of the emphasis given to the study of casein- and gelatin-hydrolyzing bacteria, many of the strains are not important in the decomposition of the more common proteins found in nature. The proteins in plant tissues need not be acted upon by the same bacteria that rapidly hydrolyze the pure substances in vitro.

Many fungi readily decompose proteins, amino acids, and other nitrogenous compounds with the liberation of considerable quantities of ammonium. Soil isolates use these materials as carbon and nitrogen sources and, with the sulfur-containing amino acids, for sulfur as well. Genera whose proteolytic enzymes have received extensive study include *Alternaria, Aspergillus, Mucor, Penicillium,* and *Rhizopus.* Fungi frequently release less ammonium than bacteria since the fungi assimilate more of the nitrogen for the purpose of cell synthesis. Without question, the fungi occupy a dominant position in proteolysis in certain soils, particularly in acid localities.

Extracellular proteolytic enzymes are likewise produced by numerous actinomycetes, and the attack by actinomycetes on vegetable proteins, serum proteins, casein, egg albumen, and gelatin in culture has been fully documented. About 15 to 70 percent of the actinomycete isolates from some soils produce protein-hydrolyzing enzymes in culture (14), and protein-cleaving enzymes are also found among thermophiles of this group. The protein-metabolizing enzymes of certain actinomycetes are released only by old cells that have begun to autolyze, and these may in reality be intracellular catalysts liberated on lysis of the hyphae. However, the slow growth habits of the actinomycetes make their role in soil proteolysis difficult to assess.

The microbiology of protein breakdown in soil is inadequately understood. It is probable that bacteria dominate in neutral or alkaline environments, but fungi and possibly actinomycetes contribute to the transformation. The key group in acid habitats is, with little question, the fungi. Anaerobic bacteria dominate the flora when O_2 is lacking. Following the initial degradation of the protein, microorganisms appear that are incapable of attacking proteins but that can utilize the metabolic wastes of the proteolytic species.

ENVIRONMENTAL INFLUENCES

The biochemical heterogeneity of the microflora bringing about nitrogen mineralization is a critical factor in determining the influence of environmental factors on the transformation. Because aerobic and anaerobic, acid-sensitive and acid-resistant, and spore-forming and non-spore-forming microorganisms function in the degradation of nitrogenous materials, at least some segment of the population is active regardless of the peculiarities of the habitat as long as microbial proliferation is possible. Consequently, mineralization is never eliminated in arable land, but the rate is markedly affected by the environment. Physical and chemical characteristics of the habitat such as moisture, pH, aeration, temperature, and the inorganic nutrient supply will govern the velocity of mineralization. The protective benefits to nitrogen compounds of clay and possibly lignin and polyphenols are similarly of importance.

The rate of production of inorganic nitrogen is closely correlated with the total nitrogen content of the soil. Sites rich in nitrogen liberate more of the inorganic ions than deficient areas in a given time interval. It is possible to approximate the amount of the element liberated during a growing season under optimal climatic conditions because of the correlation between N_i production and total nitrogen. In the northeastern United States, for example, crops that make their maximum development in the summer have made available to them each year 2 to 4 percent of the total humus nitrogen while crops growing in the cooler part of the year have 2 percent or less made available during the period of peak growth. In poorly drained land, the unsuitable water relationships depress microbial metabolism, thereby lowering the mineralization percentage. As a general rule, 1 to 4 percent may be taken to indicate the quantity of soil organic nitrogen released to plants during the growing season in temperate latitudes.

The ammonifying populations include aerobes and anaerobes, and organic nitrogen is mineralized, consequently, at moderate or at excessively high moisture levels. Ammonium is slowly formed at water levels slightly below the permanent wilting percentage, but improving the moisture status stimulates mineralization. In arid or semiarid regions and in climatic zones having wet-dry cycles, the onset of rainfall is typically associated with a rapid initiation of mineralization. In Figure 15.1 are presented typical data showing the effect of moisture level on N_i production. Although soils differ in the precise moisture values that are optimal for the conversion, the optimum for ammonification generally falls between 50 and 75 percent of the water-holding capacity. Ammonification is not eliminated by soil submergence, and the process is rapid in wet paddy fields, where the O_2 level is quite low. As a rule, soils active in aerobic mineralization also form ammonium readily in the absence of O_2, and those slow in this process in aerated conditions likewise generate ammonium slowly under anaerobiosis (34). The final inorganic product, however, is

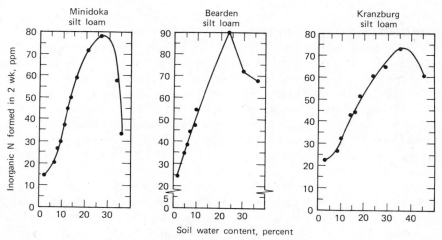

Figure 15.1. Effect of soil water content on inorganic nitrogen accumulation (30). Reproduced by permission of the Soil Science Society of America.

affected by the presence of O_2 as ammonium is converted to nitrate only in well-aerated habitats.

An as yet unexplained change occurs when soils become dry and then wet again. When remoistened, the soils that had been previously dry bring about mineralization at a more rapid rate than if they were kept continuously wet. If a number of such drying and wetting cycles occur, considerable quantities of the organic nitrogen may be degraded to the inorganic form, although the rate of the process declines in the later cycles. Furthermore, the longer is the dry period, generally the greater is the amount of inorganic nitrogen released during the subsequent wet phase. The drying-wetting cycle may make inaccessible substrates more easily available to the community or the drying may cause cell disintegration; however, none of the explanations for this anomalous flush of microbial development has convincing supporting evidence.

Mineralization is influenced by the pH of the environment. All other factors being equal, the production of inorganic nitrogen—ammonium plus nitrate—is greater in neutral than in acid environments (13), although some soils show little influence of pH on the transformation. Acidification tends to depress but does not eliminate mineralization. The liming of acid soils is commonly stimulatory as the pH is brought closer to the optimum for the active microflora. Organic nitrogen accumulates in highly acid sites, presumably because of the slow mineralization, so that a rapid release is noted when such soils are limed.

Temperature also affects the mineralization sequence as each biochemical step is catalyzed by a temperature-sensitive enzyme produced by microorganisms whose growth is in turn conditioned by temperature. Thus, at 2°C, the microflora slowly mineralizes the organic complexes, but there is no increase in ammonium or nitrate when soil is frozen. Elevation of the temperature from the freezing point to the optimum enhances the mobilization of nitrogen in proportion to the greater warmth. In contrast to most microbiological transformations, the optimum temperature for ammonification is not in the mesophilic range but rather it is above 40°C, usually between 40 and 60°C. Ammonium accumulates in composts and manure piles maintained at 65°C, demonstrating thereby the activity of thermophiles. Freezing followed by thawing, much as drying followed by wetting, often increases the rate of degradation of humus nitrogen as compared with soils not exposed to freezing; in this instance, as with the drying cycle, the reason for the effect remains obscure.

The addition of inorganic nitrogen sometimes enhances the mineralization of humus, a stimulation that can be demonstrated by measuring N_i accumulation or plant uptake of the element (4, 35). To show such an enhancement commonly requires a means for distinguishing between the added inorganic nitrogen and that formed from soil constituents, but this can be accomplished readily by labeling the former with ^{15}N. On the other hand, such amendments appear to have no influence in other soils; the stimulation, where it occurs, may result from the response of the community to the addition, the larger and more active community then being able to degrade the nitrogenous complexes more readily than the original microflora.

Cultivation of certain virgin lands causes a massive decline in their content of organic nitrogen. This promotion of mineralization is associated with and probably is dependent upon the parallel decline in organic carbon content of virgin lands that are brought into food production. So much nitrogen is released as ammonium, which is then oxidized to nitrate, that serious nitrate pollution of the underlying waters, and ultimately adjacent bodies of surface water, may occur. In some instances, thousands of kilograms of nitrogen may be mineralized per hectare in a few years. After a number of years, the decline ceases, and the level of humus nitrogen remains at a new steady-state concentration.

NUCLEIC ACID METABOLISM

Nucleic acids are second only to proteins in importance as nitrogenous substrates for the microbial flora. These nitrogen-rich substances are found in plant and animal tissues and are also concentrated within microbial protoplasm; consequently, the fate of nucleic acids has a considerable bearing upon the mineralization sequence. Plant and animal tissues and microbial cells contain two types of nucleic acids, ribonucleic acid (RNA) and deoxyribonucleic acid (DNA). Structurally, each is a polynucleotide, that is, a polymer of the structural unit known as a mononucleotide. Mononucleotides consist of a purine or pyrimidine base, a sugar, and phosphate. The sugar in RNA is ribose, and in DNA it is deoxyribose. Adenine and guanine are the purines that are found in both RNA and DNA molecules. Of the pyrimidines, cytosine is found in RNA and DNA, uracil only in the former, and thymine in the latter type of nucleic acid. Both the purine and pyrimidine bases bound in some manner, probably to a significant extent as nucleic acids, have been found in soils.

Purine base   Pyrimidine base

Nucleic acids and purines and pyrimidines or their derivatives are readily degraded when added in pure form to soil (12). The nucleic acids, purines, and pyrimidines are adsorbed by clays, however, and the substrate thus retained may be protected to a significant extent from microbial degradation.

In the decomposition of nucleic acids, the long molecules are initially converted into smaller fragments and ultimately to the individual mononucleotides. The attack is initiated by the enzymes ribonuclease and deoxyribonuclease that act on RNA and DNA, respectively, or by enzymes hydrolyzing both RNA and DNA. The latter are termed nucleases. Extracellular ribonuclease is formed by species of *Bacillus, Pseudomonas,* and *Mycobacterium* among the bacteria, and *Aspergillus, Cephalosporium, Fusarium, Mucor, Penicillium,* and *Rhizopus* among the fungi. Extracellular deoxyribonuclease is characteristic of species of *Arthrobacter, Bacillus, Clostridium,* and *Pseudomonas* of the bacteria and *Cladosporium* and *Fusarium* of the fungi. Many other genera also are able to elaborate the enzymes, moreover: In the few soils studied, from 1 to 79 percent of the isolates produce extracellular ribonuclease and up to 86 percent excrete deoxyribonuclease, so the potential for degrading the polynucleotides is indeed common. Furthermore, when the polymers are added to soil, short gram-negative rods especially of the pseudomonad type proliferate, and the substrates are soon extensively decomposed (10, 21).

The enzymatic hydrolysis of RNA and DNA by soil samples is estimated by incubation of the sample with the nucleic acid and measuring product accumulation after a short incubation period. In common with most depolymerases, the activity varies with soil type, pH, and vegetation, and it is greater in surface horizons and near plant roots.

As with other enzymes catalyzing the hydrolysis of polymers, nucleic acid-hydrolyzing enzymes may be of the endo or exo type. The former enzymes attack the polymer at many places within the nucleic acid molecule, whereas the latter act on the end of the molecule, progressively making it shorter. A few of these enzymes, however, may catalyze both an endo and an exo type of cleavage. The products of degradation by ribonucleases, deoxyribonucleases, or the enzymes acting on both RNA and DNA are often the mononucleotides, but frequently di-, tri-, or oligonucleotides accumulate; these are molecules containing two, three, or several nucleotides in a single compound. The exo enzymes typically generate the mononucleotides, and the endo enzymes commonly liberate chiefly or solely the oligomers. The oligonucleotides, however, are ultimately cleaved by other enzymes, giving rise to the mononucleotides.

By contrast with the polysaccharides such as cellulose, the monomers formed—in this instance, they are the mononucleotides—are not all identical. They may have one or another of the purine or pyrimidine bases, either ribose or deoxyribose, as well as phosphate. Regardless of the nitrogenous base in the mononucleotide, however, its fate is usually the same as the organisms decompose the molecule and, in the process, obtain energy, carbon, and nitrogen to support proliferation. After the monomer appears, the next step in the decay generally is the removal of the phosphate from the simple nucleotide to yield a substance composed of a purine or pyrimidine base still linked to the sugar. The final stage in the degradation is the separation of the purine or pyrimidine base from the sugar.

$$(\text{base-sugar-P})_n \rightarrow \text{base-sugar-P} \rightarrow \text{base-sugar} \rightarrow$$
Nucleic acid   Mononucleotide

$$\text{sugar} + \text{purine or pyrimidine base} \quad (II)$$

In the subsequent metabolism of the sugar, CO_2 is evolved and, depending on the availability of O_2, organic acids may be produced. The nitrogenous bases are most often decomposed as shown in Figure 15.2. Some of the metabolites shown as products of uric acid breakdown have been detected in soil (8) as well as in culture. *Nocardia* and bacteria of the genera *Pseudomonas, Micrococcus, Corynebacterium,* and *Clostridium* degrade the purine and pyrimidine compounds shown in the figure, but it is unlikely that they are the sole responsible agents in soil.

DECOMPOSITION OF UREA

Urea is a product of the destruction of the nitrogenous bases contained in the nucleic acids (Figure 15.2). Urea is an important fertilizer too, but it may also enter the soil with the excretions of higher animals. The position of urea as an intermediate in microbial metabolism, as an animal excretory product, and as a fertilizer makes it a key compound in the nitrogen cycle.

Urea applied to soil is very readily hydrolyzed, and much of the added urea nitrogen may be transformed to ammonium in several days. The pH of soils receiving urea rises as ammonium is produced, and the pH in the immediate vicinity of particles of urea fertilizer may reach values in excess of

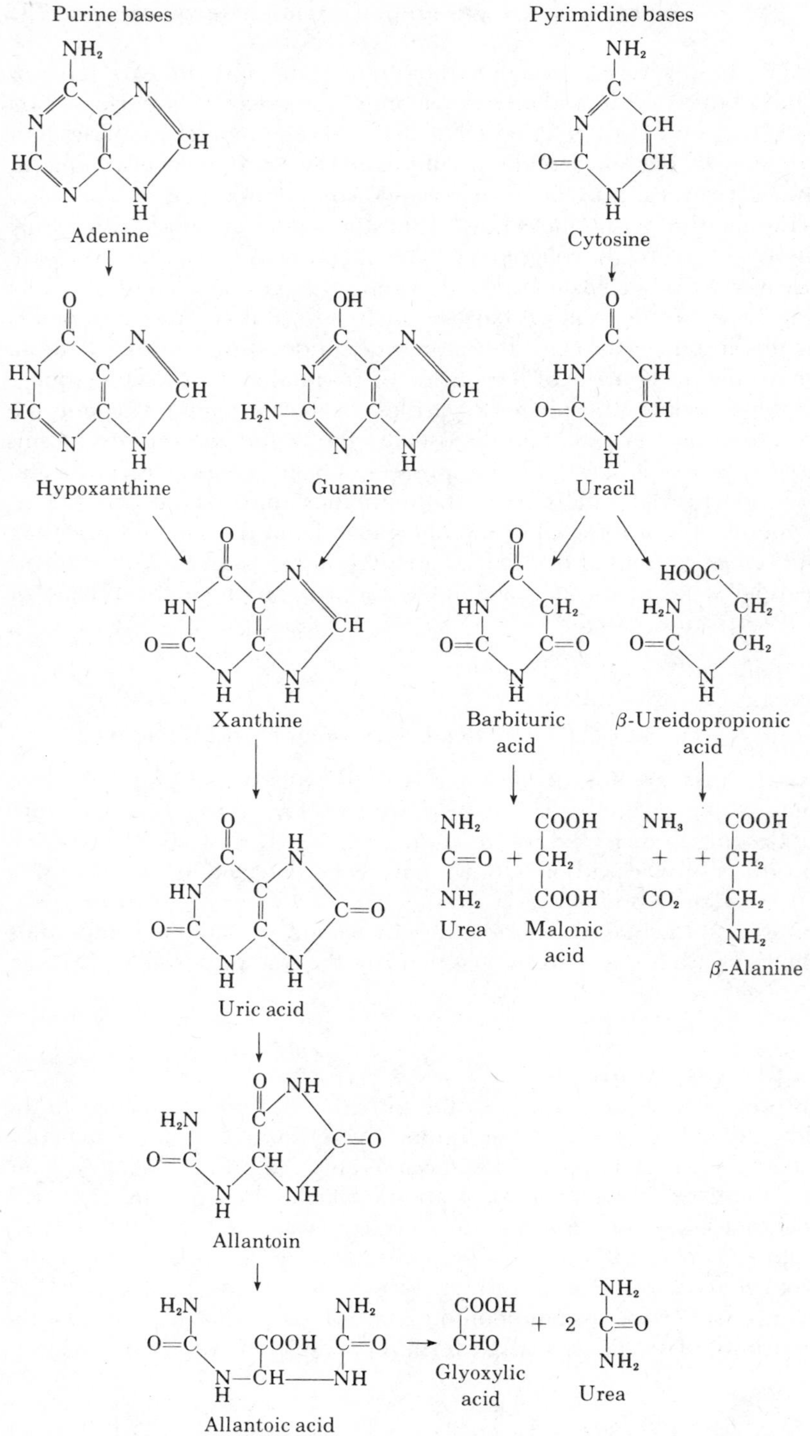

Figure 15.2. Pathway of metabolism of purines and pyrimidines.

8.0 and sometimes about pH 9.0, although the pH only a minute distance away may be near neutrality or below. In these alkaline conditions, the product is in fact ammonia and not ammonium, and much of the nitrogen from the fertilizer may be lost to the atmosphere as gaseous ammonia.

Because this initial phase of the process leads to considerable loss of a key element in fertilizers, much attention has been devoted to understanding factors affecting the loss. The results of this research have amply demonstrated that the volatilization may be considerably less than 10 percent or be as high as 70 percent of the added urea nitrogen. The loss is promoted by increasing the temperature, and it is greater if the urea is applied to the surface rather than below ground and in soils of high than low pH values. A method that has been extensively explored to reduce volatilization involves the use of inhibitors. Many compounds have been evaluated, and a few offer considerable promise (2). With time, the dramatic rise in pH is reversed, inasmuch as the ammonia-ammonium is oxidized to nitric acid.

Urea hydrolysis is enhanced with increasing temperatures, but the conversion proceeds even at values close to freezing. Moisture, the availability of O_2, and pH of the environment influence the reaction rate, but the effect of these variables on urea hydrolysis is not as marked as observed with many other microbiological processes.

Many microorganisms possess the enzyme urease, the catalyst responsible for hydrolyzing urea. In some species, urease is constitutive; in others, it is an inducible enzyme. Ammonium carbamate is an intermediate in the hydrolysis.

$$CO(NH_2)_2 + H_2O \xrightarrow{\text{urease}} H_2NCOONH_4 \rightarrow 2NH_3 + CO_2 \qquad \text{(III)}$$

The number of active organisms measured by plating varies from several thousand in the highly acid peats to greater than a million per gram in the most suitable locales. Bacteria, fungi, and actinomycetes synthesize urease and therefore can use urea as a nitrogen source for growth. Most frequently studied are species of *Bacillus, Micrococcus, Pseudomonas, Klebsiella, Corynebacterium, Clostridium,* and a diverse collection of filamentous fungi and actinomycetes. In one soil examined, for example, 32 to 69 percent of the bacteria and 58 to 100 percent of the fungi growing on agar media hydrolyzed urea (27). In other soils, from 17 to 30 percent of the bacteria had the capability of urea cleavage (17). Obligate anaerobes that hydrolyze urea are sometimes hard to obtain, and the bacteria that usually are active in the absence of air are probably the facultative anaerobes.

A small group of true bacteria have been termed *urea bacteria* not because they are necessarily the dominant organisms in the hydrolysis but because of their tolerance to high urea levels and their nutritional affinity for the compound. Two types of urea bacteria can be differentiated, cocci and spore-forming rods. Both develop well in alkaline solution and release enormous quantities of ammonia. The spore-forming rods are members of the genus *Bacillus* and are best represented by *Bacillus pasteurii* and *Bacillus freudenreichii*. By comparing counts performed with and without pasteurization at 80°C, it is possible to show that a large proportion of this group of urea decomposers exists in the spore form in soil.

Urea bacteria are isolated easily from manure or soil by enrichment in solutions containing high concentrations of urea. In pure culture, the organisms show a characteristic sensitivity to acidity, and growth proceeds only in neutral or alkaline media. The sensitivity to the hydrogen ion is apparent in field investigations as well since the urea bacteria are absent or sparse in areas of pH 4.0 to 5.5. Urea breakdown in acid environments, therefore, probably can be attributed to the nonspecific flora; indeed, in some acid forest soils, the dominant urea-cleaving members of the community may be the fungi (27).

Urease activity in soil is determined by measuring the ammonia formed in samples incubated for short periods with the substrate. It is rarely clear whether the enzyme is, in fact, apart from the cells or intracellular in these assays, although urease free of cells has been extracted from soil. The rate of hydrolysis, that is, the urease activity, varies with the vegetation and the presence of plant roots, and it declines with depth. Addition of a carbon source initiating rapid development of heterotrophs increases urease activity and the number of ureolytic organisms (Figure 15.3), but this enhancement probably merely reflects a nonspecific stimulation of bacteria and fungi, which also

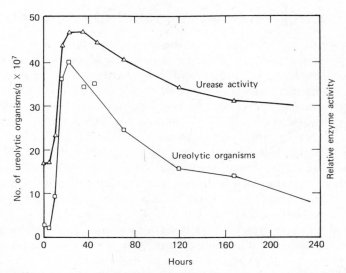

Figure 15.3. Influence of glucose addition on urease activity and ureolytic organisms in soil (25). Reproduced by permission of the Soil Science Society of America.

happen to possess the enzyme. The enzyme, to the extent that it exists apart from cells, probably is adsorbed on clay and organic colloids, to which it has a strong affinity. Complexing with soil constituents increases appreciably the resistance of urease to protein-degrading enzymes that otherwise would readily destroy the catalyst (5).

Some microorganisms metabolize urea although they possess no urease. One such means is by a reaction involving the enzymatically catalyzed combination of urea with CO_2 to yield allophanic acid.

$$CO(NH_2)_2 + CO_2 \rightarrow H_2NCONHCOOH \qquad (IV)$$

Allophanic acid is then cleaved by the organisms to yield 2 NH_3 and regenerate the CO_2. Other ureaseless heterotrophs possess additional mechanisms to make use of the nitrogen source. However, the contribution of these reactions to urea breakdown in nature still remains unsure.

NITROGEN IMMOBILIZATION

To maintain the organic matter and nutrient level in agricultural land, it is common farm practice to plow under undecayed or sometimes partially rotted crop residues. Almost invariably this leads to a decrease in the inorganic nitrogen content of the soil, a depletion that may extend for some time. An analogous nitrogen change occurs following the incorporation of pure carbohydrates, the extent of the fall being directly related to the quantity of organic material added; with time, the N_i level begins to go up again (Figure 15.4). In the decay of succulent plants, however, the level of inorganic nitrogen rises even initially. The N_i disappearance following the addition of nitrogen-poor crop residues, a process known as *nitrogen immobilization*, results in a marked depression of nitrogen uptake by the plant and a consequent decrease in yield.

Immobilization results from the microbial *assimilation* of inorganic nutrients. As new cells or hyphae are formed, not only must carbon, hydrogen, and oxygen be combined into protoplasmic complexes but so must nitrogen, phosphorus, potassium, sulfur, magnesium, and iron. Each one of these elements is thus immobilized. The microbiological removal of available ions is of agronomic importance only for the plant's macronutrients, of which nitrogen is the most prominent. In the case of nitrogen, immobilization is a consequence of the incorporation of ammonium and nitrate into proteins, nucleic acids, and other organic complexes contained within microbial cells. Immobilization is, therefore, the converse of mineralization; the latter returns microbial and plant nutrient elements to the inorganic state whereas the former combines inorganic ions into organic compounds.

Microorganisms cannot multiply nor can organic matter be decomposed unless nitrogen is assimilated into microbial protoplasm, and assimilation will take place as long as there is microbial activity. Even when a pure protein is added to soil, not all its nitrogen is mineralized; some always goes into the

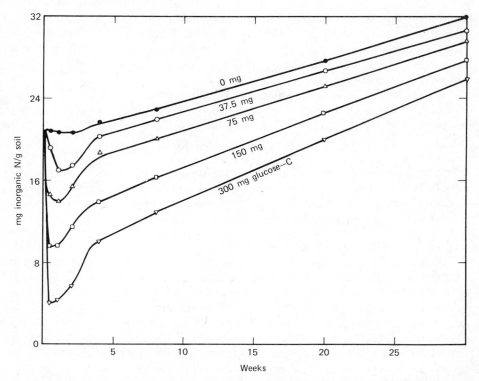

Figure 15.4. Changes in inorganic nitrogen in soil receiving various quantities of glucose (11).

biosynthesis of the cells of the microscopic inhabitants. Whenever mineralization occurs, immobilization runs counter to it; thus, a determination of the quantity of inorganic nitrogen produced or lost during incubation does not measure one or the other process but rather the net release or tie-up. The gross figure is greatly underestimated.

In the decomposition of proteins in culture, a portion of the nitrogen liberated is reassimilated by the new cells generated. The net quantity of ammonium produced is therefore the amount of protein nitrogen mineralized less the amount of nitrogen utilized in the synthesis of cellular constituents, assuming no accumulation of intermediates. The inclusion in culture media of a utilizable carbohydrate together with the organic nitrogen compounds further reduces the amount of ammonium that accumulates. The apparent retarding influence of carbohydrates on the production of ammonium from proteins and amino acids may be attributed to an assimilation of ammonium by the additional organisms appearing in the decomposition of the carbohydrate or to a microbial preference for the carbohydrate to the protein. In soil, only the former explanation applies as the mixed flora utilizes both substrates.

Ammonium salts are the most readily assimilated nitrogen sources for most bacteria, actinomycetes, and fungi. Chemically fixed or nonextractable ammonium in soil is largely unavailable, however, and only the extractable cation is used to a significant extent. Complex organic nitrogen preparations or individual amino acids are frequently incorporated in laboratory media because they provide growth factors, but the nitrogen is usually obtained from these molecules only following ammonification. Many filamentous fungi and gram-negative bacteria develop readily with either ammonium or nitrate salts in the absence of amino acids. Certain organisms cannot use nitrate, and a rare few fail to utilize ammonium as sole source of the element. Strains that use nitrate, however, invariably use ammonium so that assimilation of the former indicates a higher order of physiological development. In laboratory media containing both ions, the ammonium tends to disappear first, and the nitrate may sometimes remain until almost all the ammonium is gone. In like fashion, ammonium is preferentially utilized to nitrate in the decomposition of organic matter, and the cation is assimilated in soil almost to the exclusion of nitrate. However, much nitrate can be immobilized when ammonium levels are low (24).

A large part of the nitrogen of proteins or protein-rich compounds added to soil is recovered in the inorganic form. The added protein serves as both a carbon and a nitrogen source for the microflora, and much of the organic nitrogen is liberated as ammonium because the demand for the former element far exceeds the need for the latter. Frequently as much as three-fourths of the nitrogen of pure proteins is recovered as inorganic nitrogen in short incubation periods. Similarly, ammonium may accumulate to high levels in the immediate vicinity of nitrogen-rich plant remains (29), an accumulation that may upset the composition of the community inasmuch as low concentrations inhibit the germination of spores of many fungi (15). Should the protein be mixed with a carbohydrate, however, the population developing on the greater supply of carbon will be larger, and more nitrogen will be required for the new flora. Consequently, the supply of nitrogen in the protein will not outweigh the demand to as great an extent, and less ammonium will be released. From this point of view, accumulated ammonium reflects but a waste product of metabolism, the cation remaining behind only when not needed for microbial proliferation.

A quantification of the aforementioned observations permits characterization of the interactions between the microbial nutrient demand and the supply. Calculations of this nature are of no mere academic interest as they are used for predictive purposes in fertilizer recommendations. Data on the chemistry and physiology of microorganisms or empirical field results can be used to develop the critical relationships. Consider the decay of a typical organic material. In the process of degradation, the carbon is liberated as CO_2, the organic nitrogen as ammonium. The decomposition results in a simultaneous synthesis of additional microbial protoplasm, a process requiring the assimilation of carbon and nitrogen from either the substrate or the environment. To estimate the nitrogen needed to satisfy the demands for cell synthesis, data on the extent of carbon assimilation and the C:N ratio of the cells formed are required. As a rule, for mixed populations, 5 to 10 percent of the substrate carbon is assimilated by bacteria, 30 to 40 percent by fungi, and 15 to 30 percent by actinomycetes. The carbon content of microbial protoplasm is typically 45 to 50 percent of the dry weight, but the percentage of nitrogen varies with the age of the culture and the environment. Cells or hyphae in old cultures usually have less nitrogen than those in young cultures. As first approximations, C:N ratios of 5:1, 10:1, and 5:1 may be proposed for the cellular components of bacteria, fungi, and actinomycetes, respectively. By combining the figures for carbon assimilation and protoplasmic constitution, it can be calculated that for the decomposition of 100 units of substrate carbon it is necessary to provide 1 to 2, 3 to 4, and 3 to 6 units of nitrogen for bacteria, fungi, and actinomycetes, respectively. Taking the example of a plant product having approximately 40 percent carbon, for each 100 parts of organic matter undergoing decay, 0.4 to 0.8, 1.2 to 1.6, and 1.2 to 2.4 parts of nitrogen are needed by the bacterial, fungal, and actinomycete flora.

Straw containing 0.5 percent nitrogen and 40 percent carbon, when subjected to attack by fungi, has only 0.5 units of nitrogen to satisfy the active biological agents that require 1.2 to 1.6 units so that a deficit of 0.7 to 1.1 units of nitrogen appears in the environment. Consequently, any ammonium or nitrate present or formed will be immobilized immediately. The extent of immobilization is less for mixtures of bacteria, greater for actinomycetes. This argument, based entirely on considerations of microbial nutrition, maintains that the microflora assimilates essentially all the nitrogen contained in residues poor in protein. In protein-rich residues, the excess is liberated into the environment as ammonium, which may be subsequently oxidized to nitrate. For example, in the fungal decomposition of 100 kg of alfalfa containing 3.0 percent nitrogen and 40 percent carbon, there is a surplus, and 1.4 to 1.8 kg of nitrogen is mineralized. Similar calculations can be applied to the immobilization of any element provided the appropriate values are known.

Extension of these theoretical considerations to agricultural practice is not difficult. Illustrative of many studies are the data cited in Table 15.2, which were obtained from soil treated with various organic materials. The nitrate present after a three-month incubation period can be taken as indicative of a net mineralization or immobilization. The experiment suggests that neither a net loss nor a net gain of inorganic nitrogen occurs in soils receiving organic materials containing approximately 1.7 percent nitrogen while nitrate accumulates in soil treated with materials richer in nitrogen. In practice, organic

TABLE 15.2

Nitrate Level in Soils Incubated with Materials of Varying Nitrogen Contents (20)

Organic Amendment	Nitrogen Content of Substrate[a]	After Three Months	
		Nitrate-N in Soil	N Change
	%	mg	mg
Untreated soil	—	947	—
Dried blood	10.71	1751	+804
Clover roots	1.71	924	−23
Corn roots	0.79	511	−436
Timothy roots	0.62	398	−549
Oat roots	0.45	207	−740

[a] All materials applied at rates to give 600 mg N.

substrates containing more than 1.8 percent nitrogen, on decomposition, almost immediately increase the N_i level. Here, the mineralization rate, m, exceeds i, the immobilization rate. A temporary removal of N_i follows the application of materials with 1.2 to 1.8 percent nitrogen, but the initial period is followed by a stage in which m exceeds i so that ammonium and/or nitrate accumulates. Plant residues with less than 1.2 percent nitrogen deplete the inorganic nitrogen reserve within about one week, and the deficiency may not be alleviated in periods of several months or longer. In the last instance, crops will suffer from nitrogen starvation unless proper fertilization practices are followed.

The critical nitrogen levels are frequently expressed in terms of C:N ratios. In natural materials with approximately 40 percent carbon, the critical levels corresponding to 1.2 to 1.8 percent are C:N ratios of ca. 20 to 30:1. Wider ratios favor immobilization, narrower ratios mineralization. It must be borne in mind, however, that decomposition results in CO_2 release, the volatilization decreasing the C:N ratio of protein-deficient residues (Figure 15.5). Ultimately, when the ratio falls below the critical range, m will exceed i, and nitrogen will appear in inorganic form. The economics of crop production often does not allow for the long wait required for the slow change, and it is a common practice to fertilize in order to bring the C:N ratio of plowed-under residues to ca. 30 to 35:1 or to a final nitrogen content of 1.2 to 1.5 percent on a dry weight basis. This fertilizer is for the microflora, not for the crop.

A convenient means of expressing the deficit following the application of materials with wide C:N ratios is by the *nitrogen factor*. This factor is defined as the number of units of inorganic nitrogen immobilized for each 100 units of

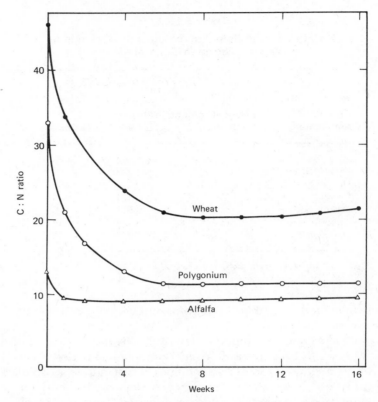

Figure 15.5. Changes in C:N ratio during the decomposition of plant remains (32).

material undergoing decomposition or, operationally, the amount of nitrogen that must be added in order to prevent a net immobilization from the environment. To determine the nitrogen factor, excess ammonium is added to the crop residue, and the increase in organic nitrogen is determined at intervals during the decay. Values for the nitrogen factor vary from as little as 0.10 or less to as high as 1.3. For mature crop remains, values for the factor usually fall between 0.5 to 1.0 when the nitrogen is applied at the beginning of the decomposition. But, if the carbonaceous matter is allowed to rot for some time prior to fertilizer application, the nitrogen factor is less because the C:N ratio of the resultant material has been narrowed.

The nitrogen factor is an equilibrium value representing the opposing forces of mineralization and immobilization. Both mineralization and immobilization take place regardless of the percentage of nitrogen in the organic matter.

In order to estimate the rate of one of two opposing transformations, isotopic tracers are used. For example, when ^{15}N-labeled ammonium is applied to soil, the ^{15}N tracer moves into the organic reservoir with immobilization while, in mineralization, the nontracer ^{14}N from the organic fraction dilutes the ^{15}N in the inorganic pool. Such experiments show that both transformations are proceeding regardless of environmental conditions. Should the immobilization rate, i, exceed the mineralization rate, m, N_i will not accumulate, and any initially present will disappear. Where i is less than m, N_i accumulates and becomes available for plant utilization. Because nitrogen is generally available to plants growing in natural ecosystems, m usually exceeds i.

Comparison of the simultaneous mineralization and immobilization is of considerable importance, for the results indicate the course of the nitrogen transformations at a given time. Typically, following the addition of materials with wide C:N ratios, for example, 70:1, i exceeds m, so that the net effect is one of inorganic nitrogen disappearance. With carbonaceous substrates of narrow C:N ratios, for example, 15:1, m exceeds i, so the inorganic nitrogen content increases.

From the pure culture standpoint, the products of assimilation are simply the nitrogenous constituents of the cells: proteins, amino sugars, nucleic acids, and the like. Because the ratios of the components vary markedly from organism to organism and with culture conditions and since neither the identities nor the ages of the cells constituting the bulk of the soil biomass are known, analyses have been conducted to ascertain the actual products in soil. These have been found to be compounds chiefly of high molecular weights, with 50 to 60 percent being as bound amino acids and about 5 to 10 percent as bound amino sugars (1, 3). The initial products very likely are the proteins and cell walls made by the heterotrophs, but the composition undoubtedly changes as the initial populations themselves lyse or are attacked by succeeding populations. With time, the nitrogen thus assimilated and then subjected to various subsequent transformations becomes quite resistant to attack, and only a part is mineralized in each following year. This is demonstrated by providing the microflora with ^{15}N-tagged inorganic salts so that the immobilized nitrogen is labeled, and the source of the nitrogen later mineralized can be traced: ^{15}N from the pool of immobilized material, ^{14}N from native organic matter. Years, decades, or possibly centuries may be required for all that the community has incorporated to once again be available for plant uptake.

The rate and extent of microbial nitrogen assimilation are intimately linked with the type of carbonaceous substances undergoing decay. In general, a stimulation of biological activity is accompanied by an enhanced immobilization as protoplasmic turnover is increased. The rate of immobilization is therefore related to the availability of the organic molecule, very rapid with readily oxidized carbohydrates, moderate with less suitable materials, and particularly slow with extensively decayed materials or resistant tissue components such as lignin. Thus, for a constant percentage of nitrogen or C:N ratio, the disappearance of inorganic nitrogen is associated with the rate of organic matter breakdown. Furthermore, the nitrogen requirement for the decomposition of lignified or resistant tissues is generally quantitatively less than for the more succulent plants because little microbial protoplasm is synthesized. Immobilization is also correlated with pH and available phosphate and is stimulated by increasing temperature, results that are not unexpected because of the qualitative and quantitative effects of pH, temperature, and phosphorus on the biochemical capacities of the microflora.

The immobilization of inorganic nitrogen has important agronomic ramifications. Plants are poor competitors with the microflora when the inorganic nitrogen level is inadequate for maximum development of both macro- and microorganisms. Immobilization accompanying soil amendment with nitrogen-poor crop residues is undesirable during the growing season since a critical nutrient is rendered unavailable. On the other hand, the same reaction may be beneficial in the autumn of the year in temperate climatic areas because nitrate and ammonium are tied up and are not lost by leaching during the winter season. The following spring, the nitrogen bound into microbial protoplasm is mineralized, at least in part, to ammonium and nitrate that can then be utilized by plants. The season of year thereby determines whether immobilization is beneficial or detrimental.

BIOCHEMISTRY OF PROTEIN DECOMPOSITION

The protein molecule is composed of a long chain of amino acids, all of which have the general type structure $H_2NCHRCOOH$ where R may be a hydrogen atom, a single methyl group, a short carbon chain, or a cyclic structure. Some twenty different amino acids are found in the protein molecule, linked together by peptide bonds (CO—NH). Peptides are composed of short chains of amino acids. The molecule of the protein and peptide has the type structure shown below.

$$
\begin{array}{ccc}
\text{H} & \text{H} & \text{H} \\
\ldots\text{NHCCONHCCONHCCO}\ldots & & \\
\text{R} & \text{R}' & \text{R}''
\end{array}
$$

Enzymes that attack and hydrolyze the peptide bonds of proteins and peptides are known as *proteases*. These are considered to be of two general types, exopeptidases, which hydrolyze peptide bonds in the vicinity of the end of the amino acid chain, and endopeptidases, which hydrolyze bonds at a distance from the terminal end of the chain.

In the decomposition of proteins and peptides, free amino and free carboxyl groups are released.

$$
\ldots\underset{R}{N}H\overset{H}{C}CONH\overset{H}{\underset{R'}{C}}CONH\overset{H}{\underset{R''}{C}}CO\ldots \xrightarrow[\text{enzyme}]{H_2O}
$$

$$
\ldots\underset{R}{N}H\overset{H}{C}COOH + H_2N\overset{H}{\underset{R'}{C}}CONH\overset{H}{\underset{R''}{C}}CO\ldots \quad (V)
$$

In the process, the proteolytic enzymes cleave the protein molecule to polypeptides, simple peptides, and finally to the free amino acids that are the end products of protease action. The reaction is a hydrolysis as the enzyme ruptures the peptide bond by the addition of water. Many microorganisms utilize polypeptides, simple peptides, and amino acids in contrast to the few genera degrading native proteins.

Because it is too large to enter into the microbial cell, the protein molecule must first be cleaved to smaller units that can be assimilated. Within the confines of the cell, the simple derivatives can serve as sources of energy, carbon, and nitrogen. To convert the protein to assimilable forms, the microorganism must excrete an extracellular protease. Demonstration of extracellular proteases in vitro is accomplished by removing the cells from the culture medium and testing the filtrate for activity. A single population may excrete several different protein-hydrolyzing enzymes. Following the degradation of the large protein molecule, the simple derivatives enter the cell for further metabolism. For the release of energy for growth, the substrate must be transformed within the cell itself.

The activity of proteolytic enzymes in soil is evaluated by incubating test samples with proteins or peptides and measuring amino acid formation in short periods of incubation. These activities are markedly affected by temperature and pH, with the optimal temperature for hydrolysis being in the thermophilic range (16). Different proteins are cleaved by these catalysts, and their activity is enhanced by the addition not only of proteins but also of sugars that bring about extensive microbial proliferation.

The amino acids liberated by proteases serve as carbon and nitrogen sources for innumerable heterotrophs, each of which may be able to utilize one or several of these compounds. The nitrogen of most amino acids is removed as ammonia prior to significant decomposition of the carbon-containing portion of the molecule, and the microorganism gets its nitrogen by assimilation of the ammonia. The common mechanisms for the initial degradation of amino acids are *deamination*, the removal of ammonia, and *decarboxylation,* in which the carboxyl is removed.

a.   Deamination by direct removal of ammonia:

$$RCH_2CHNH_2COOH \rightarrow RCH{=}CHCOOH + NH_3 \qquad (VI)$$

b.   Oxidative deamination:

$$RCHNH_2COOH + \tfrac{1}{2}O_2 \rightarrow RCOCOOH + NH_3 \qquad (VII)$$

c.   Reductive deamination:

$$RCHNH_2COOH + 2H \rightarrow RCH_2COOH + NH_3 \qquad (VIII)$$

d.   Decarboxylation:

$$RCHNH_2COOH \rightarrow RCH_2NH_2 + CO_2 \qquad (IX)$$

The amino acids produced from the proteins are mineralized at different rates. Some amino acids are resistant and others highly susceptible to decomposition. Ammonia is formed readily from some, while others have a more extended persistence in soil. After deamination, the carbon residue is attacked aerobically or anaerobically to yield CO_2 and various organic products.

The inanimate portion of soil has a profound effect on proteolysis. Because the initial degradation of the protein molecule requires the elaboration of extracellular enzymes, substances that adsorb or otherwise inactivate the biological catalysts influence the decomposition. Clay is prominent in such adsorption phenomena, and some clay minerals remove a number of enzymes from

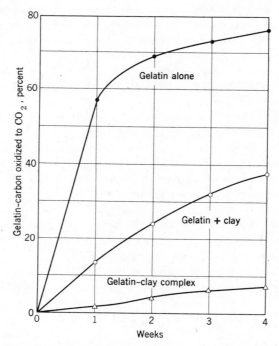

Figure 15.6. Effect of montmorillonite on gelatin decomposition in sand culture (26).

solution and thereby diminish biochemical activity. Little adsorption by kaolinite occurs above the isoelectric point while the removal is magnified at pH values below the isoelectric point of the enzyme. The former effect arises from the fact that the enzyme has a negative charge above its isoelectric point and is repelled by the negatively charged kaolinite (19, 22).

Clays may also decrease the rate of protein decomposition by adsorption of the proteinaceous substrate (Figure 15.6). Thus, substances like gelatin are quite resistant to attack when adsorbed in a monolayer on the crystal lattice of montmorillonite whereas resistance to hydrolysis is diminished when the protein is present on the clay in two or more layers (26). The clay mineral attapulgite has a similar protective role while illite has little influence. There is also an intriguing reverse effect. When a denatured protein is adsorbed by kaolinite, the rate of its hydrolysis by a proteolytic *Pseudomonas* sp. and a *Flavobacterium* sp. is greater in the presence of the colloidal material than in its absence. The greater transformation of the adsorbed compound likely results from the clay mineral functioning as a surface for the concentration of adsorbed substrate and enzyme (9).

REFERENCES

Reviews

Bartholomew, W. V. 1965. Mineralization and immobilization of nitrogen in the decomposition of plant and animal residues. In W. V. Bartholomew and F. E. Clark, eds., *Soil nitrogen*. American Society of Agronomy, Madison, Wisc., pp. 285–306.

Bremner, J. M. 1967. Nitrogenous compounds. In A. D. McLaren and G. H. Peterson, eds., *Soil biochemistry*. Marcel Dekker, New York, pp. 19–66.

Egami, F. and K. Nakamura. 1969. *Microbial ribonucleases*. Springer-Verlag, New York.

Lehman, I. R. 1971. Bacterial deoxyribonucleases. In P. D. Boyer, ed., *The enzymes*. Academic Press, New York, vol. 4, pp. 251–270.

Literature Cited

1. Allen, A. L., F. J. Stevenson, and L. T. Kurtz. 1973. *J. Environ. Qual.*, 2:120–124.
2. Bremner, J. M. and L. A. Douglas. 1973. *Soil Sci. Soc. Amer. Proc.*, 37:225–226.
3. Broadbent, F. E. 1968. In *Isotopes and radiation in soil organic-matter studies*. Intl. Atomic Energy Agency, Vienna, pp. 131–140.
4. Broadbent, F. E. and T. Nakashima. 1971. *Soil Sci. Soc. Amer. Proc.*, 35:457–460.
5. Burns, R. G., A. H. Pukite, and A. D. McLaren. 1972. *Soil Sci. Soc. Amer. Proc.*, 36:308–311.
6. Davies, R. I., C. B. Coulson, and D. A. Lewis. 1964. *J. Soil Sci.*, 15:299–309.
7. Denmead, O. T., J. R. Simpson, and J. R. Freney. 1974. *Science*, 185:609–610.
8. Durand, G. 1961. *Compt. Rend. Acad. Sci.*, 252:1687–1689.
9. Estermann, E. F. and A. D. McLaren. 1959. *J. Soil Sci.*, 10:64–78.
10. Greaves, M. P. and M. J. Wilson. 1970. *Soil Biol. Biochem.*, 2:257–268.
11. Haroda, T. and R. Hayashi. 1968. *Soil Sci. Plant Nutr.*, 14:13–19.
12. Hrubcova, M. and V. Drobnikova. 1971. *Zentr. Bakteriol.*, II, 126:713–724.

13. Ishaque, M. and A. H. Cornfield. 1972. *Plant Soil,* 37:91–95.
14. Ishizawa, S., M. Araragi, and T. Suzuki. 1969. *Soil Sci. Plant Nutr.,* 15:104–112.
15. Ko, W. H., F. K. Hora, and E. Herlicska. 1974. *Phytopathology,* 64:1398–1400.
16. Ladd, J. N. and J. H. A. Butler. 1972. *Soil Biol. Biochem.,* 4:19–30.
17. Lloyd, A. B. and M. J. Scheaffe. 1973. *Plant Soil,* 39:71–80.
18. Luebs, R. E., K. R. Davis, and A. E. Laag. 1973. *J. Environ. Qual.,* 2:137–141.
19. Lynch, D. L. and L. J. Cotnoir. 1956. *Soil Sci. Soc. Amer. Proc.,* 20:367–370.
20. Lyon, T. L., J. A. Bizzell, and B. D. Wilson. 1923. *J. Amer. Soc. Agron.,* 15:457–467.
21. Mahmoud, S. A. Z., A. M. Abdel-Hafez, M. El-Sawy, and E. A. Hanafy. 1973. *Zentr. Bakteriol.,* II, 128:196–202.
22. McLaren, A. D. 1954. *J. Phys. Chem.,* 58:129–137.
23. Mosier, A. R., C. E. Andre, and F. G. Viets, Jr., 1973. *Environ. Sci. Technol.,* 7:642–644.
24. Overrein, L. N. and F. E. Broadbent. 1967. *Trans. 8th Intl. Cong. Soil Sci.,* vol. 3, pp. 791–799.
25. Paulson, K. N. and L. T. Kurtz. 1969. *Soil Sci. Soc. Amer. Proc.,* 33:897–901.
26. Pinck, L. A., R. S. Dyal, and F. E. Allison. 1954. *Soil Sci.,* 78:109–118.
27. Roberge, M. R. and R. Knowles. 1967. *Soil Sci. Soc. Amer. Proc.,* 31:76–79.
28. Robinson, E. and R. C. Robbins. 1970. In S. F. Singer, ed., *Global effects of environmental pollution.* Reidel Publ., Dordrecht, Netherlands, pp. 50–64.
29. Smith, J. H. 1967. *Soil Sci. Soc. Amer. Proc.,* 31:377–379.
30. Stanford, G. and E. Epstein. 1974. *Soil Sci. Soc. Amer. Proc.,* 38:103–107.
31. Taylor, C. B. and A. G. Lochhead. 1938. *Can. J. Res.,* C, 16:162–173.
32. van Schreven, D. A. 1964. *Plant Soil,* 20:149–165.
33. Verma, L., J. P. Martin, and K. Haider. 1975. *Soil Sci. Soc. Amer. Proc.,* 39:279–284.
34. Waring, S. A. and J. M. Bremner. 1964. *Nature,* 201:951–952.
35. Westerman, R. L. and L. T. Kurtz. 1973. *Soil Sci. Soc. Amer. Proc.,* 37:725–727.
36. Winsor, G. W. and A. G. Pollard. 1956. *J. Sci. Food Agr.,* 7:618–624.

16
Nitrification

The termination of the reactions involved in organic nitrogen mineralization occurs at the point where ammonium is formed. This, the most reduced form of inorganic nitrogen, serves as the starting point for a process known as *nitrification*, the biological formation of nitrate or nitrite from compounds containing reduced nitrogen. The importance of the nitrifying microorganisms rests to a great degree on their capacity to produce the nitrate that is the major nitrogen source assimilated by higher plants. Nitrate is produced not only in soil but also in marine environments, manure piles, and during sewage processing, where it is the product of the final stage in rendering organic nitrogen inoffensive.

During the Napoleonic era, nitrate salts were in great demand in France for gunpowder manufacture, and with the natural source cut off by war, the nitrate was prepared in niter heaps that consisted of piles of soil mixed with manure and lime. The explanations for the phenomenon of nitrate synthesis advanced during the early nineteenth century were of a chemical nature, the product reputedly being formed by reaction of atmospheric O_2 and ammonium, with the soil considered as chemical catalyst. In contrast to this theory, Pasteur postulated that the formation of nitrate was microbiological and analogous to the conversion of alcohol to vinegar. The first experimental evidence that nitrification is biological is attributed to Schloesing and Muntz (36), who added sewage effluent to a long tube filled with sterile sand and $CaCO_3$. For 20 days the ammonium concentration in the liquid remained unaltered, but then ammonium was destroyed and nitrate appeared. Heating the column or the addition of an antiseptic eliminated the transformation, but it could be reinitiated by the application of a small quantity of garden soil. It remained for S. Winogradsky (42) to isolate the responsible agents.

Under certain conditions, two separate and distinct steps are distinguishable in nitrification. Since nitrite frequently appears during ammonium oxidation, it seemed apparent to early microbiologists that the transformation involved an initial oxidation to nitrite and a subsequent conversion of the latter to nitrate. The validity of this theory was established when two groups of bacteria were isolated and described, each catalyzing a separate portion of the reaction sequence. Because nitrite is rarely found in nature even in habitats where nitrification is proceeding rapidly, the nitrate formers must generally occur in the same environments as the ammonium oxidizers. Information on the responsible species, however, has been collected with difficulty because of their frequent overgrowth by rapidly proliferating contaminants.

ENVIRONMENTAL INFLUENCES

Physical and chemical factors affect the rate of ammonium oxidation. This fact is simply demonstrated as nitrification rates in sterile soils receiving the same inoculum differ according to soil type. The remarkable degree of sensitivity of the process to external influences is attributed in part to the great physiological similarity of the responsible species; as a result, modification of the environment often has a profound significance in governing the production of the end product. When the habitat becomes unfavorable, in acid or anaerobic conditions for example, little or no nitrate is detected, but ammonium accumulates because ammonification is less sensitive to environmental change.

Chief among the ecological influences is acidity. Several careful investigations have demonstrated a significant correlation between nitrate production and pH. In acid environments, nitrification proceeds slowly even in the presence of an adequate supply of substrate, and the responsible species are rare or totally absent at great acidities. An exact limiting pH cannot be ascertained since a variety of physicochemical factors in soil will alter any specific boundaries. Typically, the rate falls off markedly below pH 6.0 and becomes negligible below 5.0 (14), yet nitrate may occasionally be present in fields down to 4.0 and sometimes lower. Some soils nitrify at 4.5, others do not; the difference is possibly attributable to acid-adapted strains or to chemical differences in the two habitats. The acidity affects not only the transformation itself but also the microbial numbers, neutral to alkaline soils having the largest populations. The pH values for growth of these bacteria depend to some extent on the locality from which they originated. Strains derived from acid soils are frequently more tolerant of high hydrogen ion concentration than those from areas of alkaline pH, and the optimum pH for individual isolates may vary from as low as 6.6 to as high as 8.0 or sometimes higher.

Because of the microbial sensitivity to hydrogen ions, nitrification in acid soils is usually markedly enhanced by liming. At reactions near the optimum, the lime addition has little to no effect. Indeed, in extremely acid environments devoid of nitrate, liming may lead to ammonium oxidation in sites where the transformation had not previously occurred. Often the failure to nitrify is entirely a consequence of acidity, and the condition may therefore be readily alleviated by liming.

Oxygen is an obligate requirement for all species concerned, making ✓ adequate aeration essential. Where the O_2 supply is inadequate for microorganisms, there will be little ammonium oxidation, and the reaction ceases in the total absence of O_2. Because of this nutritional characteristic, soil structure will affect the accumulation of nitrate through its influence on aeration. In controlled experiments, artificial aeration stimulates the bacteria to greater activity and nitrate accumulates more rapidly, while a reduction in the partial pressure of O_2 below that found in air reduces nitrification. Nitrification occurs even in submerged soils used for growing rice, however, because O_2 diffusing from the atmosphere through the water phase keeps the upper few millimeters of soil oxygenated (46). Below the O_2-containing zone, no nitrate is synthesized.

Because moisture affects the aeration regime of soil, the water status of the microbial habitat has a marked influence on nitrate production (35). At one extreme, waterlogging limits the diffusion of O_2, and nitrification is suppressed. At the opposite pole, in arid conditions, bacterial proliferation is not retarded by the supply of gaseous nutrients but rather by the insufficiency of water, and irrigation increases both the nitrifying population and nitrate biosynthesis. The optimum moisture level varies considerably with different soils, but nitrate generally appears most readily at one-half to two-thirds the moisture-holding capacity.

The interaction of moisture, temperature, cropping, and inorganic nutrients largely makes up the seasonal effect. The dominant influence will be determined by the specific environment and cannot be predicted accurately. Furthermore, the season of the year at which nitrate is most abundant does not necessarily coincide with the time of maximal microbial activity because plant uptake, microbial immobilization, and leaching all reduce the nitrate level. In temperate zones, nitrate formation is generally most rapid in spring and autumn and slowest during the summer and winter months, but annual fluctuations of moisture and temperature can alter the seasonal effects appreciably.

Nitrification is markedly affected by temperature, and many investigations have confirmed the fact that below 5°C and above 40°C the rate is very slow. There is evidence of a slow nitrate formation almost down to the freezing point, a fact of significance to nitrogen losses in cooler climates where leaching or denitrification of the end product may occur during autumn and winter; this loss can lead to a lack of response in the spring to fall-applied nitrogenous fertilizers unless the element is tied up in some way. Increasing the temperature from the lower extremes produces a more rapid ammonium oxidation until the optimum range is attained. This range varies, presumably because of physiological differences in the dominant bacterial strains, but the optimum usually lies between 30 and 35°C.

The type of crop may have a bearing on the size and activities of the nitrifying microflora. With many crops, no preferential influences can be ascertained, but sometimes a single plant species is without effect on nitrification in one soil type while increasing it in another. A striking feature of many grasslands is the paucity of inorganic nitrogen and the greater relative abundance of ammonium than nitrate. These characteristics have led to the hypothesis that the roots of grasses, as well as those of other plants, excrete compounds deleterious to nitrification. Some evidence indeed exists that root exudates reduce the rate of nitrification (29), while other findings provide no evidence of inhibition (32). Hence, generalizations are not as yet possible. A striking effect of roots became apparent when trees and shrubs in a northern forest were cut and regrowth was prevented by herbicide application: significant quantities of nitrate accumulated in the soil and were transported to underlying waters. This flush of nitrate may be a reflection of an inhibition when the vegetation was present. Alternatively, it may merely signify that the vegetation was able to remove the nitrate as readily as it was made (27).

Nitrification may have undesirable consequences. Although plants readily assimilate nitrate, the anion is also far more susceptible to leaching than ammonium so that the nutrient may pass through the soil and out of the rooting zone. This leads to a loss of an essential element for food production and may also be responsible for water pollution because excessive nitrate in waters can be hazardous to infants and livestock or can enhance growth of aquatic plants in nearby surface waters. Nitrogen in the nitrate form may also be converted to gaseous products that are unavailable for crop use. For these reasons, a great deal of research has been devoted to finding inhibitory chemicals that might be mixed or applied together with ammonium or urea fertilizers to reduce the rate of nitrification. The compound of choice clearly must not be phytotoxic nor can it itself be an environmental pollutant. Many chemicals have been patented for this purpose: azide, chlorinated pyridines and pyrimidines, thiourea, cyanoguanidine, dicyandiamide, nitrobenzenes, and substituted formamidines, isothiocyanates, N-alkylmaleimides, pyrazoles, thiadiazoles, thiazoles, s-triazines, triazoles, and trichloroacetamides. One of the most promising, from the practical standpoint, is 2-chloro-6-(trichloromethyl)pyridine. Remarkably low concentrations of this compound inhibit the ammonium- and nitrite-oxidizing autotrophs but have no effect on heterotrophs that can participate in nitrification in culture (37). By suppressing nitrification, the amount of inorganic nitrogen lost through leaching and denitrification is diminished; thus, more fertilizer nitrogen is available for plant assimilation and consequently crop yields may be increased with lower rates of fertilizer application. For reasons not yet clear, the enhanced yield occurs in some but not all experiments (20, 26). However, the effectiveness of the various inhibitors is governed by soil type and temperature (11) as well as by other undefined environmental factors.

PRODUCTION OF NITRATE FROM VARIOUS SUBSTRATES

The soil perfusion apparatus such as that shown in Figure 16.1 is a convenient means for the study of nitrification. In normal operation of the apparatus, a metabolite is continuously percolated through a soil column, the repeated perfusion permitting direct study of the kinetics of the transformation and the effects of environmental change. The original metabolite-containing solution, after completing its passage through the column, is recirculated so that the system is complete, a single community being investigated as a biological unit. Designed originally for nitrification investigations, it has subsequently been adapted to studies of other organic and inorganic conversions.

By means of the perfusion technique, the rate of ammonium oxidation or nitrate synthesis may be conveniently estimated. When expressed as a function of time, a plot of the transformation yields a sigmoid curve similar to that obtained when increases in bacterial numbers are depicted on a linear scale. In the initial perfusion, the system becomes enriched with the nitrifying bacteria, and the rate of ammonium oxidation upon reperfusion is linear. Such an

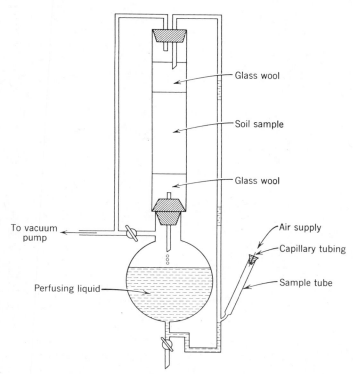

Figure 16.1. The soil perfusion apparatus.

enriched sample will, when treated with fresh ammonium, consume almost the theoretical quantity of O_2 for the complete conversion to nitrate. Alternatively, when a soil that had been initially perfused with ammonium is reperfused with nitrite, the oxidation commences immediately, demonstrating that both an ammonium- and a nitrite-oxidizing flora developed during the conversion of ammonium to nitrate.

The sigmoid relationship in linear expressions shows, when presented logarithmically, the exponential plot characteristic of bacterial growth. This is demonstrated by the results in Figure 16.2; such exponential changes indicate that the reaction in soil is bacterial as the fungi and actinomycetes would not be expected to exhibit logarithmic increases in biochemical activity under such

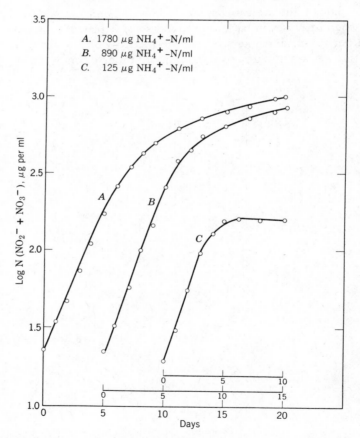

Figure 16.2. Logarithmic plot of the nitrification rate at three ammonium concentrations (40).

circumstances. The reaction rate cannot be determined precisely by ammonium disappearance because of retention of the cation by the soil's cation exchange complex, but it can be measured by the accumulation of the products of the transformation, nitrite and nitrate. Consequently, a logarithmic plot of the appearance of nitrite nitrogen plus nitrate nitrogen under controlled conditions yields a straight line from the slope of which an apparent generation time for the nitrifying bacteria can be obtained. In soils having pH values near neutrality, the apparent generation time may be about 20 hours for the nitrite oxidizers and about 30 hours for the ammonium oxidizers, but the doubling time increases in more acid environments (30). The apparent generation times for different sites undoubtedly vary greatly and depend on local conditions, but few such rates have been established.

The fixation or strong retention of ammonium by the clay fraction has a profound influence on its microbiological availability, and the tenacity of the adsorption and the rate of release directly affect ammonium metabolism. Such retained cations are oxidized by the chemoautotrophs although the type of clay and the amount of ammonium affect the percentage of the fixed chemical that is nitrified. The availability of the chemically fixed ion is low with usually less than 25 percent nitrified in periods of several months. A reasonably high percentage of the ammonium fixed by the clays of some soils, however, can be oxidized in periods of a month or somewhat longer (9).

The natural intermediate, nitrite, is readily metabolized in slightly acid, neutral, and in calcareous soils. The oxidation under most circumstances is quantitative, and the theoretical yield of nitrate is observed. As with ammonium, nitrite additions lead to an enrichment of the microflora active on the anion. If enriched soil is placed in a respirometer and treated with a fresh increment of nitrite, one-half mole of O_2 is consumed for each mole of nitrite supplied.

$$NaNO_2 + \tfrac{1}{2}O_2 \rightarrow NaNO_3 \qquad\qquad (I)$$

NITRIFICATION IN VARIOUS ENVIRONMENTS

During the growing season, agricultural crops assimilate large quantities of nitrate so that the production of this plant nutrient in soil is of considerable importance. Certain chemical properties of the microbiological habitat serve to alter the magnitude of the transformation. Acidity has been the most fully investigated. There is also evidence that nitrification is proportional to the cation exchange capacity (25). In alkali soils of high salt content, nitrate production is commonly retarded as the tolerance of the nitrifiers to salinity is not great.

The rate of nitrate production varies with the material undergoing transformation. In environments having a near neutral reaction, nitrate appears more rapidly from ammonium salts than from organic nitrogen compounds whereas nitrate is formed faster from organic materials in certain acid soils. The unexpected difference results from the increase in alkalinity during ammonification that makes the environment more favorable for the nitrifiers. The oxidation of anhydrous NH_3 or NH_4OH in acid soils is similarly more rapid than the nitrification of ammonium salts, and here too the rates parallel the differences in alkalinity following treatment. When urea is hydrolyzed by urease-producing microorganisms, the resulting ammonia increases the pH so that the formation of nitrate in acid soils is also greater from urea than from $(NH_4)_2SO_4$. No such effect of nitrogen carrier would be expected in soils whose reaction is closer to the optimal range for the nitrate-producing bacteria.

Some soils, particularly highly acid fields that are freshly limed, respond to the addition of biochemically active nitrifying inocula. Frequently, the use of fungicides largely or completely eliminates the nitrifying species so that ammonium will persist and may become potentially phytotoxic. Once the pesticide has been dissipated, the treated soil often responds to inoculation although natural contamination would ultimately overcome the deficiency.

The nitrifying activity of these organisms is of importance to plant nutrient availability. As ammonium salts are oxidized, nitric acid is formed, and the pH falls. This tends to increase the availability to plants of certain ions that are normally obtained only with difficulty, and nitrification has thus been implicated in changes in concentration of soluble potassium, phosphate, magnesium, iron, manganese, and calcium.

THE NITRIFYING BACTERIA

Considerable uncertainty still exists as to the status of many of the organisms reported to be capable of catalyzing ammonium or nitrite oxidation. This results from the difficulties in obtaining isolates whose purity is beyond reasonable doubt. Contaminants in autotrophic cultures often remain undetected because many of the offending strains are morphologically identical to the predominant organisms and because neither the autotrophs nor many of the contaminating heterotrophs develop appreciably on conventional laboratory media.

The pioneering work of S. Winogradsky established that nitrification is typically associated with the metabolism of certain chemoautotrophic bacteria. Two groups were distinguished, one deriving its energy for cell synthesis by the oxidation of ammonium, the other by the oxidation of nitrite. The nitrifying autotrophs form no endospores and vary in shape to include rods, ellipsoids, cocci, and spirilla. Despite the lack of morphological homogeneity, physiological similarity stands out, especially with respect to the energy-yielding reactions; the energy for growth is derived solely from the metabolism of ammonium or nitrite. The following genera are known to occur in soil:

I. Oxidize ammonium to nitrite
   Nitrosomonas. Ellipsoidal or short rods.
   Nitrosococcus. Spherical cells.

Nitrosospira. Spiral-shaped cells.
Nitrosolobus. Lobate and pleomorphic cells.
II. Oxidize nitrite to nitrate
   Nitrobacter. Short rods.

Of the five, only *Nitrosomonas* and *Nitrobacter* are encountered frequently, and these two are undoubtedly the major nitrifying chemoautotrophs. Only two species are currently recognized, *Nitrosomonas europaea* (Figure 16.3) and *Nitrobacter winogradskyi*. *Nitrosococcus* species are not frequently found in nitrifying enrichments and have been rarely reported. Three genera have been established in the past by their ability to form a zoogloea, *Nitrosocystis* and *Nitrosogloea* among the nitrite producers and *Nitrocystis*, a nitrate former, but the validity of these three genera has been the subject of considerable doubt. The other two genera, *Nitrosospira* and *Nitrosolobus*, are rarely found and do not seem to be common in nature. The established chemoautotrophic nitrifying genera and species are, therefore, fewer in number than might be expected in view of the significance of nitrate production for plant growth.

The nitrifying autotrophs are typically obligate in their reliance upon inorganic materials for energy; not only is organic carbon not utilizable as sole carbon source for growth, but no energy is obtained from the oxidation of any inorganic substrates other than those containing nitrogen. The carbon for cell synthesis is derived solely, or possibly largely, from CO_2, carbonates, or bicarbonates while the energy for the reduction of CO_2 is obtained by the oxidation of the inorganic nitrogen compounds. Growth does not occur in

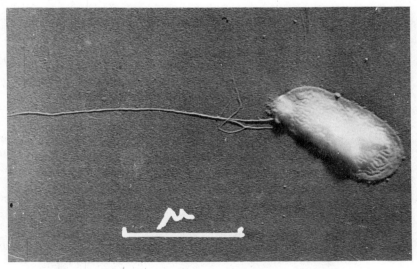

Figure 16.3. Electron photomicrograph of *Nitrosomonas europaea*.

conventional laboratory media, many of the medium ingredients even being bactericidal. However, *Nitrosomonas* and *Nitrobacter* cells are able to assimilate and metabolize certain simple organic molecules (12, 21), although it is unlikely that such assimilation is of much value for their growth in nature. Except for the peculiarities regarding the carbon and energy sources, the inorganic nutrient requirements resemble those known for heterotrophs although the nutrient demand in laboratory studies is low since the number of cells is never great. A requirement for low levels of molybdenum exists in *Nitrobacter* (17). These bacteria have no known growth factor requirements, and still all polysaccharides, structural constituents, amino acids, and vitamins found within the organisms can be produced from inorganic nutrients and CO_2. A remarkable nutritional simplicity is indicated, yet the metabolic complexity must be great since the bacteria are capable of synthesizing from inorganic materials all enzymes and other factors necessary for life.

Because of their capacity to utilize CO_2 as sole carbon source, the nitrifiers must bring about a reduction of CO_2 in order to convert it to the types of carbon compounds found in microbial protoplasm. For each molecule of CO_2 reduced to the level of carbon in the cell, that is, $(CH_2O)_n$, a fixed amount of energy is required. The driving force for the reduction is the inorganic oxidation; hence, biochemical efficiency in these organisms is conveniently expressed by the ratio of inorganic nitrogen oxidized to CO_2 carbon assimilated, an N:C ratio. For strains of *Nitrosomonas*, the ratio of ammonium nitrogen oxidized to CO_2 carbon assimilated varies from approximately 14 to 70:1. In *Nitrobacter*, the ratio of nitrite nitrogen oxidized to CO_2 carbon fixed varies from 76 to 135:1.

The reaction characterizing the chemoautotrophic bacteria of the first step in nitrification is

$$NH_4^+ + 1\tfrac{1}{2}O_2 \rightarrow NO_2^- + 2H^+ + H_2O \tag{II}$$

For equations I and II, the observed disappearance of substrate, the accumulation of product, and the utilization of O_2 agree with the theoretical values.

The microorganisms do not get all the energy potentially available but only a small proportion of it; the proportion is determined by their efficiency of energy utilization. For *Nitrosomonas*, the free energy efficiency often ranges from 5 to 14 percent while the efficiency of *Nitrobacter* has been calculated to be 5 to 10 percent. The free energy efficiency is greater in young cultures still in the logarithmic phase than in old cultures. The lower N:C ratios of the ammonium than of the nitrite oxidizers demonstrate that the former get more energy from their autotrophic reaction; that is, less nitrogen need be turned over for each cell formed.

Many microbiologists have investigated the abundance of these autotrophs in nature. The numbers of ammonium oxidizers have been found to vary from zero up to one million or more per gram of soil. The larger counts occur in soils of pH greater than about 6.0, yet many neutral or alkaline habitats have small populations. In most habitats, species of *Nitrosomonas* and *Nitrobacter* are found together; otherwise nitrite might accumulate to phytotoxic levels. Populations of both groups may be enlarged by use of ammonium salts, and values in excess of 10^7 per gram are not unknown. In temperate climates, nitrifiers are numerous during the warmth of the spring and rare during hot, dry summers and in winter; drying or freezing decreases their abundance but never entirely eliminates these bacteria.

In their early investigations of the basis for chemoautotrophy, microbiologists were perplexed by the total inability of these bacteria to use added carbonaceous nutrients, and it was not uncomforting to know that simple organic materials such as asparagine or sodium butyrate completely inhibited the nitrifiers in culture media. The inhibition by certain chemicals has been confirmed in recent investigations, but careful study has revealed that many of the effects even in culture were the result of experimental errors. A diverse group of organic compounds fails to inhibit *Nitrosomonas europaea* except when present in relatively high concentrations although the levels for inhibition are still somewhat lower than for common heterotrophs. There is no basis for the conclusion that ammonium oxidation in nature is inhibited by organic matter per se especially as the reaction occurs in environments with considerable soluble carbonaceous materials as in sewage and manure. The well-established role of carbohydrates in depressing the nitrate content of soil is not an influence on the nitrate-producing bacteria but a consequence of the depletion of the inorganic nitrogen supply by the flora requiring inorganic nutrients for carbohydrate decomposition.

Under most conditions nitrite does not accumulate in soil, and the dominant nitrogen-containing anion is nitrate. The existence and persistence of nitrite could have a marked effect on crop production because of its toxicity to plants and microorganisms. On the other hand, nitrite may accumulate under certain circumstances; in alkaline soils, for example, nitrate formation from $(NH_4)_2SO_4$ is suppressed as the rate of ammonium application is increased. This is not an inhibition of ammonium oxidation but rather the nitrite formed is not further metabolized to nitrate so that it persists as long as available ammonium is present. Considerable field evidence indicates that nitrite accumulation is the result of two factors, alkalinity and high ammonium levels. In calcareous soils, the accumulation is proportional to the rate of ammonium addition, and for constant nitrogen fertilization the effect increases with decreasing hydrogen ion concentrations. Anhydrous ammonia, a common fertilizer, may cause the pH to rise, sometimes to values of 9.0 and 9.5, and the nitrogen source together with the locally high pH frequently cause nitrite accumulations at the site of application. Even in the hydrolysis of proteinaceous matter or urea, the release of ammonium may lead to a significant although usually transitory suppression of nitrate formation by an analogous mechanism (13, 38). As the pH or the ammonium level falls with progress of nitrification, the suppression is relieved and nitrate production commences. Toxicity from the nitrite thus building up occasionally seems to be a practical problem.

The nitrite accumulation is attributed to the marked sensitivity of the *Nitrobacter* group to ammonium salts at alkaline reaction. At pH 9.5, as little as 1.4 ppm ammonium nitrogen suppresses the energy-yielding reaction of *Nitrobacter* while having no such influence on the ammonium oxidizers (2). Such observations suggest that ammonium, the natural substrate for nitrification, is a selective inhibitor in alkaline environments of the second step in nitrification by virtue of its toxicity to *Nitrobacter*. The growth of the ammonium oxidizers and the suppression of proliferation of nitrite-oxidizing bacteria in a soil that accumulates nitrite are depicted in Figure 16.4. The active principle apparently

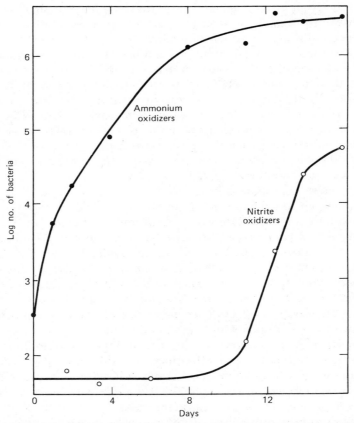

Figure 16.4. Growth of ammonium-oxidizing bacteria and inhibition of nitrite-oxidizing bacteria in a soil receiving ammonium salts (31). Reproduced by permission of the Soil Science Society of America.

is not the cation, ammonium, but the free ammonia that is favored by pH values more basic than 7.0. Once the ammonia level falls to a low level because of its oxidation or the drop in pH during nitrification, the nitrite oxidizers develop rapidly and destroy their substrate.

HETEROTROPHIC NITRIFICATION

Enrichment techniques are designed to be specific for definite physiological types, and these methods are the means whereby the autotrophs are isolated. As a result, nitrification is often taken to be a purely chemoautotrophic attribute. Heterotrophs that might be able to oxidize inorganic nitrogen compounds, as they derive no energy from the process, cannot be isolated by preferential enrichment since the necessary culture solutions would require organic substrates that would favor a variety of microorganisms growing in simple organic carbon-ammonium-salts media. Nevertheless, heterotrophic nitrification is now established as a common microbial phenomenon, in vitro at least, but each suspected heterotroph must be tested separately in pure culture in order to establish its capacity to nitrify.

A large number of heterotrophic bacteria and actinomycetes are able to generate traces of nitrite when grown in culture media containing ammonium salts. As a rule, nitrite appears only after active growth has largely ceased and in media in which the ammonium supply far exceeds that required for assimilatory purposes; that is, where the C:N ratio is low (10, 19). These many organisms do not form nitrate. The trace quantities of nitrogen oxidized are in contrast to the 2000 ppm nitrogen or more transformed by *Nitrosomonas*. Several fungi are also capable of oxidizing nitrite in culture, so that two heterotrophic populations acting together can presumably convert ammonium to nitrate. On the other hand, a few bacteria, such as strains of *Arthrobacter* (41), and fungi such as *Aspergillus flavus* (15) and related aspergilli may produce nitrate in media containing ammonium as the sole nitrogen source. In the process, the fungus produces 3-nitropropionic acid ($O_2NCH_2CH_2COOH$) and traces of nitrite (Figure 16.5), whereas the bacterium excretes large amounts of hydroxylamine.(NH_2OH), nitrite, and 1-nitrosoethanol, $CH_3CH(OH)NO$.

The nitrogenous substrates that are oxidized in vitro are not restricted to inorganic compounds, and the products likewise need not be inorganic. Nitrogen in amino acids of microbial cells is in the reduced form ($R-NH_2$), and it is also in the reduced state in purines, pyrimidines, and amides of protoplasm. However, fungi, actinomycetes, and heterotrophic bacteria can make compounds in which the nitrogen is in a more oxidized state, and to include these in the transformation, a more general definition of nitrification has been suggested: the biological conversion of nitrogen in organic and inorganic compounds from a reduced to a more oxidized state. The heterotroph does not make use of the energy released in the oxidation for growth, and indeed, often the more oxidized products do not appear until active proliferation has ceased.

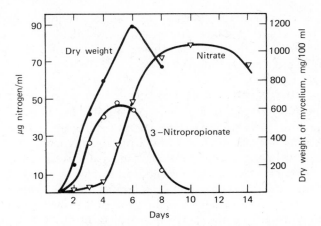

Figure 16.5. Growth and formation of 3-nitropropionic acid and nitrate by *Aspergillus flavus* in a medium containing ammonium (15).

Many microbial metabolites contain nitrogen in a partially or highly oxidized state. These include the following:

A. *N* Oxides: aspergillic acid, pulcherriminic acid.
B. *N*-Hydroxy compounds: hadacidin.
C. Hydroxamates: sideramines, mycelianamide.
D. Oximes: oximinopropionic acid.
E. *C*-Nitroso compounds: alanosine, 1-nitrosoethanol.
F. Nitrosamines: dimethylnitrosamine, *p*-methylnitrosoaminobenzaldehyde.
G. Nitrosamides: streptozotocin.
H. *C*-Nitro compounds: 3-nitropropionic acid, chloramphenicol, and others.
I. Nitramines: fragin, nitraminoacetic acid, β-nitroaminoalanine.

The type structures for these classes of molecules are given in Figure 16.6. The precursors of such metabolites in culture are not always ammonium or amino compounds, but the potential of many heterotrophs for the formation of such an array of substances—by oxidative or reductive sequences—adds a new dimension to explaining which organisms oxidize nitrogen compounds in nature. The reaction need not stop at the organic product, moreover, inasmuch as nitro compounds, oximes, and undoubtedly others from among these types of chemicals can be transformed to nitrite and sometimes nitrate.

The significance of heterotrophic nitrification is difficult to assess at this time. The evidence at present does not implicate heterotrophs in most instances of nitrite or nitrate production in soil, yet they may be prominent in some habitats. For example, certain soils contain few autotrophs and yet nitrate is synthesized, and sometimes nitrate is formed at temperatures too high for

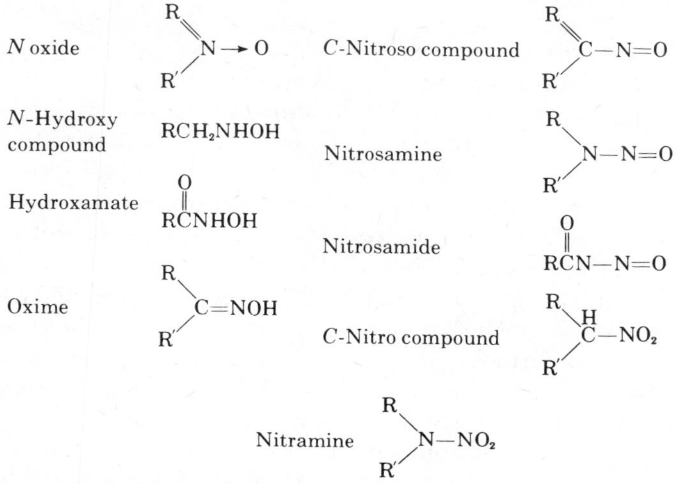

N oxide
$$\begin{matrix} R \\ \diagdown \\ \quad N \rightarrow O \\ \diagup \\ R' \end{matrix}$$

C-Nitroso compound
$$\begin{matrix} R \\ \diagdown \\ \quad C-N=O \\ \diagup \\ R' \end{matrix}$$

N-Hydroxy compound RCH_2NHOH

Nitrosamine
$$\begin{matrix} R \\ \diagdown \\ \quad N-N=O \\ \diagup \\ R' \end{matrix}$$

Hydroxamate
$$\begin{matrix} O \\ \parallel \\ RCNHOH \end{matrix}$$

Nitrosamide
$$\begin{matrix} O \\ \parallel \\ RCN-N=O \end{matrix}$$

Oxime
$$\begin{matrix} R \\ \diagdown \\ \quad C=NOH \\ \diagup \\ R' \end{matrix}$$

C-Nitro compound
$$\begin{matrix} R \quad H \\ \diagdown \\ \quad C-NO_2 \\ \diagup \\ R' \end{matrix}$$

Nitramine
$$\begin{matrix} R \\ \diagdown \\ \quad N-NO_2 \\ \diagup \\ R' \end{matrix}$$

Figure 16.6. Type structures of microbial metabolites with oxidized nitrogen.

significant replication of the chemoautotrophs (22, 33). Moreover, the results of some experiments suggest that certain inhibitors may allow for nitrate biosynthesis from humus nitrogen but not from ammonium—that is, two different populations are involved (16). Nevertheless, direct and unequivocal evidence for the heterotrophic oxidation in soil is lacking, and pure culture studies have failed to reveal any heterotroph that can nitrify as rapidly and to as great an extent as the two-genus autotrophic association. Still, the inefficiency of the heterotrophs may be compensated by their great numbers, and they may exert some influence on the rate of nitrate synthesis.

NITRATE POLLUTION

Nitrate, though an important ion in plant nutrition, may be a significant environmental pollutant. It is deemed to be undesirable because of its potential role in (a) eutrophication, (b) infant methemoglobinemia associated with the consumption of nitrate-rich waters or vegetables, (c) animal methemoglobinemia, and (d) the formation of nitrosamines. In these instances, the initiators of the problem are the nitrifying populations, the end product of their activity being often unwanted, in certain concentrations at least, in waters, food, and feed.

Recent years have witnessed a marked increase in the addition of nitrogen to soil. This increase may be attributable to the use of large quantities of chemical fertilizers for crop production, the application to fields of vast amounts of animal manure from feedlots as part of disposal operations, or the addition of digested sludge from cities and towns desiring to avoid polluting nearby waters with the sludge derived from sewage treatment plants. The chemical fertilizers—containing usually ammonium salts, anhydrous ammonia, or urea—are generally nitrified rapidly, whereas the organic nitrogen from the carbonaceous wastes are converted to nitrate more slowly. The potential fertilizer contribution to increasing nitrate levels in soils and waters must be deemed to be substantial in view of the enormous growth of the fertilizer industry in the past few decades and the need for still more industrial expansion in order to provide food for the malnourished or undernourished peoples of many of the developing countries. Just to keep pace with the growing world population requires a substantially increased input of fertilizer, but the need for a more adequate food supply in the impoverished nations must, of necessity, also be answered by greater fertilizer usage. Unfortunately, only part of the nitrogen thus added is recovered by the crop, and much of that which is not so removed is oxidized to nitrate. The nitrate is then subject either to leaching to the groundwater or to denitrification. The relative importance of the two modes of loss from soil is not yet predictable, being governed by the frequently unknown potential activities of the denitrifying populations and environmental factors governing their metabolism. As a consequence, the quantity of nitrate is sometimes surprisingly low in fields receiving considerable fertilizer, but sometimes it is remarkably high (1, 24). Nevertheless, proportionately greater quantities of fertilizer nitrogen undoubtedly will increasingly exceed the availability of substrates utilized as energy sources by the community for immobilization and denitrification, and the additional nitrate made in nitrification would then be subject to leaching to underground waters.

The feedlots so common in regions of the United States contain enormous numbers of beef cattle on small areas of land, many feedlots having tens of thousands of animals. The trend in cattle husbandry to confinement feeding and the high cost of transporting manure to farm land at some distance away have made acute the issue of nitrate pollution from feedlots. The magnitude of the problem can be visualized if one bears in mind that an average steer may excrete about 43 kg of nitrogen per year so that a feedlot of 30,000 animals may create significant pollution as the organic materials are mineralized and the resulting ammonium is oxidized. When there is sufficient precipitation for leaching, the nitrate moves through the soil profile to the groundwater, and the subterranean water then is transported laterally to enter wells used for human drinking supplies or to surface waters that have the capacity to support algal blooms. It is not uncommon for the nitrate level in the water to thus exceed the standards established to guarantee a safe supply for human consumption (18), and even in regions with many dairy cows rather than beef cattle, the nitrate may be a major contributor to the nutrient supply of lakes (39).

Lakes and rivers support a certain population of algae or sometimes rooted plants, but the biomass is usually limited by the poor supply of inorganic nutrients. However, when additional nutrients enter, algae or rooted plants may flourish to create undesirable conditions. This process is known as *eutrophication*, the enrichment of waters with nutrients. In many lakes, the element limiting development of photosynthetic organisms is phosphorus but frequently it is nitrogen, and thus nitrate entering from adjacent land areas—after passing through the groundwater—may promote modest to extensive blooms. Eutrophication occurs naturally as nutrients move from fields to inland waters, but the use of inorganic or organic nitrogen in agriculture as well as the release of nitrogen when virgin soils are brought into cultivation undoubtedly enhance the enrichment of those bodies of water in which this element limits photosynthetic populations. Only a low concentration of nitrate is necessary, moreover, and lakes with 0.3 ppm nitrate nitrogen sometimes, but not invariably, may support unwanted algal blooms, provided other nutrients are in adequate supply. Excessive growths are undesirable for many reasons: recreational uses of lakes are diminished, the cost of water treatment rises, fish may succumb as the O_2 is depleted when the algae die and are decomposed, the drinking water acquires offensive tastes and odors, and navigation of small boats may even be hindered because of extensive networks of higher plants. The relative contribution of nitrate from a land area to a particular body of water is sometimes small because of the little farming practiced, the small nitrogen inputs, or the presence of other sources of the element, but sometimes agriculture is the major contributor.

The disposal on land of solid or liquid sludge from municipalities or from large numbers of farm animals is advantageous in that it provides the vegetation with essential nutrients, often leads to greater humus levels, and eliminates a bulky waste from the municipal treatment facility or the farm. However, even this beneficial practice is fraught with hazards, one of which is the accumulation of nitrate to undesirable concentrations. In this instance, reducing the rate of sludge application decreases the amount of nitrate released from the microbial conversions so that groundwater pollution is minimized or overcome (23).

Most of the public health interest in nitrate, and hence nitrification, comes from the disease known as *methemoglobinemia*. Methemoglobinemia arises when nitrate consumed with water, food, or feed is reduced in the gastrointestinal tract to nitrite. The latter enters the bloodstream and there reacts with hemoglobin to yield methemoglobin, a change that leads to an impairment of O_2 transport in the body. This process is not of consequence in adult humans but can be serious in infants, particularly those under three months old, and in ruminants. Because infant methemoglobinemia is nearly always attributable to water and the frequency of clinical cases is directly correlated with the nitrate concentration in the water supply, the World Health Organization and various countries have established standards for the quality of water designed for human consumption. The standard is usually 10 ppm nitrate nitrogen, a level below which clinical methemoglobinemia in infants is extremely rare.

Nitrate also may be clinically important because of its presence in food or feed. In these instances, the plants that are eventually consumed assimilate much nitrate from the soil and store it in large quantities. A few species are notorious accumulators: beets, spinach, celery, and lettuce among the vegetables and corn, sorghum, sudan grass, and oats among the forage crops. Although food-induced methemoglobinemia is almost unknown among babies, active research is being conducted to find ways of reducing nitrate formation in soil under vegetables and to diminish accumulation in the crops that could present special problems (8). In livestock herds, by contrast, considerable deaths have occurred because the animals consumed feeds made from plant species that stored the ion in appreciable amounts.

Nitrosamines are not known to be a problem because of their potential formation in soil, but the potency of this class of chemicals makes an evaluation of their hazard especially important. These compounds have the following structure.

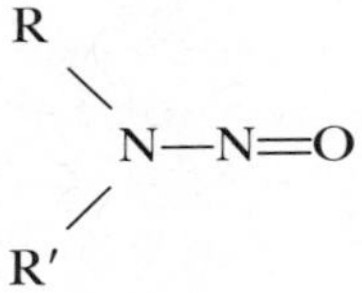

$$
\begin{array}{c}
R \\
\diagdown \\
N{-}N{=}O \\
\diagup \\
R'
\end{array}
$$

R and R' can be methyls, straight chains, rings, or other groups. Interest in these chemicals rose dramatically when it became evident that many nitrosamines are carcinogenic, mutagenic, and teratogenic; that is, they can induce cancer, mutations, or abnormalities and sometimes death in fetuses. Their formation requires the presence of a secondary amine (RNHR') and nitrite, the reaction then involving a simple condensation.

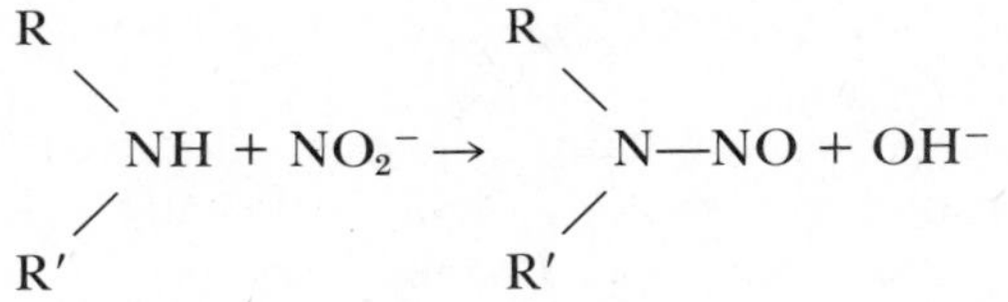

$$
\begin{array}{c}
R \\
\diagdown \\
NH + NO_2^- \rightarrow \\
\diagup \\
R'
\end{array}
\qquad
\begin{array}{c}
R \\
\diagdown \\
N{-}NO + OH^- \\
\diagup \\
R'
\end{array}
\qquad (III)
$$

This reaction can be carried out by microbial enzymes (6) or nonenzymatically by organic materials present in soil (28). Nitrite is not usually found in soil, but, as discussed above, it sometimes appears in large amounts and is constantly being generated in nitrification and nitrate reduction. Secondary amines are not known to be common soil constituents, but some pesticides are secondary amines, plant residues often contain them, and they can be produced microbiologically from other pesticides or natural products. That the concern is not entirely unfounded is supported by findings that samples of soil in the laboratory do indeed form these toxicants (7) and that plants used by humans as food can assimilate them, but to date no evidence exists that these chemicals are produced in soil under natural conditions.

BIOCHEMICAL MECHANISM

In the nitrification step catalyzed by *Nitrosomonas*, the oxidation state of nitrogen is changed from the -3 of ammonia to the $+3$ of nitrous acid through the removal of six electrons. Establishing the individual steps involved and the intermediary compounds formed has been the subject of considerable interest, but it is only in recent years that some light has been shed on this problem. Cell suspensions of *Nitrosomonas europaea* usually convert ammonium nitrogen quantitatively to nitrite, and the O_2 consumed and nitrite produced agree with the values predicted by equation II. Under certain circumstances, hydroxylamine, which is usually highly toxic, may be oxidized as readily as ammonium itself. The reaction is shown by equation IV.

$$NH_2OH + O_2 \rightarrow HNO_2 + H_2O \tag{IV}$$

To demonstrate that hydroxylamine is not only a substrate but is probably formed during nitrification inside the bacterial cells, *N. europaea* is incubated with ammonium and an inhibitor of hydroxylamine oxidation; under these conditions, hydroxylamine is excreted by the bacteria (43). Growing cultures of the organisms produce traces of nitrous oxide (N_2O), but when cells are allowed to age, both ammonium and hydroxylamine are converted to N_2O. This oxide is not an intermediate in nitrification because it is not converted to nitrite. The gas is either formed from the actual intermediate or possibly on occasion by reduction of the final product of the organism, nitrite (34, 44, 45). Small amounts of nitric oxide (NO) are also evolved. Under suitable experimental circumstances, enzyme preparations derived from the autotroph oxidize hydroxylamine to N_2O, NO, and nitrite (4, 5). These findings and related observations suggest that ammonium is converted to hydroxylamine, and the latter is then transformed to some still undefined metabolite, possibly a compound like nitroxyl (HNO). This intermediate is then oxidized to nitrite, possibly by way of NO. The latter compound may exist free, or it may be in some bound form. The N_2O might then arise nonenzymatically from the unknown intermediate.

$$NH_3 \rightarrow NH_2OH \rightarrow (HNO?) \rightarrow NO \rightarrow NO_2^- \\ \searrow \tfrac{1}{2} N_2O + \tfrac{1}{2} H_2O \tag{V}$$

Despite research in several laboratories, however, the evidence for this pathway remains somewhat equivocal.

Not equivocal, on the other hand, is the way in which *Nitrobacter* oxidizes its substrate. Members of the genus change the nitrogen oxidation state from $+3$ to $+5$, a yield of two electrons for each molecule transformed. Typical of cellular metabolism and energy-yielding reactions in biology is dehydrogenation, and dehydrogenation characterizes nitrate formation as well. Enzymes prepared from cells of this organism convert nitrite to nitrate and consume O_2 in the process. The nitrite-oxidation step is not the one that uses the O_2 because cells provided with nitrite and $^{18}O_2$ make nitrate containing almost no ^{18}O (3). The last step in the autotrophic sequence is thus visualized as involving a hydrated nitrite molecule from which hydrogens (or electrons) are removed to yield nitrate in a conversion not requiring O_2. The hydrogens (or electrons) are then passed to O_2 to give H_2O, a sequence in line with the $^{18}O_2$ experiment.

$$NO_2^- + H_2O \rightarrow H_2O \cdot NO_2^- \rightarrow NO_3^- + 2H \qquad \text{(VI)}$$

$$2H + \tfrac{1}{2}O_2 \rightarrow H_2O \qquad \text{(VII)}$$

REFERENCES

Reviews

Alexander, M. 1965. Nitrification. In W. V. Bartholomew and F. E. Clark, eds., *Soil nitrogen.* American Society of Agronomy, Madison, Wisc., pp. 307–343.

Committee on Nitrate Accumulation. 1972. *Accumulation of nitrate.* National Academy of Sciences, Washington, D.C.

Walker, N. 1975. Nitrification and nitrifying bacteria. In N. Walker, ed., *Soil microbiology: A critical review.* Halsted Press (Wiley), New York, pp. 133–146.

Wallace, W. and D. J. D. Nicholas. 1969. The biochemistry of nitrifying microorganisms. *Biol. Rev.,* 44:359–391.

Literature Cited

1. Adriano, D. C., P. F. Pratt, and F. H. Takatori. 1972. *J. Environ. Qual.,* 1:418–422.
2. Aleem, M. I. H. and M. Alexander. 1960. *Appl. Microbiol.,* 8:80–84.
3. Aleem, M. I. H., G. E. Hoch, and J. E. Varner. 1965. *Proc. Natl. Acad. Sci., U.S.,* 54:869–873.
4. Anderson, J. H. 1965. *Biochem. J.,* 95:688–698.
5. Anderson, J. H. 1965. *Biochim. Biophys. Acta,* 97:337–339.
6. Ayanaba, A., and M. Alexander. 1973. *Appl. Microbiol.,* 25:862–868.
7. Ayanaba, A., W. Verstraete, and M. Alexander. 1973. *Soil Sci. Soc. Amer. Proc.,* 37:565–568.
8. Barker, A. V., N. H. Peck, and G. E. MacDonald. 1971. *Agron. J.,* 63:126–129.
9. Blasco, M. L. and A. H. Cornfield. 1966. *J. Sci. Food Agr.,* 17:481–484.
10. Brisou, J. and H. Vargues. 1962. *Compt. Rend. Soc. Biol.,* 156:1487–1489.
11. Bundy, L. J. and J. M. Bremner. 1973. *Soil Sci. Soc. Amer. Proc.,* 37:396–398.
12. Clark, C. and E. L. Schmidt. 1967. *J. Bacteriol.,* 93:1302–1308.
13. Court, M. N., R. C. Stephen, and J. S. Waid. 1964. *J. Soil Sci.,* 15:42–48.
14. Dancer, W. S., L. A. Peterson, and G. Chesters. 1973. *Soil Sci. Soc. Amer. Proc.,* 37:67–69.
15. Doxtader, K. G. and M. Alexander. 1966. *J. Bacteriol.,* 91:1186–1191.
16. Etinger-Tulczynska, R. 1969. *J. Soil Sci.,* 20:307–317.
17. Finstein, M. S. and C. C. Delwiche. 1965. *J. Bacteriol.,* 89:123–128.
18. Gilbertson, C. B., T. M. McCalla, J. R. Ellis, O. E. Cross, and W. R. Woods. 1971. *J. Water Pollut. Control Fed.,* 43:483–493.

19. Hirsch, P., L. Overrein, and M. Alexander. 1961. *J. Bacteriol.*, 82:442–448.
20. Huber, D. M., G. A. Murray, and J. M. Crane. 1969. *Soil Sci. Soc. Amer. Proc.*, 33:975–976.
21. Ida, S. and M. Alexander. 1965. *J. Bacteriol.*, 90:151–156.
22. Ishaque, M. and A. H. Cornfield. 1974. *Trop. Agr. (Trinidad)*, 51:37–41.
23. King, L. D. and H. D. Morris. 1972. *J. Environ. Qual.*, 1:442–446.
24. Larson, K. L., J. F. Carter, and E. H. Vasey. 1971. *Agron. J.*, 63:527–528.
25. Lees, H. and J. H. Quastel. 1946. *Biochem. J.*, 40:815–823.
26. Lewis, D. C., and R. C. Stefanson. 1975. *Soil Sci.*, 119:273–279.
27. Likens, G. E., F. H. Bormann, N. M. Johnson, D. W. Fisher, and R. C. Pierce. 1970. *Ecol. Monogr.*, 40:23–47.
28. Mills, A. L. and M. Alexander. 1976. *J. Environ. Qual.*, 5:437–440.
29. Moore, D. R. E. and J. S. Waid. 1971. *Soil Biol. Biochem.*, 3:69–83.
30. Morrill, L. G. and J. E. Dawson. 1962. *J. Bacteriol.*, 83:205–206.
31. Morrill, L. G. and J. E. Dawson. 1967. *Soil Sci. Soc. Amer. Proc.*, 31:757–760.
32. Purchase, B. S. 1974. *Plant Soil*, 41:527–539.
33. Remacle, J. and A. Froment. 1972. *Oecol. Plant.*, 7:69–78.
34. Ritchie, G. A. F. and D. J. D. Nicholas. 1972. *Biochem. J.*, 126:1181–1191.
35. Sabey, B. R. 1969. *Soil Sci. Soc. Amer. Proc.*, 33:263–266.
36. Schloesing, T. and A. Muntz. 1877. *Compt. Rend. Acad. Sci.*, 84:301–303; 85:1018–1020.
37. Shattuck, G. E., Jr. and M. Alexander. 1963. *Soil Sci. Soc. Amer. Proc.*, 27:600–601.
38. Smith, J. H. 1967. *Soil Sci. Soc. Amer. Proc.*, 31:377–379.
39. Sonzogni, W. C. and G. F. Lee. 1974. *Trans. Wisc. Acad. Sci. Arts Letters*, 62:133–164.
40. Stojanovic, B. J. and M. Alexander. 1958. *Soil Sci.*, 86:208–215.
41. Verstraete, W. and M. Alexander. 1972. *J. Bacteriol.*, 110:955–961.
42. Winogradsky, S. 1890. *Ann. Inst. Pasteur*, 4:213–231, 257–275, 760–771.
43. Yoshida, T. and M. Alexander. 1964. *Can. J. Microbiol.*, 10:923–926.
44. Yoshida, T. and M. Alexander. 1970. *Soil Sci. Soc. Amer. Proc.*, 34:880–882.
45. Yoshida, T. and M. Alexander. 1971. *Soil Sci.*, 111:307–312.
46. Yoshida, T. and B. C. Padre. 1974. *Soil Sci. Plant Nutr.*, 20:241–247.

17
Denitrification

The various reactions of the nitrogen cycle transform one form of the element to another. Mineralization liberates the nitrogen in inorganic form, immobilization converts it back to an unavailable state, while nitrification changes the element from a reduced to an oxidized condition. Certain transformations of nitrogen lead to a net loss of the element from the soil through volatilization. For the purposes of crop production, nitrogen volatilization has a deleterious influence for it depletes part of the soil's reserve of an essential nutrient. The sequence of steps that results in gaseous loss is known as *denitrification*, the microbial reduction of nitrate and nitrite with the liberation of molecular nitrogen and nitrous oxide.

Denitrification is not the sole means by which microorganisms reduce nitrate and nitrite. In the utilization of the two anions as nitrogen sources for growth, microorganisms reduce them to the ammonium level. A reduction of this type serves the purpose of changing nitrogen into a form suitable for amino acid synthesis within the cell. In denitrification, on the other hand, the nitrogen is lost to the atmosphere and fails to enter the cell structure. In contrast with denitrification, which is essentially a respiratory mechanism in which nitrate replaces molecular oxygen, that is, *nitrate respiration*, the utilization of nitrate as a nutrient source may be termed *nitrate assimilation*. Both transformations involve reductive pathways, but the end products of nitrate respiration are volatilized while the products of nitrate assimilation are incorporated into cell material. From an agronomic viewpoint, nitrate assimilation differs from denitrification further in that the nitrogen remains in the soil and is not put out of the realm of potential availability as a plant nutrient.

NITROGEN LOSSES IN SOIL
When nitrate and a readily available carbohydrate are added to soil, N_2, usually N_2O (nitrous oxide), and often NO (nitric oxide) are evolved by the reduction

of the nitrate applied. These products are volatile and are lost to the atmosphere, but if the soil and the overlying gas phase are enclosed so that the volatile metabolites are still available to the community, the oxides are further reduced to N_2 (Figure 17.1). The NO that escapes is ultimately oxidized nonbiologically by O_2 in the overlying atmosphere to NO_2 (nitrogen dioxide). The evolution of nitrogenous gases is a typical result of laboratory experiments when the demand for O_2 by the microflora exceeds the supply. But the situation in practice is far more complex. The quantities of nitrate and of readily utilizable organic matter are usually not great, and the aeration status fluctuates with the moisture regime. Furthermore, since N_2 is omnipresent in nature, its release in the field is not simple to demonstrate quantitatively. Because the volatilization of nitrogen is a slow process and precise measurement of the gas evolution is difficult, nitrogen losses in the field usually are estimated by determining the changes in nitrogen content of soil with time. It should be borne in mind, however, that soils often contain in excess of 2000 kg of nitrogen per hectare, so that measurements of losses of 20 to 40 kg per hectare, an appreciable change, require sensitive analytical techniques.

To prepare a balance sheet of soil nitrogen, it is necessary to have precise data on the quantity of nitrogen entering the soil from precipitation, seed,

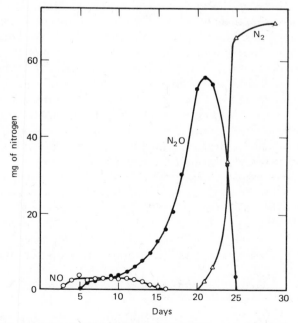

Figure 17.1. Sequence of products during denitrification in Norfolk sandy loam (8).

fertilizer, and manure as well as the quantity lost by leaching, erosion, and crop removal. If there is no net change in the soil, losses should equal gains. Should the losses exceed the gains, the nitrogen must be disappearing in some way not accounted for in the variables included in the balance sheet. Except in fields planted to legumes, a condition leading to nitrogen accretion, deficits are the rule when fertilizers are used. Frequently, only half to three-fourths of the inorganic nitrogen applied in fertilizers or formed from humus is recovered in the crop. Much of the element is removed by leaching, but an appreciable portion nevertheless disappears from the system entirely. The unaccounted-for fraction presumably is lost by volatilization.

MECHANISMS OF NITROGEN VOLATILIZATION

The foregoing discussion suggests that the liberation of gaseous nitrogen compounds may be of considerable economic importance in agriculture. The specific mechanism by which the volatilization occurs is not always readily established, but three reactions are possible: (a) nonbiological losses of ammonia, (b) chemical decomposition of nitrite, and (c) microbial denitrification leading to the liberation of N_2, N_2O, and sometimes NO.

The volatilization of free ammonia is appreciable under certain circumstances, and as much as one-fourth or occasionally more of the ammonium supplied in fertilizers or formed microbiologically may be lost in the gaseous form. In most soils, such losses are insignificant below pH 7.0, but the magnitude varies directly with increasing alkalinity. Warmer temperatures also favor the process, but losses are small in soils of high cation exchange capacity. During the rotting of manure or other nitrogen-rich organic matter near or at the soil surface, the pH rises during ammonification, and ammonia is released to the atmosphere. Should the increase in alkalinity associated with ammonification be sufficiently great, gaseous ammonia loss from surface-applied manure—as with urea—will take place even at sites in which the underlying soil is highly acid.

In those soils in which nitrite accumulates, a likely consequence at local sites with high pH values as discussed in the previous chapter, the nitrite may decompose spontaneously through a reaction with organic matter. The products of this nonenzymatic process are N_2, NO_2, and small quantities of N_2O. In the absence of O_2, NO may be evolved, but it would be oxidized readily to NO_2 once the gas came into contact with O_2. However, even in the special circumstances in which nitrite levels do rise, the microbial generation of gases may exceed the nonbiological process (3, 21). No volatilization is detected when ammonium is applied to areas too acid for nitrification because the nitrite is never synthesized.

The major mechanism of nitrogen volatilization and probably the most common means whereby N_2 and nitrogen oxides are evolved is by microbiological denitrification. In most laboratory studies, the dominant product is N_2, but these findings may sometimes be misleading because the N_2O and NO that are evolved are further reduced in the sealed and usually O_2-free incubation vessels widely employed. The oxides are thus accessible and are subject to further microbial metabolism, and N_2 then appears in large yield. Nevertheless, it is commonly held that N_2 is in fact also the chief gaseous product in nature, but this belief is not well substantiated by analytical determinations and is at times contradicted by experimental findings.

A common sequence of products involves the initial generation of nitrite from the nitrate in the soil. Nitrite levels may then be reasonably high, but they are more often low or undetectable. The nitrite then is converted to gases, but the identity of the first gas that is observed varies with the soil pH, temperature, and initial levels of nitrate and available organic matter. Often it is NO, a compound that is typically associated with denitrification in acid soils (9). The appearance of NO, if it is evolved, is generally followed by N_2O, but sometimes N_2 is the first volatile metabolite. In sealed vessels, the concentrations of N_2O and NO in the anaerobic gas phase decline, and N_2 then accumulates as the terminal product (Figure 17.2). In some instances, the dominant gas is indeed N_2 as the reduction approaches completion; in other instances, N_2O accounts for much of the nitrogen volatilized (28, 30).

The actual rate of gas evolution will vary enormously from site to site, the rate depending on temperature, moisture, O_2 status, organic matter level, and other factors. Nevertheless, although few studies of rates have been conducted,

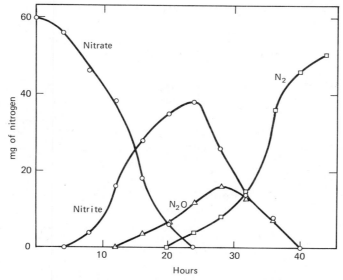

Figure 17.2. Products formed during denitrification in Melville loam, pH 7.8 (10). Reproduced by permission of the Soil Science Society of America.

the limited data available suggest that about 0.003 to 0.30 kg N_2O nitrogen and from less than 0.001 to about 0.02 kg NO nitrogen may be volatilized per hectare per day (4, 7, 15, 18). Undoubtedly the rates are low in dry or cold regions, but still larger oxide losses also may be occasionally encountered.

The major products of nitrate reduction in poorly drained land or in soils that become excessively wet appear to be gaseous rather than nitrite or ammonium. The nitrite usually is reduced almost as rapidly as it is formed. The ammonium concentration in flooded land containing or treated with nitrate increases slowly, but it never becomes too high. Moreover, not all the additional ammonium need arise from nitrate as ammonification is not greatly retarded at low partial pressures of O_2.

MICROBIOLOGY

Growth of the microorganisms concerned in denitrification is not dependent on the reduction of nitrate. Many of the responsible bacteria are active in proteolysis, ammonification, and undoubtedly in other transformations. Consequently, the presence of many denitrifying cells does not of itself indicate that conditions are suitable for denitrification. The existence of numerous organisms, on the other hand, does point to a large denitrifying potential. In this regard, it is important to emphasize the concept that the abundance of substrate-non-specific microorganisms should never be taken as a sign that any single one of their biochemical activities is prominent in the habitat from which the organisms were isolated. Thus, a large population of hemicellulose decomposers, ammonifiers, or antibiotic producers is *not* conclusive evidence for significant hemicellulose breakdown, ammonium release, or antibiotic synthesis. With substrate-specific organisms such as the nitrifying autotrophs, however, a dense population is good presumptive evidence for nitrifying activity. In the latter instance, growth of the bacteria is associated obligately with ammonium oxidation; the organisms in the former examples have a broad enzyme potential so that the reactions noted in culture media may be entirely unrelated to the activities in nature.

Arable fields contain an abundance of denitrifying microorganisms, and counts in excess of a million per gram are not uncommon in field soil. The population is typically larger in the immediate vicinity of plant roots. The potential for volatilization is therefore enormous, but conditions must be suitable for the organisms to change from aerobic respiration to a denitrifying type of metabolism. On the basis of the reservations outlined above, the existence of a large flora demonstrates only a potential for rapid nitrogen volatilization rather than an actual activity.

The capacity for true denitrification is limited to certain bacteria. Fungi and actinomycetes have not been implicated in N_2 production. Only a small number of bacterial species can bring about denitrification as strictly defined. The active species are largely limited to the genera *Pseudomonas, Bacillus,* and

TABLE 17.1
The Effect of pH on Bacterial Denitrification (31)

	pH					
	4.0	**5.0**	**6.0**	**7.0**	**8.0**	**9.0**
	ml gas evolved/hr/g cell nitrogen					
Paracoccus denitrificans	0	64	105	168	214	116
Pseudomonas denitrificans	0	15	196	138	92	0
Pseudomonas aeruginosa	0	12	218	246	251	13
Bacillus licheniformis	4	4	108	125	102	60

Paracoccus (Table 17.1) although *Thiobacillus denitrificans* and an occasional *Chromobacterium, Corynebacterium, Hyphomicrobium,* or *Serratia* species will catalyze the reduction. *Bacillus* strains, although numerous, are rarely important; their abundance is usually the result of the persistence of the endospore.

The denitrifying bacteria are aerobic, but nitrate is used as the electron acceptor for growth in the absence of O_2. Thus, the active species grow aerobically without nitrate or anaerobically in its presence. Few develop in environments devoid of both O_2 and nitrate. Most substrates used for aerobic oxidation may be attacked in the absence of O_2 in media containing nitrate; however, the organisms may gain less energy from oxidizing their carbonaceous substrates when nitrate is the electron acceptor for growth rather than O_2 (17), so that fewer cells are formed per unit of substrate oxidized when the organisms are denitrifying as compared to when they are growing in air. The physiological specialization for nitrate-dependent, anaerobic proliferation has been adapted for the isolation of the responsible agents. A common procedure is the use of anaerobic enrichment solutions containing large amounts of KNO_3. In such media, denitrifiers produce profuse quantities of gas and make the solution alkaline. Enrichments for thermophilic strains can be made by incubating the same medium at temperatures of 55 to 65°C. The thermophiles are almost invariably aerobic, spore-forming bacteria classified in the genus *Bacillus*.

Several chemoautotrophs are capable of reducing nitrate to molecular nitrogen. *Paracoccus denitrificans*, a facultative autotroph, develops in air or anaerobically with either organic compounds or H_2 as source of energy and O_2 or nitrate as electron acceptor. *Thiobacillus denitrificans* is a sulfur-oxidizing chemoautotroph that differs from other thiobacilli by its ability to proliferate anaerobically providing nitrate is available. For *T. denitrificans*, the energy source in these circumstances is sulfide, elemental sulfur, or thiosulfate, all of which are oxidized to sulfate. The nitrate is converted to N_2, but NO and N_2O are also sometimes evolved (2).

$$5S + 6KNO_3 + 2H_2O \rightarrow 3N_2 + K_2SO_4 + 4KHSO_4 \qquad (I)$$
$$5K_2S_2O_3 + 8KNO_3 + H_2O \rightarrow 4N_2 + 9K_2SO_4 + H_2SO_4 \qquad (II)$$

T. denitrificans will similarly grow well as an aerobe.

Most denitrifiers convert nitrate all the way to N_2. They use nitrate, nitrite, NO, or N_2O as electron acceptors for proliferation, reducing each to N_2. Some bacteria, however, carry out incomplete reductions. *Corynebacterium nephredii*, for example, is able to reduce only nitrate, nitrite, and NO, but the compound at the end of the sequence with this bacterium is N_2O rather than N_2 (25). Furthermore, a few fungi and bacteria that are not known to denitrify—that is, that do not grow anaerobically with nitrate as the alternate to O_2 as an electron acceptor—will still synthesize N_2O from nitrate and nitrite, and, as pointed out above, *Nitrosomonas* can liberate traces of N_2O during nitrification (32).

Many other microorganisms reduce nitrate. The reduction is required either in the process of protein synthesis or as a substitute for the reduction of O_2 in conventional aerobic metabolism. In the former instance, any microorganism that utilizes nitrate as a nitrogenous nutrient is able to reduce it to ammonium as it is ammonium that enters into the organic combination necessary for protein synthesis within the cell. The latter reaction, on the other hand, is found in bacteria normally considered to be aerobic but that replace the O_2 by nitrate under anaerobiosis; the products here, however, are not gaseous but nitrite or ammonium. It is also not uncommon to find nitrite in culture solutions of bacteria, fungi, and actinomycetes that reduce nitrate to ammonium. Generally, however, the nitrite produced from nitrate disappears almost as readily as it is formed, but occasional strains will accumulate nitrite.

There are thus three microbiological reactions of nitrate: (*a*) a complete reduction to ammonium frequently with the transitory appearance of nitrite, (*b*) an incomplete reduction and an accumulation of nitrite in the medium, and (*c*) a reduction to nitrite followed by the evolution of gaseous compounds, that is, denitrification. Regardless of their other physiological characteristics, microorganisms that use nitrate as a nitrogen source carry out reaction *a*. Some cultures are incapable of the complete reduction and must be supplied with ammonium or other reduced nitrogen compounds for growth to proceed; thus, a number of denitrifying bacteria require ammonium for growth even as they actively reduce nitrate (31). A major difference among these organisms is the relative amount of organic nitrogen synthesized: bacteria using nitrate for assimilatory purposes convert the anion chiefly to cell constituents, whereas those using it as both a nitrogen source for biosynthesis and an electron acceptor make little organic nitrogen but consume much nitrate. This is evident in soil (1) as well as in culture.

ENVIRONMENTAL INFLUENCES

The restriction of the capacity for active nitrogen volatilization to the bacteria and, furthermore, to only a few genera suggests that the magnitude and rate of denitrification are markedly affected by the environment. The responsible organisms bear some physiological kinship to one another so that changes in the habitat can appreciably stimulate or largely eliminate these economically important populations. Chief among the environmental influences are the nature and amount of organic matter, aeration, the moisture status, acidity, and temperature.

The rate of denitrification is far more slow in soils low in carbon than in land that is rich in organic matter. In soils of the former type and to a lesser extent of the latter type, nitrogen volatilization is enhanced by the addition of carbonaceous materials. The effectiveness of organic nutrients in promoting denitrification in waterlogged soils is proportional to their availability. Readily decomposable compounds such as simple sugars or organic acids are oxidized quickly and are more stimulatory than the less readily fermentable straws or grasses, which possess a greater degree of effectiveness in bringing about nitrate reduction than sawdust and lignin preparations. Well-rotted plant or animal residues, as expected, have only a minor influence, particularly when compared with the fresh materials. In contrast to the trends in waterlogged habitats, the addition of organic substances to well-drained soils diminishes the nitrogen losses, the conserving action resulting from an immobilization of inorganic nitrogen (5, 27).

Oxygen availability is another of the critical environmental determinants. Aeration affects the transformation in two apparently contrasting ways: on the one hand, denitrification proceeds only when the O_2 supply is insufficient to satisfy the microbiological demand; at the same time, O_2 is necessary for the formation of nitrite and nitrate, which are essential for denitrification. The demonstration that nitrogen is volatilized from soils maintained at normal O_2 tensions indicates that the pores and interstices in the profile are never entirely oxygenated. Anaerobic microenvironments exist at microscopic sites in well-drained soils whenever the biological O_2 demand exceeds the supply. Needless to say, decreasing the partial pressure of O_2 enhances the denitrification of added nitrate. The need for O_2 to create the substrate for the denitrifiers and the requirement for anaerobiosis to allow for the reductive sequence to proceed are reflected in the frequently large nitrogen losses at sites undergoing cycles in which O_2 is alternately available and then unavailable (24). Similarly, appreciable losses may be encountered in submerged soils used for rice cultivation: the ammonium is oxidized to nitrate in the thin O_2-containing surface layer, and the nitrate is converted to gaseous products of denitrification as it diffuses into the underlying anaerobic zone (33).

The inhibition of denitrification by O_2 may be the result of one or more effects on the responsible populations. Because the active species are aerobes, the suppression cannot be attributable to some detrimental influence of O_2 on their growth. In some instances, the suppression of nitrate reduction may merely reflect the preferential use of oxygen as the electron acceptor. On the other hand, the inhibition may sometimes result from a detrimental effect on the enzymes bringing about nitrate and nitrite reduction, or O_2 may even repress formation of the necessary denitrifying enzymes so that the catalysts required for the conversion never appear in the cells (23).

In well-drained soils, nitrogen volatilization is related to the moisture content. Denitrification of added nitrate is appreciable at high water levels and in localities having improper drainage. No losses usually occur at moisture levels below 60 percent of the water-holding capacity regardless of the carbohydrate supply, nitrate concentration, or pH. Above this figure, the rate and magnitude of denitrification are correlated directly with the moisture regime (Figure 17.3). The effect of water is attributed to its role in governing the diffusion of O_2 to sites of microbiological activity.

Important also to the active microflora is the pH of the environment. Many of the bacteria that bring about denitrification are sensitive to high hydrogen ion concentrations, and hence various acid soils contain a sparse denitrifying population. These ecological observations agree with studies of denitrification in pure culture (31). In other soils, on the other hand, denitrification is still rapid at pH values of 4.7 (13), so that a pH range applicable to nitrogen volatilization from all soils cannot be clearly defined.

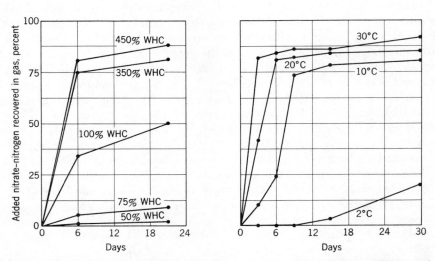

Figure 17.3. Effect of moisture, expressed as water-holding capacity (WHC), and temperature on denitrification in soil receiving glucose (5).

Acidity governs not only the rate of denitrification but also the relative abundance of the various gases. The liberation of N_2O is pronounced in environments whose pH is below about 6.0 to 6.5, and frequently N_2O makes up more than half of the nitrogenous gases evolved from acid habitats. Similarly, NO only appears in significant quantities when the pH is low. At neutral or slightly acid reaction, N_2O may be the first gas detectable, but it is reduced microbiologically so that N_2 tends to be the dominant product above pH 6. The differences in gas composition associated with pH may be largely a result of the acid sensitivity of the enzyme system concerned in N_2O reduction. From the viewpoint of biochemical mechanism, the appearance of N_2O prior to N_2 is of considerable interest.

Nitrous oxide release in soil or in culture is conditioned further by the nitrate concentration. The relative proportion of the gas is greatest at high nitrate levels, that is, when the concentration of the electron acceptor (the nitrogen salt) exceeds that of the electron donor (decomposable organic matter). When additional donors are made available, the N_2O is reduced to N_2.

Denitrification is markedly affected by temperature. The transformation proceeds slowly at 2°C, but increasing the temperature enhances the rate of biological loss. The optimum for the reaction is at 25°C and above (Figure 17.3). The transformation is still rapid at elevated temperatures and will proceed to about 60 to 65° but not at 70°C. The rapid release of N_2 in the more elevated ranges suggests an active thermophilic flora, a fact verified by the ease of obtaining enrichment cultures of thermophilic denitrifiers. The low temperature effects have considerable economic importance in the temperate zone because the progress of nitrate reduction at 10°C or below indicates the possibility of nitrogen volatilization during the colder part of the year when plants are not assimilating the nitrate formed by nitrification.

In flooded soils receiving ammonium and supporting a crop of rice, the presence of plants serves to reduce the extent of nitrogen volatilization (6). In other kinds of agriculture, too, cropping serves to reduce losses through denitrification. This retardation in cropped land suggests that the vegetation is competing with the microflora for the available supply of inorganic nitrogen, thereby reducing the amount of nitrate acted on microbiologically. By contrast, when the nitrate supply is adequate, plants promote the microbial release of volatile nitrogenous products (Figure 17.4) and also alter the relative amounts of N_2 and N_2O that are liberated. The enhancement may be a result of the roots providing exudates that support proliferation of the denitrifying bacteria, the diminished O_2 tension in the root region, or both.

DENITRIFICATION AND ENVIRONMENTAL POLLUTION

The attention given to nitrate pollution arising from farm practices has led to numerous assessments of denitrification in the profile and means of promoting it. On the one hand, denitrification in the root zone is deemed to be detrimental

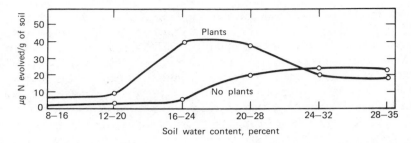

Figure 17.4. Denitrification in soil with different water content in the presence and absence of wheat plants (29).

because it removes an essential plant nutrient, and indeed chemical inhibitors to reduce this loss in the surface horizons have been sought. On the other hand, nitrate below the region occupied by the root system is no longer agriculturally useful, and it may at times pose a substantial environmental hazard. Hence, interest has also focused on microbial destruction of nitrate at lower soil depths. Commonly, the bulk of the microbial activity is in the surface horizon where the supply of readily available organic matter and the community are large (12). However, where drainage is poor and the level of easily utilized carbonaceous nutrients is reasonably high at subsurfaces sites, denitrification may be appreciable below the area occupied by actively growing roots (14), thereby reducing nitrate movement to surface waters. An interesting means that has been suggested to overcome the deficiency of energy sources, thereby possibly minimizing nitrate pollution from water passing through the profile, involves the addition of elemental sulfur. This amendment stimulates the population of *T. denitrificans*, which oxidizes the sulfur and reduces the nitrate to volatile products (19).

The increasing use of land for the disposal of sewage and animal wastes has also led to a dramatic rise in interest in denitrification, particularly inasmuch as this transformation, following mineralization and nitrification, makes the nitrogenous components of the wastes environmentally acceptable. The practical side of these investigations is directed to determining the quantity of sewage, sludge, feedlot manure, or other wastes that can be applied without seriously affecting the community's capacity to continue its degrading activities. If the wastes are applied at the proper rates and the nitrate ultimately produced coexists with available carbon, considerable denitrification may occur (16, 20). Thus, land spreading, if properly managed, can be a means of both removing wastes from an unwanted site and also of eliminating part of the nitrogen.

A surprising outcome of the interest in environmental problems has been the increasing body of evidence that the microflora plays a key role in the atmospheric cycles of nitrogen oxides and, because of that role, indirectly in exposing life on earth to harmful ultraviolet radiation. Despite the considerable and totally justifiable outcry about the NO and NO_2 released to the atmosphere by the combustion of coal, gasoline, natural gas, and other fuels by industry and the automobile, current data suggest that microbial release of these atmospheric pollutants by far exceeds that resulting from human actions, possibly being some 15 times as great. In this instance, the microorganisms in soil—and possibly in the sea—generate NO, which is spontaneously oxidized in the atmosphere to NO_2. Still greater in amount, however, is the N_2O that is generated microbiologically (26).

The N_2O is generally deemed to be innocuous at the low concentrations found at ground level, but atmospheric scientists have discovered that this product of denitrification does in fact have a major ecological influence, an effect associated with ozone (O_3) in the upper atmosphere. Ozone is formed by a photochemical reaction at considerable altitudes.

$$O_2 \rightarrow 2O \xrightarrow{2O_2} 2O_3 \tag{III}$$

Although present in the upper atmosphere in small amounts, O_3 is the only effective barrier protecting living organisms from the detrimental action of ultraviolet light at wavelengths less than 300 nm. In the absence of the O_3 shield, the incidence of skin cancer would be extremely high, and the ultraviolet radiation might even diminish plant growth. The denitrifiers enter the picture because the N_2O they synthesize diffuses to the stratosphere where it is oxidized to NO,

$$N_2O + O \rightarrow 2NO \tag{IV}$$

and indeed this reaction may be the chief source of NO in the stratosphere. The NO, and the NO_2 formed from it, destroys some of the O_3 so essential as a barrier to the damaging radiation. The reactions destroying O_3 and leading to NO_2 are catalytic (11).

$$NO + O_3 \rightarrow NO_2 + O_2 \tag{V}$$

$$O_3 \xrightarrow{\text{light}} O_2 + O \tag{VI}$$

$$NO_2 + O \rightarrow NO + O_2 \tag{VII}$$

Thus, the apparently harmless soil bacteria are critical, albeit indirectly, for the cycle of O_3 in the stratosphere, and ultimately they diminish the protection against harmful radiation. Without this metabolic process, more O_3 would likely exist above the earth, and the intensity of radiation reaching the earth's surface would consequently be lower. It has even been proposed, moreover, that the continuing increase in the use of nitrogenous fertilizers—a necessary trend in order to improve the nutritional level of the impoverished peoples in much of the world as well as to provide food for an ever-growing human population—will lead to more nitrate as the fertilizer nitrogen is oxidized, and the nitrate would be reduced to yield, in part, more N_2O. This N_2O, it is further argued, would cause more O_3 destruction and thus bring about additional skin cancer and other dire consequences. Although claims for a significant change in the abundance of atmospheric N_2O are unsubstantiated, the potential importance of such an alteration undoubtedly will lead to additional research.

BIOCHEMISTRY OF NITRATE REDUCTION

In the complete, aerobic oxidation of carbohydrates, the final products are CO_2 and water. With glucose, for example, the reaction is expressed as follows:

$$C_6H_{12}O_6 + 6O_2 \rightarrow 6CO_2 + 6H_2O \qquad \text{(VIII)}$$

In the aerobic metabolism of heterotrophs, the reducing power from the carbohydrate, signified by H, is used to form water from O_2. In heterotrophic nitrate respiration, the reducing power is coupled with oxidized states of nitrogen.

$$2NO_3^- + 10H \rightarrow N_2 + 4H_2O + 2OH^- \qquad \text{(IX)}$$
$$2NO_2^- + 6H \rightarrow N_2 + 2H_2O + 2OH^- \qquad \text{(X)}$$
$$N_2O + 2H \rightarrow N_2 + H_2O \qquad \text{(XI)}$$

For each two molecules of nitrate or nitrite, one molecule of N_2 is produced. In analogous fashion, the thiobacillus energy-yielding reaction, which is normally linked to O_2, may be associated in *T. denitrificans* with nitrate (equations I and II).

This explanation implies that denitrification is not anaerobic respiration in its usual, restricted sense but rather that it has a basic aerobic character with nitrate replacing the O_2, not in terms of nitrate as a "source of oxygen" but as an electron or hydrogen acceptor. In other words, the reducing power that normally is dissipated by reaction with O_2 is now dissipated at the expense of nitrate. That this is the case is supported by observations that the true denitrifying bacteria are by and large aerobes, proliferating in O_2-free circumstances only when nitrate is the alternate electron acceptor. The role of nitrate as electron acceptor for growth may also be assumed by nitrite or N_2O, both of which permit otherwise-aerobic cultures to develop anaerobically.

The suitability of nitrite for N_2 formation and its utilization in ammonium formation suggest that it is an intermediate in both denitrification and nitrate assimilation. However, the enzyme involved in reducing nitrate to nitrite in the first step of denitrification may be different from the enzyme catalyzing the same reaction in nitrate assimilation. On the basis of current knowledge, the biochemical mechanism proposed in Figure 17.5 seems to be the likeliest pathway by which microorganisms denitrify and assimilate nitrate nitrogen. In the denitrification sequence, nitrate is first reduced to nitrite, which is then transformed to NO. The NO is converted to N_2 with N_2O as intermediate.

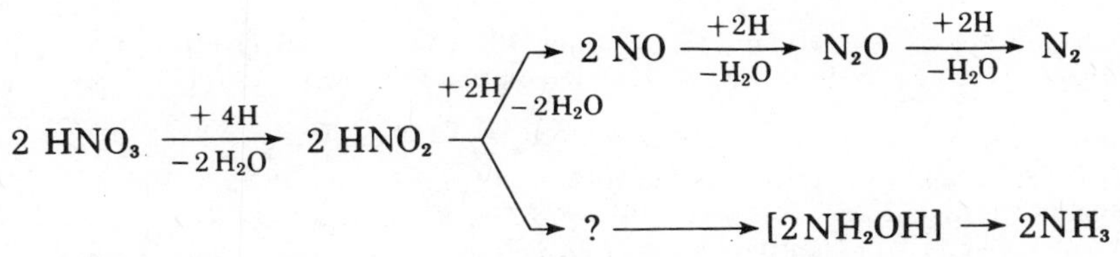

$$2\ HNO_3 \xrightarrow[-2\,H_2O]{+\,4H} 2\ HNO_2 \left\langle \begin{array}{l} \xrightarrow[-2\,H_2O]{+2H} 2\ NO \xrightarrow[-H_2O]{+2H} N_2O \xrightarrow[-H_2O]{+2H} N_2 \\[2ex] \xrightarrow{} ? \xrightarrow{} [2\,NH_2OH] \rightarrow 2\,NH_3 \end{array} \right.$$

Figure 17.5. Biochemical pathway of nitrate reduction and denitrification.

By contrast, the nitrite that is generated in the assimilatory process is converted not to NO and ultimately N_2 but rather to ammonium. How the latter process occurs remains uncertain. Although many nitrate reducers also transform hydroxylamine to ammonium, the evidence that hydroxylamine is the actual intermediate remains tenuous. In some organisms, the enzyme reducing nitrite to ammonium also reduces hydroxylamine, and in fact even enzymes reducing certain nonnitrogenous substrates may occasionally convert hydroxylamine to ammonium; hence, the action on this substrate may be fortuitous and unrelated to the assimilatory process.

The results of investigations of individual species in vitro lend further weight to the proposed mechanism. The capacity for N_2O and NO production is characteristic of most if not all denitrifiers. In addition, enzyme preparations made from the bacterial cells convert nitrite to NO, NO to N_2O, and N_2O to N_2 or catalyze the reduction of nitrite to NO, N_2O, and N_2 (2, 22). The responsible enzymes are termed the nitrate, nitrite, nitric oxide, and nitrous oxide reductases. Molybdenum is required for nitrate reductase, an observation that explains the poor growth of microorganisms in molybdenum-deficient media containing nitrate as the sole source of nitrogen.

REFERENCES

Reviews

Allison, F. E. 1966. The fate of nitrogen applied to soil. *Advan. Agron.,* 18:219–258.

Broadbent, F. E. and F. E. Clark. 1965. Denitrification. In W. V. Bartholomew and F. E. Clark, eds., *Soil nitrogen.* American Society of Agronomy, Madison, Wisc., pp. 344–359.

Hewitt, E. J. 1975. Assimilatory nitrate-nitrite reduction. *Annu. Rev. Plant Physiol.,* 26:73–100.

Payne, W. J. 1973. Reduction of nitrogenous oxides by microorganisms. *Bacteriol. Rev.,* 37:409–452.

Literature Cited

1. Ardakani, M. S., L. W. Belser, and A. D. McLaren. 1975. *Soil Sci. Soc. Amer. Proc.,* 39:290–294.
2. Baldensperger, J. and J.-L. Garcia. 1975. *Arch. Microbiol.,* 103:31–36.

3. Bollag, J.-M., S. Drzymala, and L. T. Kardos. 1973. *Soil Sci.,* 116:44–50.

4. Borisova, N. N., S. N. Burtseva, V. N. Rodionov, and O. L. Kirpaneva. 1972. *Soviet Soil Sci.,* pp. 540–546.

5. Bremner, J. M. and K. Shaw. 1958. *J. Agr. Sci.,* 51:40–52.

6. Broadbent, F. E. and M. E. Tusneem. 1971. *Soil Sci. Soc. Amer. Proc.,* 35:922–926.

7. Burford, J. R. and R. C. Stefanson. 1973. *Soil Biol. Biochem.,* 5:133–141.

8. Cady, F. B. and W. V. Bartholomew. 1960. *Soil Sci. Soc. Amer. Proc.,* 24:477–482.

9. Cady, F. B. and W. V. Bartholomew. 1963. *Soil Sci. Soc. Amer. Proc.,* 27:546–549.

10. Cooper, G. S. and R. L. Smith. 1963. *Soil Sci. Soc. Amer. Proc.,* 27:659–662.

11. Crutzen, P. J. 1974. *Ambio,* 3:201–210.

12. Dubey, H. D. and R. H. Fox. 1974. *Soil Sci. Soc. Amer. Proc.,* 38:917–920.

13. Ekpete, D. M. and A. H. Cornfield. 1965. *Nature,* 208:1200.

14. Gambrell, R. P., J. W. Gilliam, and S. B. Weed. 1975. *J. Environ. Qual.,* 4:311–316.

15. Kim, C. M. 1973. *Soil Biol. Biochem.,* 5:163–166.

16. King, L. D. 1973. *J. Environ. Qual.,* 2:356–358.

17. Koike, I. and A. Hattori. 1975. *J. Gen. Microbiol.,* 88:1–10.

18. Makarov, B. N. 1969. *Soviet Soil Sci.,* pp. 20–25.

19. Mann, L. D., D. D. Focht, H. A. Joseph, and L. H. Stolzy. 1972. *J. Environ. Qual.,* 1:329–332.

20. Meek, B. D., A. J. Mackenzie, T. J. Donovan, and W. F. Spencer. 1974. *J. Environ. Qual.,* 3:253–258.

21. Nelson, D. W. and J. M. Bremner. 1970. *Soil Biol. Biochem.,* 2:203–215.

22. Payne, W. J., P. S. Riley, and C. D. Cox, Jr. 1971. *J. Bacteriol.,* 106:356–361.

23. Pichinoty, F. 1965. *Ann. Inst. Pasteur, Suppl.* 3, pp. 248–255.

24. Reddy, K. R. and W. H. Patrick, Jr. 1975. *Soil Biol. Biochem.,* 7:87–94.

25. Renner, E. D. and G. E. Becker. 1970. *J. Bacteriol.,* 101:821–826.

26. Robinson, E. and R. C. Robbins. 1970. *J. Air Pollut. Control Assoc.,* 20:303–306.

27. Stanford, G., R. A. Vander Pol, and S. Dziena. 1975. *Soil Sci. Soc. Amer. Proc.,* 39:284–289.

28. Stefanson, R. C. 1972. *Aust. J. Soil Res.,* 10:183–195.

29. Stefanson, R. C. 1972. *Plant Soil,* 37:113–127.

30. Stefanson, R. C. 1972. *Plant Soil,* 37:129–140.

31. Valera, C. L. and M. Alexander. 1961. *Plant Soil,* 15:268–280.

32. Yoshida, T. and M. Alexander. 1970. *Soil Sci. Soc. Amer. Proc.,* 34:880–882.

33. Yoshida, T. and B. C. Padre. 1974. *Soil Sci. Plant Nutr.,* 20:241–247.

18
Nitrogen Fixation: Nonsymbiotic

Nitrogen is removed from soil by leaching, denitrification, and through cropping. Intensive agriculture accentuates the drain on the limited terrestrial supply of this critical element. Recent years have witnessed a remarkable expansion of the fertilizer industry, yet only a portion of the agricultural need for nitrogen comes from chemical fertilizers. Precipitation may add several kilograms per hectare each year as ammonium or nitrate, and the electrical discharges of the atmosphere return a small quantity of fixed nitrogen to the soil. Biological fixation, on the other hand, tends to right the nitrogen balance so that, despite the growth of the fertilizer industry, it is still necessary to stimulate the responsible microbial agencies to return gaseous nitrogen to the soil.

MICROBIOLOGY
Biological N_2 fixation is brought about by free-living bacteria or blue-green algae, which make use of N_2 by nonsymbiotic means, and by symbiotic associations composed of a microorganism and a higher plant. The agricultur-ally important legume-*Rhizobium* symbiosis and the association of microorgan-isms with certain nodulated nonlegumes will be discussed in the next chapter. The establishment of an isolate as capable of using molecular N_2 is often considered an easy task, for microbial growth in a medium free of nitrogenous ingredients presumably should of itself provide proof of the ability of an organism to assimilate N_2. Often, quantitative determinations during growth of a culture show slight increases in the total nitrogen content of the medium. On the basis of such criteria, claims have been made for nitrogen fixation by the following: a variety of actinomycetes; strains of the fungal genera *Aspergillus, Botrytis, Cladosporium, Mucor, Penicillium,* and *Phoma*; many yeasts; and species of several bacterial genera. Growth in nitrogen-free media, however, is rarely an

acceptable criterion because considerable difficulty is encountered in purifying culture solutions of the element. Furthermore, volatile nitrogenous compounds occur in the air, particularly of laboratories, and they can be selectively absorbed by microbial cells, resulting in an apparent but fictitious N_2 fixation.

To overcome the shortcomings inherent in qualitative experiments, a number of more critical techniques have been developed. Biological N_2 fixation is frequently put on a quantitative basis by measurement of the changes in the total nitrogen content of the culture or samples of the natural environment by the Kjeldahl procedure, but this useful approach is not reliable for the detection of small nitrogen gains. Gasometric measurement of N_2 disappearance has been proposed, but the analysis is laborious. Two techniques are especially suitable for conditions in which the nitrogen gains are slight. One involves the use of the stable isotope ^{15}N. Should a suspect organism be capable of fixation, it will incorporate ^{15}N into protoplasmic combination when exposed to ^{15}N-labeled N_2. The isotopic measurement is sufficiently sensitive to detect fixation of trace quantities of the gas with remarkable accuracy, but this procedure is not too widely used because ^{15}N-tagged N_2 is expensive and measurement of this stable isotope usually requires a mass spectrometer, an instrument that is both expensive and uncommon in biological and soils laboratories. The second technique entails determination of the microbial formation of ethylene (H_2C=CH_2) from acetylene (HC≡CH). This method is based on the finding that microorganisms that reduce N_2, which has a triple bond in the molecule, also can reduce acetylene, also a molecule with a triple bond. However, by contrast with the reduction of N_2—the metabolism of which yields ammonia, as will be considered below—the reduction of acetylene yields only ethylene.

$$N{\equiv}N \rightarrow 2\ NH_3 \qquad \qquad \text{(I)}$$
$$HC{\equiv}CH \rightarrow H_2C{\equiv}CH_2 \qquad \qquad \text{(II)}$$

This method has the advantage of being both sensitive and requiring an inexpensive substrate, acetylene. The ethylene produced is usually determined by means of a gas chromatograph, which is a more widely used instrument than the mass spectrometer, but the reaction can be measured by other means, too (17).

The procedures now available to verify the capacity of an organism to assimilate N_2 have resulted in the confirmation of this activity in many species in which the reaction was suspected. Conversely, various organisms that develop surprisingly well in media not receiving nitrogen compounds are now known to be merely scavengers of the element from contaminated medium ingredients or ammonia or nitrogen oxides in laboratory air. At present, strains of the following have been verified to be capable of the transformation:

a. Aerobic bacteria: *Azomonas, Azotobacter, Beijerinckia, Derxia, Methylomonas, Mycobacterium, Spirillum.*

b. Facultatively anaerobic bacteria: *Bacillus, Enterobacter, Klebsiella.*
c. Anaerobic bacteria: *Clostridium, Desulfotomaculum, Desulfovibrio.*
d. Photosynthetic bacteria: *Rhodomicrobium, Rhodopseudomonas,* and *Rhodospi*_
lum of the nonsulfur purple bacteria; *Chromatium* and *Ectothiorhodospira* of the purple sulfur bacteria; *Chlorobium* of the green sulfur bacteria.
e. Blue-green algae: *Anabaena, Anabaenopsis, Aulosira, Calothrix, Cylindrospermum, Fischerella, Gloeocapsa, Hapalosiphon, Lyngbya, Mastigocladus, Nostoc, Oscillatoria, Plectonema, Scytonema, Stigonema, Tolypothrix, Trichodesmium, Westiellopsis* (Table 18.1).

Among the facultatively and obligately anaerobic bacteria as well as in the genera *Methylomonas, Mycobacterium,* and *Spirillum*, the capacity to assimilate N_2 is not a characteristic of all species and often not of all strains of a single species.

Neither the ecological niche nor the economic or geochemical significance of many of the bacteria has been established. The most intensively investigated representative of the group is *Azotobacter*, but the vast literature concerned with the genus should not be taken to indicate that it is the most important in nature. All azotobacters are strict aerobes, and they apparently possess the highest known respiratory rate. These bacteria utilize few nitrogenous compounds—N_2, ammonium, nitrate, nitrite, urea, and an occasional organic nitrogen-containing molecule. Members of the genus are mesophilic, and their optimum temperature is near 30°C. When strains of *Azotobacter* utilize the gaseous form of the element, nitrogen gains may exceed 1.0 mg of nitrogen per milliliter of culture medium. The efficiency, measured as N_2 fixed per unit of sugar decomposed, is often quite low, from 5 to 20 mg of N_2 fixed per gram of sugar oxidized, but figures in excess of 30 mg have sometimes been recorded. *A. chroococcum* is generally the most frequently encountered azotobacter in soils of temperate regions. Azotobacter densities typically vary from nil to several thousand per

TABLE 18.1

Fixation of Molecular Nitrogen in Culture Media

Microorganism	Conditions	Incubation Period, Days	Nitrogen Fixed, μg N/ml
Azotobacter vinelandii	Aerobic	3	1050
Cylindrospermum cylindrica	Aerobic, light	55	52
Klebsiella pneumoniae	Anaerobic	2	60
Clostridium butyricum	Anaerobic	10	136
Chlorobium sp.	Anaerobic, light	5	20
Rhodospirillum sp.	Anaerobic, light	10	76

gram, and it is uncommon to find counts exceeding 10^3 and indeed rare to note values in excess of 10^4 per gram in soils of the temperate zone and many of those in the tropics. On the basis of the paucity of numbers at such sites, it is likely that this easily isolated and thoroughly studied organism is generally not an important contributor to N_2 fixation. On the other hand, there are localities where the population is particularly large, sometimes the numbers exceeding 10^6 per gram of soil in regions of northern Africa (1), and the roots of certain tropical grasses may teem with azotobacters; under these special circumstances, the genus may contribute substantially to nitrogen gains.

Beijerinckia species also are aerobic N_2 fixers, but they grow well in acid conditions and sometimes develop at a reaction as low as pH 3. *Beijerinckia* species are commonly found in tropical soils, and strains have been isolated in South America, Asia, northern Australia, and tropical areas of Africa, their numbers ranging from a few to several thousand per gram. *Beijerinckia* is rarely detected in soils of the temperate zone (6). *Derxia* is similarly an aerobic, acid-tolerant genus, and it will grow in media from pH 5 to 9. Strains of *Derxia* have been detected in soils of Asia and Brazil, but the group has still received scant attention.

The dominant anaerobes using N_2 are members of the genus *Clostridium*. The population of N_2-fixing clostridia in arable land may be as small as 10^2 or as large as 10^5 cells per gram. Under some conditions, more than 10^6 clostridia per gram have been detected. The dominant N_2-utilizing species appear to be *C. pasteurianum, C. butyricum,* and *C. acetobutylicum* (33), but other species have been found to carry out the same biochemical process. Clostridia proliferate when organic matter is added, and they often are numerous around plant roots. In contrast with the azotobacters, which they usually outnumber, the clostridia are found at sites of pH 5.0, and they are still capable of growth at pH 9.0. Under suitable conditions, N_2 incorporation is appreciable, and up to 180 μg of nitrogen is fixed per milliliter of culture medium. The efficiency is low, and only 2 to 10 mg of nitrogen is assimilated per gram of carbohydrate consumed.

Blue-green algae are not uncommon in well-drained fields, but their numbers are far greater in flooded soils. Many blue-green algae of soil origin grow in culture solutions devoid of fixed nitrogen compounds and effect an increase in the nitrogen content of the medium. Not all blue greens and representatives of none of the other algal classes can utilize N_2, however. Nitrogen fixation by the active strains is stimulated by increasing light intensities, but excessive light is inhibitory. The transformation is invariably slow, and gains of 30 to 115 μg of nitrogen per milliliter of solution require periods of $1\frac{1}{2}$ to 2 months for most isolates of *Aulosira, Anabaenopsis, Anabaena, Cylindrospermum, Nostoc,* and *Tolypothrix*. The fixation, therefore, is much less rapid than in the azotobacters or clostridia. Several of the algae grow slowly in pure culture in the dark provided that sugar is present to serve as a carbon and energy source for heterotrophic metabolism, but the nitrogen gains are correspondingly small. The dark fixation of N_2 by blue-green algae is probably of no ecological significance. In the light, however, blue-green algae and the legume-*Rhizobium* symbiosis have a marked advantage over the heterotrophic N_2 users—their ability to develop without relying on the limited supply of available carbohydrates in soil.

In addition to vegetative cells, certain filamentous blue-green algae—such as *Anabaena, Cylindrospermum,* and *Mastigocladus*—possess structures known as heterocysts. Under the light microscope, the thick-walled heterocyst is distinctive because it gives the illusion of being an empty cell; however, this cell has a special function in the filamentous species because it is here that active N_2 fixation occurs. From the heterocysts, the nitrogen is transferred readily to adjacent vegetative cells. In some filamentous species that bear heterocysts, it is possible that vegetative cells also can metabolize N_2, but further research is required to confirm this possibility. However, certain filamentous blue-green algae that produce no heterocysts (*Lyngbya, Oscillatoria, Phormidium,* and *Plectonema*) and a unicellular, nonheterocystous form like *Gloeocapsa* also can make use of N_2.

Provided they contain a blue-green alga, lichens may fix N_2. Of the two organisms that live in symbiosis and make up the lichen, the alga and the fungus, it is unquestionably the former that makes use of the gaseous nitrogen source. Lichens—which are classified as if they were separate organisms—such as *Collema, Lichina, Peltigera,* and *Stereocaulon* thus assimilate N_2, a process that may aid in their growth in nitrogen-poor coastal habitats, the subarctic, arid regions, and on glacial drifts. The blue-greens are similarly able to assimilate N_2 when living in symbiosis with some liverworts, a pteridophyte (*Azolla*), an angiosperm (*Gunnera*), and plants classified among the Cycadaceae.

The photosynthetic N_2-fixing bacteria are divided into three groups: nonsulfur purple bacteria, purple sulfur bacteria, and green sulfur bacteria. All are affected favorably by light but are inhibited by O_2. The rate of N_2 assimilation by the photoautotrophic bacteria is quite slow, and periods of weeks are required for gains of 100 μg of nitrogen per milliliter of culture medium. The nonsulfur purple bacteria are found in flooded soils and in ditches, lake muds, or sea bottoms, but they are essentially absent from farm or forest land. When soils in the tropics are waterlogged, however, these bacteria sometimes flourish. Thus, flooded fields of Asia planted to rice may sometimes show the presence of 10 to 10^5 cells of nonsulfur purple bacteria or from essentially nil to 10^3 cells of purple sulfur bacteria per gram, and sometimes higher counts are even recorded (26).

Little is known about the ecology or frequency of the other N_2-fixing bacteria. Many isolates of *Klebsiella* catalyze the reaction, but the abundance of this genus in nature is largely unexplored. Studies of individual soils, however, have disclosed the presence of from 20 to 18,000 *Klebsiella* and less than 10^3

Enterobacter and *Bacillus* cells per gram potentially able to utilize N_2 (28). Sometimes, *Bacillus polymyxa* may be the chief N_2 fixer, as in parts of the Alaskan tundra (22).

FACTORS AFFECTING NITROGEN FIXATION

A number of environmental factors govern the rate and magnitude of nonsymbiotic N_2 fixation, and the transformation is markedly affected by the physical and chemical characteristics of the habitat. Considerable information has been obtained from investigations of pure cultures, and extension to the field is sometimes possible.

Microorganisms that assimilate N_2 have the ability to utilize ammonium and sometimes nitrate and other combined forms of nitrogen. In fact, ammonium salts are used preferentially and often at a greater rate than molecular nitrogen so that the presence of ammonium, in effect, inhibits the fixation; that is, the bacteria or algae use the nitrogen salt rather than N_2 from the atmosphere. Ammonium or compounds converted to it, for example, urea or nitrate, are most effective in the inhibition of fixation in culture, but a number of amino acids have a less marked deterring influence. In soil, as in culture, ammonium and nitrate at moderate or high concentrations inhibit N_2 fixation; however, as the level of free ammonium or nitrate falls to a low concentration through microbial assimilation of the element, N_2 utilization resumes (25). It seems likely that many soils may support the fixation when nitrate or available ammonium is present at low levels.

Many inorganic nutrients are necessary for the development of the microorganisms, but only a select few are specifically implicated in the metabolism of N_2, that is, they are indispensable for N_2-linked proliferation. Some are required in lesser amounts for growth on ammonium salts or other fixed nitrogenous compounds. Molybdenum, iron, calcium, and cobalt are critical for the fixation reaction. Molybdenum is required for the metabolism of N_2, but microorganisms will similarly not use nitrate unless molybdate is present although the molybdenum requirement for nitrate utilization is less than for N_2 fixation. Growth on ammonium salts, however, proceeds rapidly in the absence of added molybdenum. For at least some organisms, vanadium will replace molybdenum, but it is never fully as effective. In like manner, iron salts are implicated in the N_2 metabolism of *Azotobacter, Clostridium*, algae, and *Klebsiella*, but the specific requirement for N_2 metabolism is often difficult to establish because iron is required, to a lesser extent, for growth on fixed compounds of nitrogen. The not-infrequent observations of stimulation of azotobacter by humus or soil decoctions result from the use of media deficient in iron or molybdenum, or both. A requirement for calcium has been demonstrated during N_2 assimilation by blue-green algae and some species of *Azotobacter*, but the calcium can sometimes be replaced by strontium. Calcium is also required for growth on fixed nitrogen compounds, but the need is less than for N_2-

grown cultures (15). Similarly, organisms making use of N_2 must have cobalt available to them, although a lesser concentration of this element may be essential for growth on combined nitrogen (19). A role for cobalt in the process has been established for *Azotobacter, Beijerinckia, Clostridium,* and several algal genera.

The availability of energy sources is a major factor limiting the rate and extent of N_2 assimilation by heterotrophic populations. Thus, the addition of simple sugars, cellulose, straw, or plant residues with wide C:N ratios often markedly enhances the transformation either in aerobic or anaerobic circumstances. The extent of nitrogen gain is related to the quantity of the carbon source added and the prevailing temperature (Table 18.2), and from less than 1.0 to about 30 mg N_2 may be assimilated per gram of carbon source by the soil populations (34). Because the active species must compete with other populations for the energy sources, the more competitive is the active organism, the greater will be the nitrogen gains.

The prevailing pH has a profound influence on the abundance of these organisms. For example, *Azotobacter* is characteristically sensitive to high hydrogen ion concentrations. Ecological investigations show that, despite the wide distribution of bacteria of the genus, many soils contain none or an insignificant number. Their absence is associated directly with pH. As a rule, environments more acid than pH 6.0 are free of the organism or contain very few azotobacter cells. Similarly, the bacteria generally will neither grow nor fix N_2 in culture media having a pH below 6.0, but an occasional variant will tolerate greater hydrogen ion concentrations. *Beijerinckia* species do not possess the acid sensitivity of the azotobacters, and they develop and fix N_2 from pH 3 to 9. Blue-green algae, however, develop poorly in media and are sparse in soils more acid than approximately pH 6.0. In a typical investigation, N_2-using blue-green algae were found to be rare in acid forest soils, a paucity that suggests these organisms may not add nitrogen to established forests in regions of low

TABLE 18.2

Effect of Glucose Concentration and Soil Temperature on Nitrogen Fixation in 14 days (16)

Glucose Added %	N_2 Fixed, mg/kg of Soil		
	15°C	25°C	35°C
0.00	2.0	0.9	1.3
0.04	3.7	0.7	1.3
0.20	4.4	1.7	1.6
1.00	34.3	33.5	12.4

pH values (21). The acid tolerance of *Clostridium* falls between *Azotobacter* and *Beijerinckia*.

The rate of fixation is frequently determined by soil moisture. Gains are insignificant when little water is available, but the rate and magnitude of the process increase as moisture becomes more abundant. Sometimes the activity is especially great at or near field capacity and occasionally under waterlogged conditions, the optimum water level varying with the soil and the quantity of available organic matter (23, 36). As with other microbiological transformations, the changes associated with excessive moisture are intimately linked with the shift from aerobiosis to anaerobiosis, and this reaction sequence likewise is affected by the O_2 status of the environment. With most soils having adequate moisture, the rate is higher when the soil is anaerobic than when O_2 is present, but some soils exhibit greater activity with O_2 in the overlying atmosphere or the transformation is unaffected by the availability of O_2. It has also been proposed that waterlogging promotes N_2 gains because complex organic nutrients are decomposed to simple products in either the aerobic top portion of the flooded soil or in the underlying anaerobic zone, and the simple products then diffuse to the adjacent microhabitat where they are metabolized by the N_2 fixers (31, 36).

Temperature also has a profound influence on N_2 metabolism. Little activity is evident at low temperature, and warming promotes the microbial uptake of the gas. The process takes place at moderate temperatures but ceases a few degrees above the optimum temperature (Figure 18.1). In some regions of the northern temperate zone, fixation occurs even during the winter. In at least certain of these northern soils, the fixation is not heterotrophic but instead results from indigenous algae or lichens containing blue-green symbionts, and these photosynthetic organisms or symbioses may still be active during parts of the winter when the temperature is somewhat below 0°C (14).

NITROGEN GAINS IN SOIL

Assessments of nitrogen increases in nature were difficult to make in the past because the analyst had to measure what undoubtedly was a small change of nitrogen in a soil containing a large reserve of the element. The first significant advance in assays involving soils in the laboratory or field was the introduction of isotopes, the use of which provides a specific procedure for detecting changes of small magnitude. In experiments designed to this end, soil samples are exposed to ^{15}N-tagged N_2, and the soil is analyzed for its ^{15}N content after suitable incubation periods. The quantity of isotope recovered in the sample is a measure of the N_2 incorporation. The second major breakthrough was the acetylene reduction technique, in which measurements are made of the rate or quantity of ethylene formed in a soil incubated with acetylene. If the capacity of microorganisms to reduce both N_2 and acetylene is not limiting the rates of

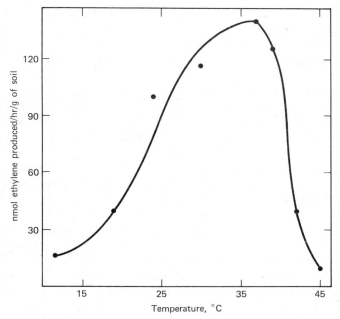

Figure 18.1. Effect of soil temperature on N_2 fixation (7). The reaction was measured by determining the rate of ethylene formation from acetylene incubated with the soil.

reaction, one may compare the two sequences with H being taken as the equivalent of reducing power.

$$N_2 + 6H \rightarrow 2\,NH_3 \qquad\qquad (III)$$
$$3C_2H_2 + 6H \rightarrow 3\,C_2H_4 \qquad\qquad (IV)$$

Thus, when equations III and IV are written so that the rates of microbial production of the H needed for the two sequences are equal, a ratio of $3C_2H_2:1N_2$ is obtained. This theoretical ratio is frequently approached in pure culture or soil and is sometimes attained. In actual assays of soil, however, the observed ratio is at times as low as 0.75 or as high as 4.5 or occasionally greater so that caution needs to be observed in using the acetylene reduction technique as a means for the quantitative measurement of N_2 fixation rates.

The biological return of nitrogen to soil requires the presence of a suitably large population to catalyze the transformation and environmental conditions conducive to the activities of the responsible population. In the early research, when sensitive analytical methods were unavailable and emphasis was placed on *Azotobacter* as the chief agent of fixation, considerable skepticism was expressed as to the significance of nonsymbiotic N_2 fixation in agricultural practice. There were just too few azotobacters in fields used for most kinds of commercial agriculture. The more recent research disclosing the many other groups of organisms able to utilize N_2, in culture at any rate, has led to reassessments of the practical importance of the transformation, and fortunately the two assay methods described above have given means for obtaining the requisite information.

Three conditions must be satisfied for heterotrophic fixation to be appreciable: (a) the population must be large, (b) the rate of new cell formation needs to be rapid because N_2 incorporation is a growth-linked process, and (c) the bulk of the nitrogen in cells of the active species must come from the atmosphere. It can be calculated that some 10^7 bacterial cells per gram are necessary for the fixation of 2 kg of nitrogen per hectare. Should the growing season provide as many as 200 favorable days for microbial metabolism, a turnover or formation of 50,000 cells per gram per day are required for the 2-kg nitrogen increase.

Representative measurements of nitrogen fixation are presented in Table 18.3. These data should be interpreted with care because many of the experiments were conducted with small samples in the laboratory or involved short incubation periods, and then the results were corrected to a hectare and long-term bases. Moreover, in those studies making use of the acetylene reduction technique, the precise factor needed to convert the values so obtained to nitrogen assimilation rates were not always established. Nevertheless, the tabulated figures show that the process may be appreciable in some areas and of little agricultural significance in others. The marked influence of region is also apparent, and season also has a profound impact. In those localities where the

TABLE 18.3
Nitrogen Fixation in Various Soils

Environment	Time Period	N_2 Fixation Rate, kg/ha	Reference
California soils	Year	2.1–4.8	42
Hawaiian pastures	Year	0–32	27
Soil following burning	Year	23	20
Swedish soil: algae	Year	15–51	18
Philippines: flooded soil	Year	10–55	30
Arctic soil	Year	0.03–3.8	43
Desert crust	Year	10–100[a]	40
Canadian grassland	Growing season	1–2	44
Alaska: lichens, algae	Summer	0.15	3
Desert crust	Day	1.68[a]	29
Ivory Coast grassland	Day	0.025	4

[a] At the specific site colonized by the algae or lichens.

gains are but a few kilograms per hectare per year, the process is probably not important in intensive agriculture, but it still will have an appreciable impact on the nitrogen cycle of unfertilized fields. The reason for the low figures for heterotrophic fixation is probably the scarcity of available carbon sources, but low moisture or temperature may also be implicated.

Grasslands appear generally to have low activity (4, 44). On the other hand, some mountain meadows are moist for much of the time, bear a thick sod mat, and contain more nitrogen than nearby dry soils. These meadows can be active in N_2 fixation, part of which is of algal and part of bacterial origin (35). The deliberate burning of woodlands to control vegetation and litter accumulation is one means of promoting this transformation (20). In numerous localities where the blue-greens flourish, as in flooded soils planted to rice or where moisture is otherwise adequate for such photoautotrophs, much N_2 can be assimilated; the functioning of this light-promoted reaction of the algae may explain why rice has been grown for so long in regions of Asia despite the use of little or no fertilizer or manure.

Algal assimilation of N_2 has been confirmed in numerous habitats. In desert or semiarid regions, the blue-greens or lichens containing them become active following the occasional rains, and considerable quantities of N_2 may be incorporated by the patchy growths arising when the moisture supply favors the photosynthetic populations (29, 40). Part of the N_2 thus fixed may be released as algal excretions, and the rest will be liberated as the cells are decomposed, the element then being mineralized and made usable for higher plants. Lichens such as *Peltigera*, which contain blue-green symbionts, and free-living genera such as *Nostoc* may also bring about nitrogen accretion in localities they colonize in the arctic and subarctic (3, 43), and either lichens or bacteria associated with them bring about fixation even on the surfaces of rocks, an activity that probably allows for the establishment of pioneer plant species.

N_2 FIXATION ASSOCIATED WITH ROOTS

If the capacity of heterotrophs to make use of N_2 is indeed limited because of the small quantities of carbon available to them, it seems quite plausible to believe that roots of plants excreting simple organic molecules might sustain large populations of N_2-utilizing bacteria. This possibility seems particularly likely inasmuch as certain grasses and crop species, especially among tropical plants, are known to grow surprisingly well with no additions of fertilizer nitrogen. The multiplication of bacteria able to metabolize N_2 has long been suspected as the reason for the surprisingly good growth, but it was not until the introduction of sensitive procedures for measuring the process that this view was confirmed.

Considerable evidence now exists to show that a loose association exists between roots of certain species of higher plants and heterotrophic, nonsymbiotic N_2-fixing bacteria. These bacteria differ from *Rhizobium* in that they are not localized in specialized nodules or protuberances on the roots, instead they grow on the root surface and make use of carbonaceous exudates to satisfy their energy demands. The bacteria are members of three genera: *Azotobacter*, especially *A. paspali*; *Beijerinckia*; and *Spirillum*. *A. paspali* is most notably associated with the tropical grass *Paspalum notatum*, but it is common on only certain varieties of *P. notatum* and is rare on the roots of other varieties and on other species of *Paspalum* as well as dissimilar grasses (11). *Beijerinckia*, by contrast, is often especially abundant on the roots of sugar cane growing in parts of Brazil (10), but it has been found on additional plants as well. *Spirillum* is often present on the roots of corn, wheat, sorghum, *Digitaria decumbens*, *Panicum maximum*, and *Melinis multiflora*. The bacteria are somehow linked closely with the roots inasmuch as gentle washings do not dislodge the N_2-metabolizing activity.

By means of the acetylene reduction technique, N_2 fixation has been shown to occur on the roots of corn, wheat, millet, sorghum, rice, and species of the following genera of higher plants: *Andropogon, Anthriscus, Brachiaria, Convolvulus, Cynodon, Cyperus, Digitaria, Eleusine, Heracleum, Hyparrhenia, Melinus, Panicum, Paspalum, Pennisetum, Rumex, Stachys,* and *Viola*, and extrapolation from the rate of ethylene formation in the assay suggests that from 0.002 to 1.1 kilograms of N_2 may be fixed per hectare each day, depending on the plant. Although not in all instances are the bacteria well characterized, they usually seem to be strains of *A. paspali, Beijerinckia*, and *Spirillum* that have roots as their habitat. The results of acetylene reduction assays of roots of a number of species are presented in Figure 18.2. The numerical values represent high or maximum figures and are not necessarily those for most varieties; nevertheless, the high values are important, especially for cereal crops like corn or rice and forage grasses, because they signify that breeding can lead to varieties capable of supporting large populations and impressive fixation. Tests involving the use of ^{15}N-tagged N_2, moreover, confirm that the element can be rapidly translocated from the bacteria to the plant tissues, at least in sugar cane (39). Some evidence also exists that N_2 may be fixed by bacteria living in conjunction with the roots of conifers (37).

Extrapolation from experiments with plants growing in controlled conditions in pots, especially for only short periods, to natural conditions is fraught with danger, yet such extrapolations are needed until reliable field methods are available. Calculations of these sorts, which are derived from measurements of acetylene rather than ^{15}N-N_2 metabolism, suggest that *A. paspali* on *P. notatum* roots incorporates from 15 to 93 kg/ha/yr, *Beijerinckia* on sugar cane may assimilate 50 kg/ha/yr (12), a *Pennisetum purpureum* stand might fix a maximum of 1 kg/ha/day (9), and bacteria on the most active varieties of corn may fix 2.4 kg/ha/day (45). Since varieties of many of these crops and forage grasses obviously show symptoms of nitrogen deficiency when growing naturally, the

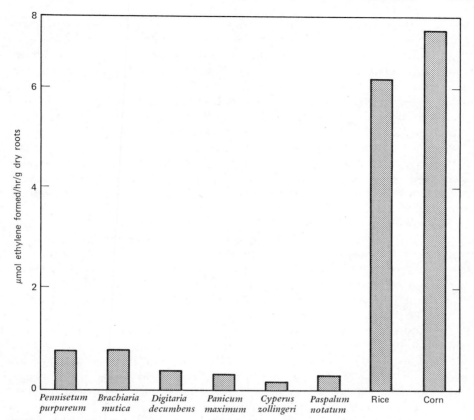

Figure 18.2. Maximum rates of acetylene reduction by roots of varieties of cereals and forage grasses (9, 13, 45).

breeding for the best genotypes and an understanding of the factors conducive to greatest activity should be of immense benefit to food and forage production.

INOCULATION

The frequent paucity of free-living N_2-fixing organisms in farm land and the obvious need for nitrogen supplementation in crop production have prompted innumerable attempts to stimulate the fixation. Not infrequently, inoculant preparations containing *Azotobacter* or blue-green algae are added to soil or seed in the hope of favoring the process. Among the crops claimed to benefit from azotobacter inoculation are corn, oats, wheat, barley, sugar beets, potatoes, cabbage, tomatoes, carrots, and cotton. The yield increases are often reported to be about 10 percent, but greater stimulations are occasionally claimed. In many trials of these same bacterial preparations, no responses are noted (32).

Careful investigations using chemical and isotopic techniques have shown that heavy inoculation with azotobacter cultures fails to affect the incorporation of N_2 into soil. Frequently the introduced bacteria rapidly die out. Nevertheless, plant responses are sometimes evident, and thus explanations must be sought elsewhere than in N_2 metabolism. Among the more plausible alternative hypotheses are that species of *Azotobacter* produce compounds detrimental to pathogens or that act as plant-growth regulators, and indeed azotobacters do synthesize stimulatory compounds such as gibberellins, cytokinins, and indole-acetic acid. In some instances, the benefit may result not from the bacteria proliferating on the developing roots or in the soil but instead from substances they liberated during their original growth in culture and that are still present in the inoculum preparation (5). Furthermore, careful analysis of the field experiments that have been performed is difficult because of the scarcity of statistical evaluations. It is possible that the azotobacter effects are not real and may be accounted for by the normal variability of field experimentation. *Clostridium* has sometimes been the organism employed as a soil inoculum for experimental purposes. The anaerobes on occasion enhance growth; often they do not (30, 38).

By contrast with inocula of heterotrophs, which must compete with other populations for a limited supply of readily available carbon, algae have an adequate supply of carbon and energy, as CO_2 and light. Both the light and the moisture the algae require are abundant in flooded rice fields in the tropics, and inoculation with these organisms has attracted much interest to promote rice production in the developing countries where fertilizers are expensive or in short supply. Species of *Aulosira, Calothrix, Cylindrospermum, Nostoc,* and *Tolypothrix* have been used for this purpose. Rice yield increases have indeed been observed in several of the field tests (2, 46) as well as in many greenhouse experiments, but again convincing evidence that N_2 fixation underlies the rice response is wanting.

BIOCHEMICAL MECHANISM

Microorganisms that assimilate molecular nitrogen are biologically unique because they utilize a gas considered to be relatively inert. The specific enzyme that combines with and activates the commonly nonreactive gas is called *nitrogenase*. The partial pressure of N_2 at which the fixation is proceeding at half maximal velocity is approximately 0.02 atmospheres for *Azotobacter, Rhodospirillum,* and *Nostoc*, 0.03 atmospheres for *Clostridium*, and 0.05 atmospheres for red clover. The most rapid N_2 incorporation takes place at somewhat higher concentrations of the gas, and the rate is appreciably slower below 0.02 and 0.05 atmospheres.

The use of accurately regulated gas mixtures for certain types of experiments resulted in an entirely unexpected discovery: H_2 prevented N_2 assimilation in many free-living N_2-fixing microorganisms. Inhibition by H_2 is specific for the assimilation of N_2, and the utilization of combined nitrogen compounds is unaffected. The inhibition by H_2 is competitive; that is, the effect depends on the relative concentrations of H_2 and N_2. Thus, the influence of a fixed amount of H_2 is lessened as the N_2 concentration is increased. Another anomaly is that all microorganisms that assimilate N_2 also metabolize H_2, the former by means of nitrogenase, the latter by the enzyme *hydrogenase*. Hydrogenase is capable of activating H_2 for the reduction of a number of substances or of liberating H_2 from reduced compounds. In equation V, R represents the compound that is reduced by hydrogenase.

$$H_2 + R \xrightleftharpoons{\text{hydrogenase}} RH_2 \tag{V}$$

The inhibition of N_2 fixation by ammonium in soil is readily explained from careful laboratory experiments on the biochemistry of the process. It has already been stated that bacteria and algae utilize ammonium rather than N_2, at least at the ammonium concentrations common in culture media. Detailed inquiry has established that ammonium salts inhibit nitrogenase synthesis so that the organism is in fact incapable of metabolizing N_2. However, the cells assimilate the combined form of nitrogen as they grow, thereby reducing its concentration, and then as the inhibitor falls to a very low level or disappears entirely, nitrogenase is synthesized and growth resumes (24, 41). Nitrate and organic nitrogen compounds may act similarly, but only if they are converted to ammonium by the cells. The very fact that ammonium is generated in the organism from N_2 and is at the same time an inhibitor of nitrogenase synthesis signifies that the ammonium must either be converted rapidly to organic nitrogen compounds or be removed from the intracellular site of N_2 metabolism. In addition, ammonium added to cultures of some but not all N_2-fixing bacteria actually inhibits the functioning of nitrogenase already present in the cells.

In view of the need of the aerobic bacteria for O_2 and the evolution of O_2 by the algae, the discovery that O_2 inhibits nitrogenase activity was startling. This inhibition applies to the enzyme system in anaerobes, aerobes, and O_2-evolving algae, yet fixation occurs in O_2-containing soils and cultures. Clearly, an intracellular mechanism exists in the aerobes to protect nitrogenase from inactivation by O_2. It is not clear as yet how aerobes exclude this gas from the cellular site of nitrogenase function, but the protection may be associated with special structural devices in some species.

The actual conversion of N_2 to combined nitrogen involves a reduction of the substrate to ammonium (or ammonia) as indicated in equation III. This simple enzymatic reaction, which proceeds at room temperature and normal atmospheric pressures, gives the same products as is made in factories manufacturing much of the world's fertilizers, yet the ease of the enzymatic synthesis has yet to be achieved with nonbiological catalysts.

Considerable effort has been directed to determining what intermediates are formed during the enzymatic reduction of N_2 to ammonia, with prime emphasis being placed on diimide (HN$=$NH) and hydrazine (H$_2$N—NH$_2$). However, no free intermediates have been detected. It is widely believed, nevertheless, that these or structurally related molecules are indeed made during the reaction, but they are retained by the enzyme that makes them and are reduced so that neither compound appears free. The reaction may thus be envisioned as involving nitrogenase (written here as E) combining with N_2, and the enzyme-substrate complex being reduced sequentially with the ultimate release of ammonia (8).

$$N\equiv N + E \rightarrow E-N=NH \rightarrow E-NH-NH_2 \rightarrow E + 2NH_3$$

The ammonia (ammonium) itself does not accumulate in the cell, although a few species may excrete it; rather it is incorporated into organic forms by combining with an organic acid to give rise to an amino acid. Often the organic acid is α-ketoglutaric acid, and the product would then be the amino acid, glutamic acid. The ammonia may also combine with organic molecules to yield alanine or glutamine.

Nitrogenase is unique in that it is made up of not one but two proteins, both of which are required for N_2 to be assimilated. One of the two contains molybdenum in the molecule and has a molecular weight of about 200,000. The second has no molybdenum, and its molecular weight is in the vicinity of 60,000. Both molecules contain iron. The specific need of these organisms for molybdenum and iron is probably a consequence of the presence of the elements in the two protein components of nitrogenase. Bearing in mind the diverse species capable of utilizing N_2, it is remarkable that all contain similar proteins, suggesting thereby a similar if not identical reaction sequence. In addition to the two proteins, moreover, nitrogenase activity requires the presence of ATP as a source of energy and a reductant to allow the reaction to proceed.

REFERENCES

Reviews

Benemann, J. R. and R. C. Valentine. 1972. The pathways of nitrogen fixation. *Advan. Microbial Physiol.*, 8:59–104.

Dalton, H. 1974. Fixation of dinitrogen by free-living microorganisms. *CRC Crit. Rev. Microbiol.*, 3:183–220.

Dilworth, M. 1974. Dinitrogen fixation. *Annu. Rev. Plant Physiol.*, 25:81–114.

Jurgensen, M. F. 1973. Relationship between nonsymbiotic nitrogen fixation and soil nutrient status. *J. Soil Sci.,* 24:512–522.

Mishustin, E. N. and V. K. Shil'nikova. 1972. *Biological fixation of atmospheric nitrogen.* Penna. State Univ. Press, University Park, Pa.

Postgate, J. R., ed. 1971. *The chemistry and biochemistry of nitrogen fixation.* Plenum Press, New York.

Quispel, A., ed. 1974. *The biology of nitrogen fixation.* North-Holland Publishing Co., Amsterdam.

Stewart, W. D. P. 1973. Nitrogen fixation by photosynthetic microorganisms. *Annu. Rev. Microbiol.,* 27:283–316.

Zumft, W. G. and L. E. Mortenson. 1975. The nitrogen-fixing complex of bacteria. *Biochim. Biophys. Acta,* 416:1–52.

Literature Cited

1. Abd-El-Malek, Y. 1971. In T. A. Lie and E. G. Mulder, eds., *Biological nitrogen fixation in natural and agricultural habitats.* Nijhoff, The Hague, pp. 423–442.
2. Aiyer, R. S., S. Salahudeen, and G. S. Venkataraman. 1972. *Indian J. Agr. Sci.,* 42:380–383.
3. Alexander, V. and D. M. Schell. 1973. *Arctic Alp. Res.,* 5:77–88.
4. Balandreau, J. and G. Villemin. 1973. *Rev. D'Ecol. Biol. Sol,* 10:25–33.
5. Barea, J. M. and M. E. Brown. 1974. *J. Appl. Bacteriol.,* 37:583–593.
6. Becking, J. H. 1974. *Soil Sci.,* 118:196–212.
7. Brouzes, R. and R. Knowles. 1973. *Soil Biol. Biochem.,* 5:223–229.
8. Dalton, H. and L. E. Mortenson. 1972. *Bacteriol. Rev.,* 36:231–260.
9. Day, J. M., M. C. P. Neves, and J. Döbereiner. 1975. *Soil Biol. Biochem.,* 7:107–112.
10. Döbereiner, J. 1968. *Pesq. Agropec. Bras.,* 3:1–6.
11. Döbereiner, J. 1970. *Zent. Bakteriol.,* II, 124:224–230.
12. Döbereiner, J., J. M. Day, and P. J. Dart. 1973. *Pesq. Agropec. Bras.,* 8:153–157.
13. Dommergues, Y., J. Balandreau, G. Rinaudo, and P. Weinhard. 1973. *Soil Biol. Biochem.,* 5:83–89.
14. Englund, B. and H. Meyerson. 1974. *Oikos,* 25:283–288.
15. Evans, H. J. and M. Kliewer. 1964. *Ann. N.Y. Acad. Sci.,* 112:735–755.
16. Fehr, P. I., P. C. Pang, R. A. Hedlin, and C. M. Cho. 1972. *Agron. J.,* 64:251–254.
17. Hardy, R. W. F., R. C. Burns, and R. D. Holsten. 1973. *Soil Biol. Biochem.,* 5:47–81.
18. Henriksson, E. 1971. In T. A. Lie and E. G. Mulder, eds., *Biological nitrogen fixation in natural and agricultural habitats.* Nijhoff, The Hague, pp. 415–419.
19. Jakobsons, A., E. A. Zell, and P. W. Wilson. 1962. *Arch. Mikrobiol.,* 41:1–10.
20. Jorgensen, J. R. and C. G. Wells. 1971. *Soil Sci. Soc. Amer. Proc.,* 35:806–810.
21. Jurgensen, M. F. and C. B. Davey. 1968. *Can. J. Microbiol.,* 14:1179–1183.
22. Jurgensen, M. F. and C. B. Davey. 1971. *Plant Soil,* 34:341–356.
23. Kalininskaya, T. A., Yu. M. Miller, and I. T. Kultyshkina. 1974. *Izv. Akad. Nauk SSSR, Ser. Biol.,* pp. 927–930.
24. Kleiner, D. 1974. *Arch. Microbiol.,* 101:153–159.
25. Knowles, R. and D. Denike. 1974. *Soil Biol. Biochem.,* 6:353–358.
26. Kobayashi, M., E. Takahashi, and K. Kawaguchi. 1967. *Soil Sci.,* 104:113–118.
27. Koch, B. L. and J. Oya. 1974. *Soil Biol. Biochem.,* 6:363–367.
28. Line, M. A. and M. W. Loutit. 1973. *New Zeal. J. Agr. Res.,* 16:87–94.
29. Macgregor, A. N. and D. E. Johnson. 1971. *Soil Sci. Soc. Amer. Proc.,* 35:843–844.
30. MacRae, I. C. and T. F. Castro. 1967. *Soil Sci.,* 103:277–280.
31. Magdoff, F. R. and D. R. Bouldin. 1970. *Plant Soil,* 33:49–61.
32. Mishustin, E. N. and A. N. Naumova. 1962. *Mikrobiologiya,* 31:543–555.
33. Mishustin, E. N. and V. T. Yemtsev. 1973. *Soil Biol. Biochem.,* 5:97–107.

34. O'Toole, P. and R. Knowles. 1973. *Soil Biol. Biochem.*, 5:789–797.
35. Porter, L. K. and A. R. Grable. 1969. *Agron. J.*, 61:521–523.
36. Rice, W. A., E. A. Paul, and L. R. Wetter. 1967. *Can. J. Microbiol.*, 13:829–836.
37. Richards, B. N. 1973. *Soil Biol. Biochem.*, 5:149–152.
38. Rovira, A. D. 1963. *Plant Soil*, 19:304–314.
39. Ruschel, A. P., Y. Henis, and E. Salati. 1975. *Soil Biol. Biochem.*, 7:181–182.
40. Rychert, R. C. and J. Skujins. 1974. *Soil Sci. Soc. Amer. Proc.*, 38:768–771.
41. Seto, B. and L. E. Mortenson. 1974. *J. Bacteriol.*, 120:822–830.
42. Steyn, P. L. and C. C. Delwiche. 1970. *Environ. Sci. Technol.*, 4:1122–1128.
43. Stutz, R. C. and L. C. Bliss. 1975. *Can. J. Bot.*, 53:1387–1399.
44. Vlassak, K., E. A. Paul, and R. E. Harris. 1973. *Plant Soil*, 38:637–649.
45. von Bülow, J. F. W. and J. Döbereiner. 1975. *Proc. Natl. Acad. Sci., U.S.*, 72:2389–2393.
46. Watanabe, A. and Y. Yamamoto. 1971. In T. A. Lie and E. G. Mulder, eds., *Biological nitrogen fixation in natural and agricultural habitats*. Nijhoff, The Hague, pp. 403–413.

19
Nitrogen Fixation: Symbiotic

The continuous drain on the nitrogen resources of the soil and the necessity for higher crop yields have led to an ever-increasing emphasis on means of conserving the limited supply of the element. Because only a fraction of the total agricultural need for nitrogen comes from synthetic and natural fertilizers, the remaining portion must be satisfied from the soil reserves and through the biological fixation of atmospheric N_2. It has been pointed out in the previous chapter that a number of free-living microorganisms can assimilate molecular nitrogen, but no higher plant or animal has the needed enzyme to catalyze the reaction. In certain instances, however, a symbiosis can become established in which one of the more prominent effects of the association is the acquisition of nitrogen from the atmosphere. Two members are required for the association—a plant and a microorganism. The classical example of such a symbiosis is that between leguminous plants and bacteria of the genus *Rhizobium*. The seat of the symbiosis is within the nodules that appear on the plant roots (Figure 19.1).

Legumes, the most important plant group concerned in symbiotic N_2 fixation, are dicotyledonous plants of the family Leguminosae. The 13,000 or more species, of which only about 200 are cultivated by man, are divided into three subfamilies. The largest of the three is Papilionoideae. In this subfamily are found *Trifolium, Melilotus, Medicago, Lotus, Phaseolus, Dalea, Crotalaria, Vicia, Vigna, Pisum,* and *Lathyrus*. A smaller number of genera are classified in the Caesalpinioideae while Mimosoideae is the smallest subfamily of legumes.

THE MICROSYMBIONT

Members of the genus *Rhizobium*, on infection of the appropriate legume, can cause the formation of nodules and participate in the symbiotic acquisition of N_2. The bacteria are gram-negative, non-spore-forming, aerobic rods, 0.5 to 0.9

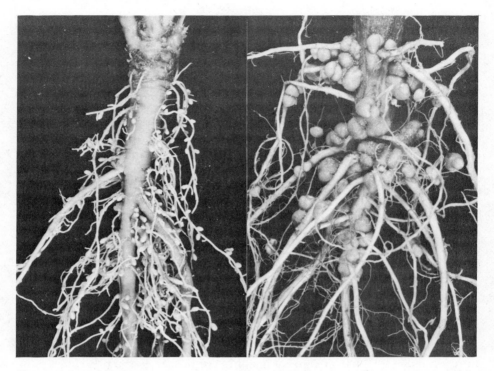

Figure 19.1. Well-nodulated legume roots. Left, red clover; right, soybeans. (Courtesy of J. C. Burton.)

μm wide and 1.2 to 3.0 μm long. Representatives of this genus are typically motile. Several carbohydrates are utilized, sometimes with the accumulation of acid but never of gas. The rhizobia are quite similar to *Agrobacterium radiobacter*, a bacterium that differs from the root nodule bacteria in certain minor cultural traits and in its inability to infect legume roots. The means of differentiating species of *Rhizobium* from related bacteria is highly unsatisfactory because it relies on the ability of the organisms to nodulate test plants. Delineation of a microbial group on the basis of a particular ecological relationship is undesirable as a bacterium living free in the soil usually is not examined for its ability to infect or nodulate given hosts. Moreover, a classification scheme of this type is inadequate because an individual isolate cannot be excluded from the genus *Rhizobium* until its ability to nodulate all leguminous species has been ascertained.

Nevertheless, the agronomic significance of the rhizobia dictates that some usable diagnostic system be developed, pragmatic though it be. Clear differences between strains of the root nodule bacteria are apparent, but standard laboratory tests for their differentiation into species are rare. Speciation within the genus is based entirely, for the present at least, on host specificity since the bacteria are limited in the plant groups they infect. The characteristic on which the classification is based is the capacity of an isolate to invade roots of a restricted number of plant species in addition to the legume from which the microorganism was obtained. Because of the limited number of hosts, so-called *cross-inoculation* groups have been established.

A cross-inoculation group refers to a collection of leguminous species that develop nodules when exposed to bacteria obtained from the nodules of any member of that particular plant group. Consequently, a single cross-inoculation group ideally includes all host species that are infected by an individual bacterial strain. More than 20 cross-inoculation groups have been established, only 7 of which have achieved prominence, and no more than 6 have been sufficiently well delineated for the responsible bacterium to have attained species status. The accepted classification scheme based on cross-inoculation groupings is outlined in Table 19.1. Many legumes of agricultural importance as well as noncultivated plants, however, are not nodulated by bacteria of the six major types. Birdsfoot trefoil (*Lotus corniculatus*), black locust (*Robinia pseudoacacia*), hemp sesbania (*Sesbania exaltata*), and others that do not fit into the established

TABLE 19.1

Cross-Inoculation Groups and *Rhizobium*-Legume Associations

Cross-Inoculation Group	*Rhizobium* Species	Host Genera	Legumes Included
Alfalfa group	R. meliloti	*Medicago*	Alfalfa
		Melilotus	Sweet clover
		Trigonella	Fenugreek
Clover group	R. trifolii	*Trifolium*	Clovers
Pea group	R. leguminosarum	*Pisum*	Pea
		Vicia	Vetch
		Lathyrus	Sweetpea
		Lens	Lentil
Bean group	R. phaseoli	*Phaseolus*	Beans
Lupine group	R. lupini	*Lupinus*	Lupines
		Ornithopus	Serradella
Soybean group	R. japonicum	*Glycine*	Soybean
Cowpea group	—	*Vigna*	Cowpea
		Lespedeza	Lespedeza
		Crotalaria	Crotalaria
		Pueraria	Kudzu
		Arachis	Peanut
		Phaseolus	Lima bean

categories require distinctly different bacterial strains. Only a small percentage of the leguminous species reported in the botanical literature are included within the six defined cross-inoculation groupings.

The six species of *Rhizobium* are not entirely distinct. For example, the soybean and cowpea bacterial groups, commonly considered to be separate, contain many similar bacterial strains, and organisms isolated from soybean nodules frequently infect cowpeas and vice versa. These results suggest that at least some of the cowpea rhizobia may be varieties of *R. japonicum*. Moreover, certain *R. lupini* strains bear a degree of similarity to the cowpea-soybean type. A reasonably clear distinction can be made between the rhizobia that may have generation times of two to four hours and produce acid in culture media and those commonly with generation times of six to eight hours that create alkaline conditions in culture. The former group includes *R. leguminosarum*, *R. meliloti*, *R. phaseoli*, and *R. trifolii*, and the latter includes *R. japonicum* and *R. lupini* (40).

The validity of the cross-inoculation system has not gone unchallenged because many legumes are nodulated by rhizobia of other host-bacterial groups. The bacterial strains that invade legumes outside of their particular class and plants that are thus infected are examples of a phenomenon termed *symbiotic promiscuity*. Occasionally, one host is infected by microorganisms normally classified in a number of different plant-bacterium groups. The cross-inoculation classes are therefore not entirely adequate for the description of the nodulating performance of many root nodule organisms. The instances of symbiosis outside of the established cross-inoculation classes often are incapable of fixing N_2. Moreover, the agricultural significance of the groups still remains a key feature of the established taxonomic system, and, until a new scheme is proposed, one is forced to rely on the existing criteria that form the basis for sound legume inoculation in farm practice.

Rhizobia grow readily in culture media containing a carbon source such as mannitol or glucose, ammonium or nitrate to supply the required nitrogen, and several inorganic salts. In addition to the organic carbon source, one or several B vitamins are often needed by the microorganisms. The vitamins that are required include biotin, thiamine, pantothenic acid (17), and sometimes riboflavin. Of some practical significance is the fact the bacteria either require or their growth is stimulated by cobalt, although only minute quantities completely satisfy their needs (29). Until recently it was believed that none of the bacteria in culture solution utilizes N_2, but it is now evident that at least some strains will fix N_2 apart from the plant (32). It is quite likely that additional strains will also be found to have this capacity. Nevertheless, it seems probable that rhizobia are unable to carry out the process alone in nature to a significant degree, so the association can still be deemed a true symbiosis, from an ecological and practical standpoint at least.

Of particular importance to the development of the symbiotic relationship is the presence of a large population of rhizobia. Because there are no selective media for the nodule organisms, a common procedure for their enumeration is the inoculation of tenfold dilutions from the test soil into pots of sterile soil or nutrient media seeded with the particular legume. After a suitable incubation period, the roots are examined, and the highest dilution giving rise to nodules is taken as an indication of the population density. When this counting method is used, the numbers of *R. japonicum, R. meliloti,* and *R. trifolii* are found to vary from as few as 10 to as many as 10^6 per gram depending on the season of year, cropping history, and management practices. The organisms are sometimes more numerous in fields having a rotation containing the specific legume than in fields in which the legume is absent. The rhizobia are frequently more abundant in proximity to roots than at some distance from the plant, but often this stimulation is just as pronounced in the vicinity of nonlegumes as in the proximity of roots the bacterium can potentially infect (37). The microorganisms are native to soils with no recent history of the macrosymbiont, but their frequency generally rises if the host plant is present; nevertheless, it is not uncommon to find no correlation between their abundance and the number of years since the last crop of the appropriate macrosymbiont (43).

Several methods in addition to plant testing are employed for studies of the ecology of the rhizobia. The use of suitable hosts to show the presence of the bacteria in soil dilutions is an excellent counting procedure, but it is both laborious and time-consuming. Immunofluorescent microscopy has also been developed for investigations of the ecology of the group, this procedure allowing for the direct visualization of rhizobia in soil (5). An entirely different approach involves the use of antibiotic-resistant mutants derived from the rhizobia of particular interest. These mutants tolerate such high concentrations of antibiotics that, when soils containing the test strains are plated on agar containing the toxic chemical, few soil bacteria grow to form colonies. The mutant, however, is readily enumerated even when its population has fallen considerably (9).

Assessments of survival of the nodule inhabitants have been facilitated by these various procedures. When added to moist soil in large numbers, the populations that have been examined do not decline readily, although the rate of viability loss is promoted by increasing temperature (Figure 19.2). Similarly, many strains persist well in dry soil (9) provided the temperature is not too high or the pH too low. Certain groups of root nodule bacteria do not survive long, hot summers when the soil is dry. Under such circumstances, the loss in viability may be so marked that legumes fail owing to poor nodulation, despite the presence of abundant nodules on the plants the year before (8). The dying associated with dry environments exposed to high temperatures is greatly affected by the kind of clay, certain of these colloids preventing the marked loss of mortality of cells resulting from dry heat (30).

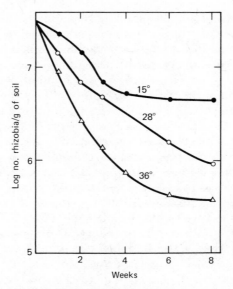

Figure 19.2. Effect of temperature on survival of *Rhizobium meliloti* in a loamy fine sand (9). Reproduced by permission of the Soil Society of America.

THE NODULATION PROCESS

The relationship between the formation of nodules and N_2 assimilation was first demonstrated in 1888 by Hellriegel and Wilfarth, but more than half a century elapsed before studies using ^{15}N-N_2 provided unequivocal proof that nodules are the seat of the fixation reaction. The localization of the N_2-metabolizing enzymes in the modified root tissue suggests that nodulation and the associated biochemical processes are of prime importance to the well-being of leguminous crops.

In the development of the nodular structure, the initial step appears to involve the release into the root zone of plant excretion products stimulatory to bacteria. The growth-promoting chemicals do not specifically benefit the organism potentially able to induce nodulation (33), their effect probably being one associated with stimulating a variety of microbial types. The rhizobia may then aggregate at distinct sites adjacent to the root. Recent evidence suggests that polysaccharides on the surface of the invasive bacteria are involved in or cause a binding of these cells to constituents on the surface of plants they can potentially nodulate. Little or no adhesion is evident between rhizobia and plants of heterologous cross-inoculation groups (6, 11).

In most legumes, the invasion occurs through the root hair, which, in the presence of suitable bacteria, undergoes a deformation or curling under the influence of some microbial product. The curling or deformation of the hairs, though characteristic of legumes and not of the nonlegumes so far examined, is induced on a single plant species by dissimilar *Rhizobium* strains so the host specificity typical of the symbiosis is not linked with this phase of the nodulation process. Indoleacetic acid was once believed to be the inducer of curling, but its role in the process is now in doubt or the compound is only one of the contributors to the morphological change. On the other hand, rhizobium synthesizes and excretes one or more products provoking deformation of legume root hairs. These compounds probably include a nucleic acid and a polysaccharide or protein (36). The deformed hairs are then penetrated in the first phase of the actual infection, although some plant species are invaded in entirely different ways. In this stage, the root-hair wall invaginates, and the invagination continues to develop into a tubelike structure.

Only a small proportion of the invaded root hairs develop nodules, usually less than 5 percent of the infections ultimately resulting in nodules. Following the microbial penetration into the root hair, a hyphalike *infection thread* is formed. In the narrow infection tube, typically surrounded by a wall of cellulose synthesized by the host, the bacterial population is never too dense, but the microorganisms can be seen easily under the microscope (Figure 19.3A). The wall of the thread seems to be continuous with the wall of the root hair itself. It is not clear what physiological changes bring about continued extension of the tube bearing the rhizobia, but it is at this point that microbial production of auxins or plant-growth regulators may play a role. Finally, the thread branches into the central portions of the developing nodule, and the bacteria ultimately are released into their symbiont's cytoplasm, there to multiply. Shortly prior to or immediately following the release, a period of rapid cell division takes place in the host's cells. The final structure consists of a central core containing the rhizobia and a surrounding cortical area in which is found the plant vascular system. A curious feature of the plant cells in the central portion of the nodule is their possession of twice the chromosome number characteristic of the host. The doubling of the chromosome number occurs in the nodules of polyploid as well as diploid legumes. The disomatic tissue probably originates from disomatic cells of the uninfected roots that, on the approach of the invading bacteria, are stimulated to multiplication. In the nodule, the microorganisms are seen to be enclosed in a membrane derived from cells of the host, and the rhizobia may there multiply so that four to six bacteria may be found in a single membrane-enclosed entity (Figure 19.3B). Most nodules are likely derived from a single strain of *Rhizobium* proliferating outside the root, but sometimes a nodule may originate from cells of more than one strain (27).

Significant differences are evident among legumes in the morphology of the nodules. Red and white clovers have club-shaped and lobed structures, the nodules of alfalfa are more branched and longer, while those of cowpea, peanut, and lima bean exhibit a spherical shape. In some plants, velvet bean for example, the nodules may approach the size of a baseball whereas the nodules

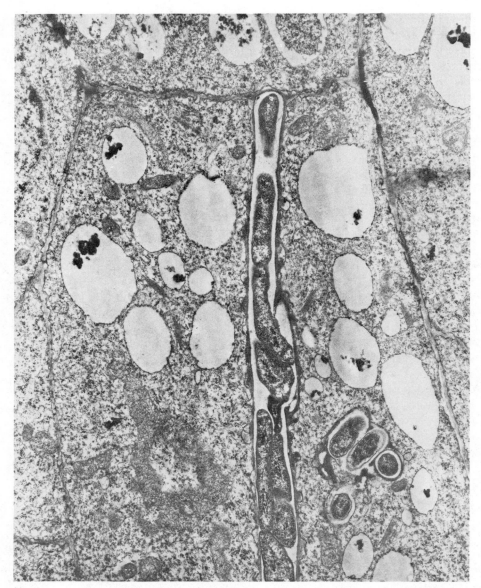

Figure 19.3 (A) Infection thread extending across a plant cell. Note bacteria in the thread, 10,500×.

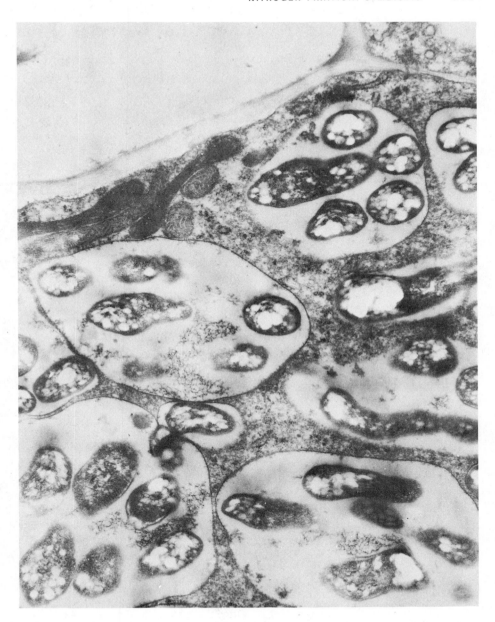

Figure 19.3 (B) Bacteroids in a membrane envelope contained in a cell of a mature soybean nodule (16), 23,000×.

of other legumes are no more than several millimeters in diameter. Legumes with fibrous roots frequently have a greater abundance of nodules than plants with well-formed tap roots, and plants bearing large nodules often have only a few whereas roots with smaller structures have them in greater numbers.

Nodules are not found on all of the genera and species of Leguminosae. The apparent failure to develop the symbiosis is most evident in the subfamily Caesalpinioideae in which about three-fourths of the species critically examined contain no nodular structures. Yet only some 1300 of a total of approximately 13,000 species of Leguminosae have been investigated to determine the presence of root nodules. The large percentage of Caesalpinioideae species that fail to develop root infections may be indicative of a degree of physiological primitiveness in which the capacity for symbiosis has never developed. However, in addition to the innate biochemical or genetic traits of the host precluding infection is the possibility that the ecological investigations have not been sufficiently intensive.

Once liberated from the infection thread into the root cytoplasm, the rhizobia may assume a peculiar morphology, a cellular form that has been termed the *bacteroid*. In morphology, the bacteroids found within the nodule are swollen and irregular, frequently appearing in star, clubbed, or branched shapes. Bacteroids vary in size and shape, and those formed by *R. leguminosarum*, for example, are distinctly different from those of *R. trifolii*. Many authorities do not restrict the term bacteroid to the peculiarly shaped microorganisms found in the nodules of many legumes, and they use the word to designate the bacteria present in all types of nodules, whether the rhizobia exhibit the odd morphology or not.

Not all rhizobia are capable of invading leguminous plants so that *infectiveness*, the ability of a strain to nodulate a given host, is of considerable economic importance. Among infective strains, moreover, the capacity of nodule bacteria to bring about N_2 fixation in conjunction with the plant varies greatly. The relative capacity of the plant-bacterial association, once established, to assimilate molecular nitrogen is known as its *effectiveness*. Many strains of *Rhizobium* are highly effective while others are largely or completely ineffective. A strain that does not permit fixation at a rate sufficient to meet the demands of the host is either partially effective or totally ineffective. Consequently, the mere presence of nodules is not a guarantee that a leguminous crop can benefit from gaseous nitrogen. The biochemical reasons why ineffective strains are unable to assimilate N_2 when located in the nodular structure have not so far been established (15).

Ineffective root nodule bacteria produce a greater number of nodules than effective cultures, but the nodules are smaller in size and tend to be more widely distributed over the root system. On the other hand, a single microbial strain may be ineffective or partially effective on one host yet be associated with active N_2 fixation on another legume variety or species. Furthermore, bacterial strains that appear effective on certain hosts may approach parasitism on others. It is therefore not possible to conclude that a given strain is effective or ineffective in absolute terms.

The environment has a marked influence on the development of the symbiotic association. Environmental effects are exerted largely through alterations in the physiology of the host. Inadequate levels of many inorganic nutrients alter nodule development in one of several manners, but such alterations are usually reflections of abnormal plant growth. Nodulation generally takes place at all soil temperatures tolerated by the free-living plant, but nodule abundance is reduced at the cooler and warmer extremes. Day length and light intensity also affect the number of nodules. Shading tends to depress nodule weights whereas high but not excessive light intensities and high CO_2 levels increase nodule numbers. An opposite effect is noted following nitrogen additions; that is, the nodule number and weight are reduced. The influence of day length, light intensity, nitrogen, and CO_2 supply may all be interpreted in terms of the internal concentration of carbohydrates; abundant light and CO_2, which increase the plant's carbohydrate storage, favor nodule production whereas nitrogen depresses the internal carbohydrate supply and has a retarding influence on nodulation at the same time. The same facts, however, may be used to support a carbohydrate:nitrogen ratio theory for nodulation. Whether the nitrogen or carbohydrate supply is the more critical has not yet been resolved.

FIXATION OF NITROGEN

The nitrogen status of soil under legumes is governed by cultural practices. When the crop is turned under, the full nitrogen gain is realized. When the legume is grazed or fed on the farm and the manure returned, the gain is not as great, yet it is appreciable. Removal of the aboveground portion of the crop, on the other hand, leads to little increase in nitrogen (Table 19.2). In the latter practice, however, the fact that the nitrogen content of the soil does not decrease, or increases somewhat, despite the removal of large quantities of proteinaceous material is of great merit, particularly when the depleting influence of grasses or cereals and the need for expensive fertilizers are taken into consideration. Furthermore, the results cited in Table 19.2 demonstrate that frequently the nitrogen gain is essentially as great in grass-legume mixtures as in the pure legume stand, the alfalfa or birdsfoot trefoil apparently assimilating the same quantity of N_2 in a pure or a mixed stand.

The fixation of N_2 by legumes that are effectively nodulated is appreciable. An idea of the magnitude of the gains under optimal conditions can be obtained from the following figures compiled from a number of field trials in the United States.

TABLE 19.2

Nitrogen Fixation and Nitrogen Gains in Soils of Ithaca, New York

Crop	Nitrogen Harvested in Crop in 4 yr	Nitrogen Balance in Soil[a]	
		4 yr	Avg/yr
	kg/ha	kg/ha	kg/ha
Timothy	157	45	11
Brome grass	132	43	11
Timothy + N[b]	480	−47	−37
Brome + N[b]	535	−69	−17
Timothy + trefoil	778	823	206
Brome + Alfalfa	1169	1169	292
Birdsfoot trefoil	809	876	219
Alfalfa	1146	1168	292

[a] Net gain or loss of nitrogen taking into account the nitrogen removed by the crop, nitrogen changes in the soil, and fertilizer added.
[b] Received a total of 695 kg of nitrogen. All other plots received 22 kg of fertilizer nitrogen.

	kg N_2 fixed/ha/year
Alfalfa	125–335
Red clover	85–190
Pea	80–150
Soybean	65–115
Cowpea	65–130
Vetch	90–155

Alfalfa, clovers, and the lupines are among the vigorous N_2 fixers while peanuts, beans, and peas are notably poor. Under normal farm conditions with well-nodulated roots and assuming favorable meteorological conditions, figures of 100 kg per hectare per year or higher are not uncommon in temperate regions. In a well-managed pasture, the biologically catalyzed N_2 incorporation varies from 100 to 200 kg per year on each hectare. Gains of this magnitude provide sufficient amounts of the element to satisfy the needs for rapid plant development. In addition, the beneficial effect of leguminous crops may persist for three or more years as measured by the improved yields of grasses or cereals in land previously growing legumes. The leguminous tree known as black locust (*Robinia pseudoacacia*) can bring about a slow but appreciable gain of nitrogen in its underlying soil. In addition, wild nodulated legumes contribute to the cycle of nitrogen on the earth's surface, but their significance is difficult to assess.

In order to obtain maximal benefits from the activities of the root nodule bacteria, one cannot usually rely on spontaneous infection by the indigenous soil microflora. Many localities contain few fully effective rhizobia, and it is not uncommon to observe as many as 25 percent of the bacteria in a given field to have a low degree of effectiveness, 50 percent to have moderate ability, and only 25 percent to be fully effective. Because of the large indigenous population of rhizobia that are not fully effective in N_2 fixation, it is not surprising that supplemental inoculation with selected bacterial strains commonly results in highly significant agronomic responses.

Several procedures have been developed to insure beneficial nodular associations. One of the most primitive is the application of soil obtained from a field previously cropped to the particular legume. Techniques of this sort are unsound because ineffective rhizobia, plant pathogens, and innumerable weed seeds may be applied together with the active microorganisms. A better method is the use of effective nodules obtained from plants of the same species. The nodules are homogenized, suspended in water, and applied to the seed at planting time. Methods involving the use of solid or liquid carriers have met with more widespread acceptance. In the original technique, the seeds were coated with a suspension of the bacteria that had been cultured in liquid media or on agar slants. In recent years, however, solid-base carriers have replaced the early inoculants. For the preparation of solid-base inoculants, a heavy suspension of rhizobia grown in liquid culture in large fermentors is mixed with a carrier of moist humus, finely ground peat, or a peat-charcoal mixture. In the farmer's hands, the seeds are wet with water and the inoculant mixed with the moist seed immediately prior to sowing.

Inoculation is recommended the first time a field is planted to a new legume species, and responses to the supplemental bacteria frequently are quite marked in these circumstances. A twofold or greater increase in dry weight yield resulting from inoculation is not uncommon. On the other hand, where the indigenous rhizobia are numerous and effective, the response measured in terms of yield or nitrogen uptake may be slight or lacking; the absence of a stimulation is often noted in fields previously cropped to effectively and abundantly nodulated hosts of the same cross-inoculation group. But, in the absence of clear evidence to the contrary, the recommended practice is to inoculate.

EXCRETION OF NITROGEN

Forage grasses or cereals grown together with legumes frequently contain more nitrogen than the corresponding crops grown alone. The beneficial effect of mixed cropping is often attributed to the excretion of nitrogen by legumes, the roots actively liberating the nitrogen obtained from the atmosphere. Benefits derived from mixed cropping have been observed for corn grown with soybeans or cowpeas, cereals with field peas or vetch, and pasture grasses with clover. The stimulation is evident in a study in which the yield of shelled corn was 3080 kg/ha when the corn was interplanted with *Phaseolus aureus* but only 1790 kg when the legume was absent. The enhancement was not evident if 45 kg of fertilizer nitrogen was applied per hectare (1).

The first clear demonstration of the subterranean transfer of nitrogen was made by Lipman (28). He grew oats in a pot of sand placed within a second, larger container of sand that was planted to field peas. The walls of the inner container were porous in order to permit the transfer of products through the nitrogen-free rooting media. Where nutrient diffusion was possible, the oats in the inner pot grew rapidly, had a green color, and appeared sturdy. In parallel containers in which the walls of the inner pot were nonporous to prevent the movement of products, the oats were stunted and exhibited symptoms of nitrogen deficiency. Many microbiologists have more recently confirmed the phenomenon of excretion, and in the modern experiments, the plants are grown in the absence of natural microbial communities and the excretions are characterized by various chromatographic techniques.

As a rule, probably only small amounts of nitrogen are released by legume roots (Figure 19.4), and thus only a portion of the nitrogen needed by adjacent plants can be provided by this mechanism. On the other hand, when the plants become senescent, are killed, or the aboveground portions are cut, considerable nitrogen may be released from the roots, and part of this nitrogen may be transferred to and assimilated by plants unable to utilize N_2 (35, 44). The precise mechanisms of nitrogen transfer and the basis for the benefit arising from mixed cropping still remain uncertain, however. Part of the nitrogen may indeed be derived from the active excretion of amino acids or other compounds by the legume. A portion of the additional nitrogen assimilated by a nonlegume grown together with a legume may arise from microbial decomposition of the sloughed-off root and nodular tissue of the legume. In some instances, however, little nitrogen transfer or no benefit to the associated plant can be detected.

N_2 FIXATION BY NONLEGUMINOUS PLANTS

Several genera of nonleguminous Angiosperms possess, at some stage in their life cycles, nodules on their roots. Nodulated nonlegumes confirmed as being able to make use of N_2 are represented by the following:

Family	Genus
Betulaceae	*Alnus*
Myricaceae	*Myrica*
	Comptonia

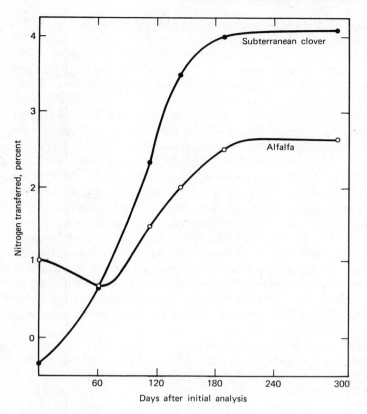

Figure 19.4. Transfer of nitrogen from legumes to grass (35).

Elaeagnaceae	*Elaeagnus*
	Hippophaë
	Shepherdia
Rhamnaceae	*Ceanothus*
	Discaria
Coriariaceae	*Coriaria*
Rosaceae	*Cercocarpus*
	Dryas
	Purshia
Casuarinaceae	*Casuarina*

Not all genera in these families and not all species in the 13 genera fix N_2. In certain environments, species of these genera may be more abundant than the legumes so that the capacity of the nodules formed by these genera to assimilate N_2 has considerable ecological significance.

The role of the nodule in such nonlegumes is not difficult to establish. The plant's growth in nutrient solutions containing low concentrations of nitrogen is distinctly poor, and nitrogen deficiencies are evident. However, the addition to the rooting medium of a preparation of ground nodules results in the production of nodules on the test plants. The nodulated hosts then develop with no deficiency symptoms, reach a greater height, and appear more vigorous than uninoculated controls. A chemical analysis for total nitrogen, exposure of the infected hosts to ^{15}N-labeled N_2, or tests with the acetylene reduction technique provide the final proof of the existence of N_2-metabolizing enzymes.

One of the most intensively studied of the group is the alder tree, a species of *Alnus*, which assimilates N_2 at a rate adequate to permit rapid development in nitrogen-deficient soils. The tree does not require nodules for growth provided that a nitrogen salt is supplied, but field-grown alders typically possess nodules that may approach the size of tennis balls, and chemical analysis of soil under alders often shows an increase in the nitrogen content. When treated with aqueous extracts of crushed nodules from field alder, greenhouse-grown plants develop nodules in two to three weeks. The optimum pH for fixation is in the vicinity of pH 5.5 to 6.0 although growth in nitrogen-free solutions is vigorous from pH 4.2 to 7.0.

As with the legumes, the extent of nitrogen gain by such Angiosperms varies enormously with soil type, climatic conditions, and plant age. The reports of nitrogen gains are 12-200 kg/ha per year for *Alnus* species and 27-179 kg/ha per year for *Hippophaë* species. Tests of individual locations suggest the fixation of 58 kg/ha for *Casuarina*, 60 kg/ha for *Ceanothus*, but only about 10 kg/ha per year for *Myrica*.

The ability of these hosts to satisfy their nitrogen demand from the air has major ecological consequences. In some instances, as with species of *Myrica*, colonization and good growth are possible in sand dunes containing little combined nitrogen. Sometimes the soil becomes distinctly enriched with nitrogen; for example, at one site, the soil under *Alnus rugosa* had 0.46 percent nitrogen, while the surrounding field only contained 0.24 percent (42). Similarly, *Alnus glutinosa* may be beneficial to the growth of adjacent, nonnodulated trees (12).

Despite reports to the contrary, isolation of the microbiological agents responsible for the nodules has not been adequately substantiated. Moreover, no known *Rhizobium* strain will invade and cause nodulation of any of such nonlegumes. Because of the inability to obtain cultures of the responsible organisms, successful infections are brought about by removing nodules from field-grown plants, crushing the tissue, and applying the preparations to test seedlings. Cross inoculation has been recorded for *Hippophaë, Elaeagnus,* and *Shepherdia,* the three genera of Elaeagnaceae, suggesting that the infecting organisms are similar or identical. Cross inoculation refers here to the fact that crushed nodules prepared from one genus will induce nodulation of a second. Among the other genera, however, there is no cross inoculation, indicating thereby that the infective agents differ from one another.

Electron and light microscopy of nodule sections have provided evidence that the microorganisms possess hyphae with diameters usually of 0.3–0.5 or 0.5–0.8 μm, a filamentous structure typical of the actinomycetes. Based on this characteristic morphology, the organisms have been placed in a genus of actinomycetes, *Frankia*. The species of this genus, the members of which have not been grown in culture, have been delineated on the basis of host specificity.

In addition to plant-microorganism associations in certain of the Angiosperms, a number of Gymnosperms possess nodulelike structures. Representatives of *Podocarpus* and *Cycas*, some of which are quite common in certain areas of the world, are included in the group of nodulated Gymnosperms. The microsymbiont in *Podocarpus* may be a fungus, although the characterization still remains equivocal, and nodules of this conifer can utilize N_2. In some Gymnosperms, as with *Macrozamia* and *Encephalartos*, the microbial partner in the N_2-fixing root nodule is a blue-green alga, and these algae—species of *Nostoc* or *Anabaena*—will also bring about fixation apart from the plant. Even water ferns of the genus *Azolla* contain N_2-using blue-green algal associates.

A number of plants, usually from the tropics, have nodulelike structures on their leaves. Such anatomical swellings have been reported in species of *Pavetta, Psychotria, Grumilea,* and *Ardisia*. From nodules of different *Psychotria* species, nitrogenase-containing isolates of *Klebsiella* and *Chromobacterium* have been obtained, but no direct evidence exists to show that plants bearing these leaf nodules gain nitrogen from the atmosphere.

ENVIRONMENTAL INFLUENCES

The amount of N_2 acquired by legumes through the *Rhizobium* symbiosis has been the subject of considerable interest because of the agricultural significance of the process, and only the highlights can be considered here. The major environmental factors governing the fixation are the type of legume, the effectiveness of the bacteria, the inorganic or mineralizable nitrogen content of the soil, the level of available phosphorus and potassium, pH, and the presence in usable form of a number of secondary nutrients. Climatic and seasonal factors have a great influence on nitrogen gains, but these will not be considered in the present discussion as such factors generally affect more the physiology of the host than the symbiotic association.

Because fixation serves as a means of obtaining nitrogen required for growth, it is not surprising that simple, inorganic nitrogen compounds inhibit N_2 fixation. This is evident in Figure 19.5, which shows that not only do nitrogen fertilizers depress the uptake of N_2 but that soil nitrogen—undoubtedly inorganic forms—is also readily assimilated by nodulated soybeans. Nodule

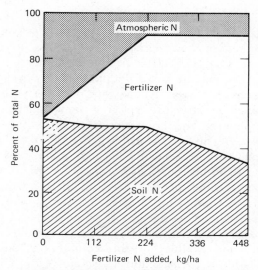

Figure 19.5. Contribution of atmospheric, soil, and fertilizer nitrogen to growth of soybeans (19). Reproduced by permission of the American Society of Agronomy.

weight and numbers are diminished at relatively high nitrate or ammonium levels, but low concentrations of inorganic nitrogen salts often enhance nodulation. The depression in nodule numbers and fixation by high levels of inorganic nitrogen and the occasional stimulation of nodulation by low levels occur with both the legume and the nonlegume associations. Measurement of the effect of fertilizers is facilitated by the use of the isotope, ^{15}N. Thus, if $(^{15}NH_4)_2SO_4$ is applied to nitrogen-free solution cultures of plants acquiring ^{14}N-N_2, the relative amount of ^{15}N and ^{14}N in the plant tissue is a measure of the quantity of nitrogen obtained from the ammonium source and the quantity from the air. By this means, it has been shown that the percentage of nitrogen derived from the atmosphere is inversely related to the nitrogen fertilization rate. It may be expected, however, that the presence of a grass will reverse the inhibition to some extent by its competition for the supply of inorganic nitrogen.

Phosphorus and potassium through their role as essential macronutrients exert a direct influence on nitrogen gains and legume yields. Responses of the N_2-fixing and the nodulation mechanisms to phosphate and potash fertilization are associated with the vigor and well-being of the host rather than reflections of a specific stimulation of the symbiosis per se.

The growth of the legume is also governed by soil acidity, and yield, nitrogen content, and nodulation of forage legumes often respond markedly to liming. The sensitivity to pH must be considered in terms of effects on the macrosymbiont, the microsymbiont, and on the symbiotic interaction of the two. With most legumes and nonlegumes, nodule formation takes place in a narrower range of hydrogen ion concentrations than plant growth. In many of the legumes of economic importance, the infection does not occur much below about pH 5.0 while, in pure culture, the responsible *Rhizobium* strain exhibits a similar sensitivity to low pH. The macrosymbiont, on the other hand, will occasionally grow at pH 4.0 or below in culture solutions provided with nitrate. Soybean and its infective bacterium are notable exceptions, and nodules are formed in highly acid environments.

The inhibition in acid soil frequently is not simply an effect of the hydrogen ion concentration. Toxicity resulting from iron or aluminum is most pronounced at low pH, and one or both of these substances may be the cause of poor legume stands. There is some evidence that calcium deficiency is important in the effects of acidity on nitrogen fixation, and insufficient calcium may decrease crop yields and diminish nodule weights.

Particularly dramatic is the influence of molybdenum in acid soils. Though the total molybdenum content may be high, the availability of the element for plant nutrition is affected markedly by pH. Little is obtained from acid soils because of chemical reactions that alter the availability of the element. Organisms that obtain their nitrogen from ammonium have a very small need for molybdenum, but this is not true for N_2-utilizing agents, either nonsymbiotic or symbiotic. For the incorporation of N_2 into fixed forms, molybdenum must be in abundant supply, and its low order of availability in acid environments is consequently a frequent cause of failures in legume seedling establishment or of low crop yields. The deficiency may be simply remedied, however, by the application of small quantities of molybdenum salts. How often the poor legume yields in acid soils result from an effect of hydrogen ions, metal toxicity, or molybdenum deficiency is not known, but some indication may be found in the remarkable stimulation by molybdenum of alfalfa, several of the clovers, and birdsfoot trefoil growing in acid soils. Molybdenum is thus necessary for symbiotic N_2 fixation, a requirement that holds for both the legume and the nonlegume associations.

Cobalt also markedly stimulates N_2 utilization by nodulated legumes and those few nonlegumes that have so far been tested. Cobalt is a component of vitamin B_{12}, and the element is found in nodules in the form of compounds containing this vitamin. No evidence of a significant cobalt requirement is usually apparent when the plants are growing on combined nitrogen. In culture, rhizobia similarly need cobalt for growth, and their cells also have compounds containing vitamin B_{12} (14); this suggests that the stimulation of fixation by cobalt reflects merely an enhancement of the proliferation and metabolism of the microsymbiont in its habitat within the root. In certain soils deficient in the element, the addition of cobalt will improve legume growth and

N_2 assimilation, with as little as 150 g/ha of a cobalt salt frequently producing a marked response.

Since the area immediately adjacent to the leguminous root system is the site of origin of the infecting rhizobia, it is not surprising that the associated microflora has an influence on the development of the symbiosis. Occasional inoculation failures may result from a microbiological competition that suppresses the desired microsymbiont and prevents initiation of the infection. Alternatively, members of the microflora may exert a beneficial influence by providing growth factors, removing toxic metabolites, or by immobilizing the inorganic nitrogen and thereby enhancing nodule formation.

When large numbers of the rhizobia are introduced into soil, as occurs at the time of planting inoculated seed, their abundance soon falls. This decline is correlated with a parallel rise in the density of protozoa, a group of predators capable of avidly feeding on strains of *Rhizobium*. It is quite likely that the protozoa are indeed responsible for the demise of many of the cells in large *Rhizobium* populations that occasionally become introduced into soil. However, many of the potential prey bacteria survive and are not eliminated after the initial drop in bacterial abundance, despite the numerous protozoa that have arisen. Apparently, the microscopic animals are effective feeders when their prey cells are common, but their feeding rate declines as the prey bacteria become more sparse (10).

Bdellovibrio strains attacking root nodule organisms are widespread, and they are able to parasitize and decimate large populations of the *Rhizobium*. However, much of the population decline when the rhizobia are added to soil is not the result of parasitism by bdellovibrios, and it is unlikely that these minute vibrios significantly affect the potential host bacteria at the numbers typically found in nature (21). Species of *Rhizobium* are also susceptible to attack and lysis by bacteriophages. Bacterial viruses specific for the soybean bacteria are found in the roots and in nodules of the respective macrosymbionts and also in land supporting these legumes (22). Continuous cropping of certain legumes, particularly alfalfa and clover, occasionally results in poor plant vigor and low yields. The condition, known as alfalfa or clover sickness, has been attributed to the deleterious effects of bacteriophage on *Rhizobium* and, therefore, on the fixation of N_2. However, it seems unlikely that bacteriophages cause the sickness as nodules actively assimilating N_2 contain the virus and bacteriophage abundance is not correlated with the incidence of the sickness.

The rhizobia are also susceptible, in culture at any rate, to toxins produced by other groups of organisms, but the frequency or activities of these toxin formers have not been convincingly correlated with the abundance of *Rhizobium*. Nevertheless, toxins harmful to *Rhizobium trifolii* have been extracted from soil, and it is possible that the heterotrophs producing such toxins occasionally flourish and synthesize the inhibitory agents, which may in turn have an influence on N_2 metabolism and the establishment of leguminous crops (7, 18).

BIOCHEMISTRY

Many of the recent advances in the biochemistry of symbiotic N_2 fixation have been achieved with the aid of ^{15}N and the acetylene reduction technique. The lack of a suitable radioactive isotope for biological experimentation with nitrogen makes the use of a stable isotope obligatory for all tracer studies, but for experiments in which the fate of nitrogen is not being sought, the acetylene reduction procedure is quite suitable. Either one of these methods can be employed to show that N_2 reduction in the legume nodule is solely associated with the bacteria. For example, the nodule tissue can be crushed, homogenized, and separated into a microbial fraction and a fraction containing plant constituents. The metabolism of N_2 is then confirmed to be a property of the rhizobia present in their unique habitat. Furthermore, although the plant-bacterial symbiosis is aerobic, O_2 inhibits the process of fixation by the microbial suspensions obtained from the nodule (3). The finding that it is the microsymbiont that carries out the transformation preceded by several years the discovery that at least several strains of *Rhizobium* use N_2 in culture.

One of the dramatic outcomes of the use of ^{15}N-labeled N_2 is the tracing of the compounds formed as intermediates in the conversion of the gaseous form of the element into cellular proteins. Thus, nodules metabolizing ^{15}N-N_2 convert the substrate to ^{15}N-NH_3, nearly all of the newly acquired nitrogen being in the form of NH_3 in short incubation periods (2). The short periods are necessary because the nitrogen thus fixed is readily converted to other products, all invariably organic. It is still not certain whether the NH_3 is converted to organic metabolites by cells of the micro- or the macrosymbiont. However, it is clear that NH_3 is combined with organic molecules to give rise to amino acids, like glutamic acid, and the amide known as glutamine (20, 24).

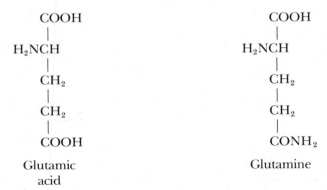

The nitrogen is then transferred to other amino acids, some of which may be transported out of the nodule. The reduction of N_2 requires considerable energy and the assimilation of NH_3 involves organic acids so it is not surprising that simple organic molecules enhance N_2 fixation by rhizobia collected from nodules. Under normal conditions, products of photosynthesis move from the leaves to the roots, and these compounds provide the energy for the reaction. However, fixation does proceed in the dark, although often at a slow rate, so that energy sources stored in the root must also be mobilized for this purpose.

When the nodules of alder are exposed to ^{15}N-N_2, the highest label is found in citrulline while glutamic acid is the second most active constituent. *Myrica gale*, on the other hand, incorporates nitrogen from N_2 most readily into the amide nitrogen of glutamine. Since the nitrogen in citrulline as well as the amide of glutamine are readily generated from ammonia, it is likely that ammonia is the inorganic precursor for the incorporation of nitrogen into organic combination in nonlegumes as well as in legumes (25, 26).

Once fixed, nitrogen is transported rapidly from the nodule to the remainder of the plant. After several hours, particularly in young legumes, the tagged nitrogen is detectable in the above-ground tissue. During the latter stages of development, more than 90 percent of the N_2 fixed is found above the soil level. The compounds involved in the translocation are as yet unknown.

The fixation of molecular nitrogen is affected greatly by the composition of the gas phase. The rate of the symbiotic incorporation varies directly with the quantity of N_2 in the atmosphere until, at concentrations greater than 15 percent, fixation is independent of the partial pressure of N_2. Further, the responsible enzyme system in red clover functions at half maximal velocity when the gas phase contains 5 percent N_2 (38). Optimal for N_2 fixation by red clover is a gas phase containing 10 to 40 percent O_2, and the rate of utilization declines at O_2 levels above 40 or below 10 percent. It is apparent from the data cited that the quantities of O_2 and N_2 present in the atmosphere are favorable for the transformation in legumes.

Nitrogenase can be obtained free of *Rhizobium* by rupturing the isolated bacteriods by mechanical means, and the enzyme is able to reduce N_2 in the absence of O_2. Nitrogenase must be provided with ATP and a reducing agent in order to function. As with free-living species that are capable of bringing about the reaction, nitrogenase of the legume bacteroids contains two protein components, one with both iron and molybdenum in the molecule and a molecular weight of about 200,000, a second with iron but no molybdenum and a molecular weight in the vicinity of 65,000 (45). The former component exists even in the bacteroids of plants grown on ammonium salts; that is, in tissues having little N_2-fixing activity (4).

Nodules of legumes actively metabolizing N_2 are distinctly red in color because of the presence of an iron-containing substance known as *leghemoglobin*. In contrast, nodules produced by ineffective rhizobia have neither leghemoglobin nor red pigmentation. The degree of effectiveness, moreover, usually can be correlated with the amount of leghemoglobin in the nodular tissue (Figure 19.6). Consequently, it is possible by visual examination of the nodule contents

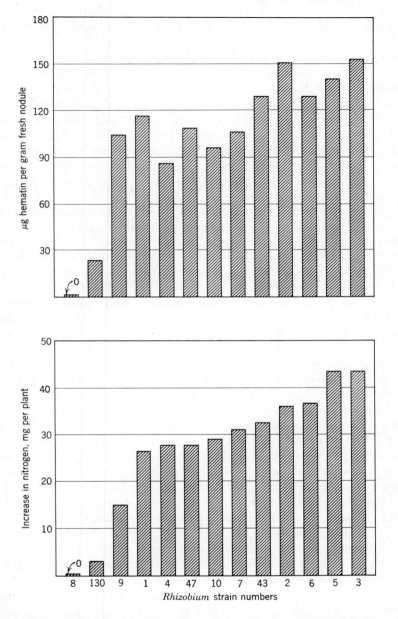

Figure 19.6. N₂ fixation and hematin content of pea plants inoculated with rhizobia of varying effectiveness (41).

to ascertain the relative capacity for active N_2 fixation or, by spectrophotometric determination of leghemoglobin concentration, to approximate bacterial efficiency. The leghemoglobin is localized outside of the bacterium in the nodule and appears to be situated between the microbial cell and the plant-derived membrane surrounding the rhizobia. Leghemoglobins in nodules formed on the same plant species by different *Rhizobium* strains are identical, whereas that in nodules of dissimilar legumes produced by the same bacterial strain are different; hence, it is the host that is the genetic determinant of the synthesis of this novel substance (13). Nevertheless, the microorganism is essential, in some unknown manner, for provoking its host to initiate the biosynthetic process.

Because of its relationship to the N_2 metabolism of Leguminosae, leghemoglobin has been the subject of considerable biochemical investigation. The bacteroids are aerobic and avidly consume O_2 that becomes available to them in their peculiar locale, and leghemoglobin promotes their O_2 utilization and, probably indirectly, thereby favors the fixation reaction. This may be taken as an indication that leghemoglobin serves in the symbiosis to facilitate the movement of O_2 through the O_2-poor host tissue to the zone immediately around the bacteria (46).

Growth of the rhizobium and its reduction of N_2 place a considerable demand on the CO_2 carbon assimilated during photosynthesis by the host. The products of photosynthesis must be present in adequate supply to maintain the activity of the nodule residents, and these products are evidently oxidized soon after their arrival in the nodule by the microorganism (23). In one trial with *Pisum sativum* (garden pea), for example, 32 percent of the carbon acquired by the shoot in short-term experiments was translocated to the nodules; 16 percent of this carbon was used for growth, 37 percent in respiration, and the rest was returned to the above-ground portions as amino compounds synthesized as a consequence of N_2 fixation (31). This demand for energy means that part of the carbon gained in photosynthesis must be diverted to the symbiotic association, a diversion that would reduce the yield of crops as compared with the same plant population growing on ammonium. On the other hand, a crop utilizing nitrate must reduce nitrate to the ammonium form, and this too entails the use of some of the energy and dry matter gained during photosynthesis.

A limitation imposed on research on the association has been the necessity of using the entire plant for critical investigations of the transformation. One way of partly overcoming this limitation is to utilize excised roots, which nodulate when provided with nutrients either from a surrounding medium or through the base of the roots. Use of this technique has permitted demonstration that the host produces a soluble substance concerned in nodule genesis. In the absence of this compound, which has yet to be identified, seedlings normally capable of being nodulated will grow through a suspension of an infective *Rhizobium* without showing signs of infection (34, 39).

REFERENCES

Reviews

Becking, J. H. 1970. Plant-endophyte symbiosis in nonleguminous plants. *Plant Soil,* 32:611–654.

Bergersen, F. J. 1973. Symbiotic nitrogen fixation by legumes. In G. W. Butler and R. W. Bailey, eds., *Chemistry and biochemistry of herbage.* Academic Press, London, vol. 2, pp. 189–226.

Dixon, R. O. D. 1969. Rhizobia (with particular reference to relationships with host plants). *Annu. Rev. Microbiol.,* 23:137–158.

Quispel, A., ed. 1974. *The biology of nitrogen fixation.* North-Holland Publishing Co., Amsterdam.

Stewart, W. D. P. 1966. *Nitrogen fixation in plants.* Athlone Press, University of London.

Vincent, J. M. 1970. *A manual for the practical study of root-nodule bacteria.* Blackwell Scientific Publications, Oxford.

Literature Cited

1. Agboola, A. A. and A. A. A. Fayemi. 1972. *Agron. J.,* 64:409–412.
2. Bergersen, F. J. 1965. *Aust. J. Biol. Sci.,* 18:1–9.
3. Bergersen, F. J. and G. L. Turner. 1968. *J. Gen. Microbiol.,* 53:205–220.
4. Bishop, P. E., H. J. Evans, R. M. Daniel, and R. O. Hampton. 1975. *Biochim. Biophys. Acta,* 381:248–256.
5. Bohlool, B. B. and E. L. Schmidt. 1973. *Soil Sci. Soc. Amer. Proc.,* 37:561–564.
6. Bohlool, B. B. and E. L. Schmidt. 1974. *Science,* 185:269–271.
7. Chatel, D. L. and C. A. Parker. 1972. *Soil Biol. Biochem.,* 4:289–294.
8. Chatel, D. L. and C. A. Parker. 1973. *Soil Biol. Biochem.,* 5:415–423.
9. Danso, S. K. A. and M. Alexander. 1974. *Soil Sci. Soc. Amer. Proc.,* 38:86–89.
10. Danso, S. K. A., S. O. Keya, and M. Alexander. 1975. *Can. J. Microbiol.,* 21:884–895.
11. Dazzo, F. B. and D. H. Hubbell. 1975. *Appl. Microbiol.,* 30:1017–1033.
12. Delver, P. and A. Post. 1968. *Plant Soil,* 28:325–336.
13. Dilworth, M. J. 1969. *Biochim. Biophys. Acta,* 184:432–441.
14. Evans, H. J. and M. Kliewer. 1964. *Ann. N.Y. Acad. Sci.,* 112:735–755.
15. Francis, A. J. and M. Alexander. 1974. *Soil Sci.,* 118:31–37.
16. Goodchild, D. J. and F. J. Bergersen. 1966. *J. Bacteriol.,* 92:204–213.
17. Graham, P. H. 1963. *J. Gen. Microbiol.,* 30:245–248.
18. Holland, A. A. and C. A. Parker. 1966. *Plant Soil,* 25:329–340.
19. Johnson, J. W., L. F. Welch, and L. T. Kurtz. 1975. *J. Environ. Qual.,* 4:303–306.
20. Kennedy, I. R. 1966. *Biochim. Biophys. Acta,* 130:295–303.
21. Keya, S. O. and M. Alexander. 1975. *Soil Biol. Biochem.,* 7:231–237.
22. Kowalski, M., G. E. Ham, L. R. Frederick, and I. C. Anderson. 1974. *Soil Sci.,* 118:221–228.
23. Lawrie, A. C. and C. T. Wheeler. 1973. *New Phytol.,* 72:1341–1348.
24. Lawrie, A. C. and C. T. Wheeler. 1975. *New Phytol.,* 74:429–436.
25. Leaf, G., I. C. Gardner, and G. Bond. 1958. *J. Exptl. Bot.,* 9:320–331.
26. Leaf, G., I. C. Gardner, and G. Bond. 1959. *Biochem. J.,* 72:662–667.
27. Lindemann, W. C., E. L. Schmidt, and G. E. Ham. 1974. *Soil Sci.,* 118:274–279.
28. Lipman, J. G. 1910. *J. Agr. Sci.,* 3:297–300.
29. Lowe, R. H. and H. J. Evans. 1962. *J. Bacteriol.,* 83:210–211.
30. Marshall, K. C. 1964. *Aust. J. Agr. Res.,* 15:273–281.
31. Minchin, F. R. and J. S. Pate. 1973. *J. Exptl. Bot.,* 24:259–271.

32. Pagan, J. D., J. J. Child, W. R. Snowcroft, and A. H. Gibson. 1975. *Nature*, 256:406–407.
33. Peters, R. J. and M. Alexander. 1966. *Soil Sci.*, 102:380–387.
34. Schaffer, A. G. and M. Alexander. 1967. *Plant Physiol.*, 42:563–567.
35. Simpson, J. R. 1965. *Aust. J. Agr. Res.*, 16:915–926.
36. Solheim, B. and J. Raa. 1973. *J. Gen. Microbiol.*, 77:241–247.
37. Tuzimura, K., I. Watanabe, and J. F. Shi. 1966. *Soil Sci. Plant Nutr.*, 12:99–106.
38. Umbreit, W. W. and P. W. Wilson. 1939. *Trans. 3rd Comm., Intl. Soc. Soil Sci.*, New Brunswick, A:29–31.
39. Valera, C. L. and M. Alexander. 1965. *J. Bacteriol.*, 89:1134–1139.
40. Vincent, J. M. 1974. In A. Quispel, ed., *The biology of nitrogen fixation.* North-Holland Publishing Co., Amsterdam, pp. 265–341.
41. Virtanen, A. I., J. Erkama, and H. Linkola. 1947. *Acta Chem. Scand.*, 1:861–870.
42. Voigt, G. K. and G. L. Steucek. 1969. *Soil Sci. Soc. Amer. Proc.*, 33:946–949.
43. Weaver, R. W., L. R. Frederick, and L. C. Dumenil. 1972. *Soil Sci.*, 114:137–141.
44. Whitney, A. S. and Y. Kanehiro. 1967. *Agron. J.*, 59:585–588.
45. Whiting, M. J. and M. J. Dilworth, 1974. *Biochim. Biophys. Acta,* 371:337–351.
46. Wittenberg, J. B., F. J. Bergersen, C. A. Appleby, and G. L. Turner. 1974. *J. Biol. Chem.,* 249:4057–4066.

MINERAL TRANSFORMATIONS

20
Microbial Transformations of Phosphorus

Phosphorus is found in soil, plants, and in microorganisms in a number of organic and inorganic compounds. It is second only to nitrogen as an inorganic nutrient required by both plants and microorganisms, its major physiological role being in certain essential steps in the accumulation and release of energy during cellular metabolism. This element may be added to soil in the form of chemical fertilizers, or it may be incorporated as leaf litter, plant residues, or animal remains. Thus, phosphorus occupies a critical position both in plant growth and in the biology of soil.

Microorganisms bring about a number of transformations of the element. These include (*a*) altering the solubility of inorganic compounds of phosphorus, (*b*) mineralizing organic compounds with the release of inorganic phosphate, (*c*) converting the inorganic, available anion into cell components, an immobilization process analogous to that occurring with nitrogen, and (*d*) bringing about an oxidation or reduction of inorganic phosphorus compounds. Particularly important to the phosphorus cycle in nature are the microbial mineralization and immobilization reactions. By the continual interplay of the latter two processes, the availability of phosphorus, like that of nitrogen, is governed to no small degree. A general cycle of phosphorus may be visualized as shown in Figure 20.1.

CHEMISTRY OF SOIL PHOSPHORUS

The chief source of organic phosphorus compounds entering the soil is the vast quantity of vegetation that undergoes decay. Agricultural crops commonly contain 0.05 to 0.50 percent phosphorus in their tissues. In plants, this element is found in several compounds or groups of substances: phytin, phospholipids, nucleic acids, phosphorylated sugars, coenzymes, and related compounds.

333

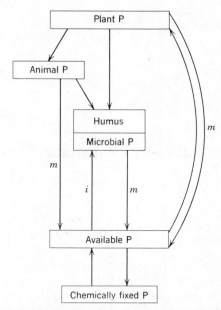

Figure 20.1. A simplified phosphorus cycle. The letters *m* and *i* denote mineralization and immobilization of phosphorus.

Phosphorus may also be present, especially in vacuoles and internal buffers, as inorganic orthophosphate.

In contrast with nitrogen and sulfur, for which the ions assimilated, nitrate or sulfate, are reduced within the cell to the amino (—NH_2) or sulfhydryl (—SH) functional groups, the plant does not reduce phosphate; this ion enters into organic combination largely unaltered. Thus, the phosphorus in phytin, phospholipids, and nucleic acids is found as phosphate. Phytin is the calcium-magnesium salt of phytic acid, the latter term being synonymous with inositol hexaphosphate.

phytic acid

Inositol phosphates may have one, two, three, four, five, or six phosphorus atoms per inositol unit, and these several compounds have been found in soils and living organisms and are formed enzymatically. Phospholipids are compounds in which phosphate is combined with a lipid. In the phosphatides, a class of phospholipids that includes lecithin and cephalin, the phosphate is esterified with a nitrogenous base. Lecithin, for example, is made up of glycerol, fatty acids, phosphate, and choline.

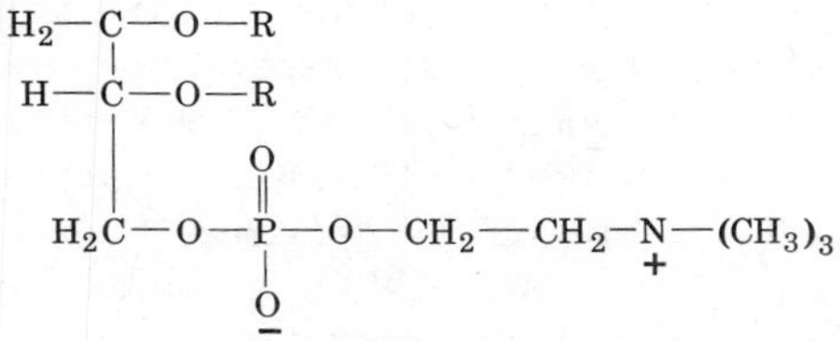

$$
\begin{array}{l}
H_2\!-\!C\!-\!O\!-\!R \\
\quad\ | \\
H\!-\!C\!-\!O\!-\!R \\
\quad\ | \\
\quad\ \ \ \ \overset{O}{\overset{\|}{}} \\
H_2C\!-\!O\!-\!P\!-\!O\!-\!CH_2\!-\!CH_2\!-\!\overset{+}{N}\!-\!(CH_3)_3 \\
\quad\ \ \ \ | \\
\quad\ \ \ \ \underset{-}{O}
\end{array}
$$

lecithin–type compounds
(R: fatty acid)

As discussed above, the nucleic acids RNA and DNA consist of a number of purine and pyrimidine bases, a pentose sugar, and phosphate.

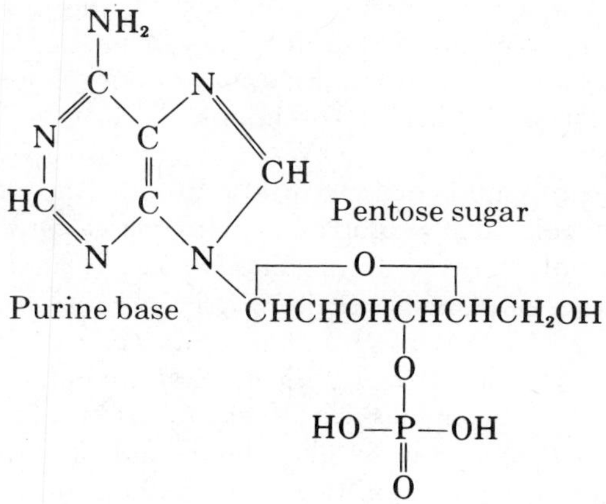

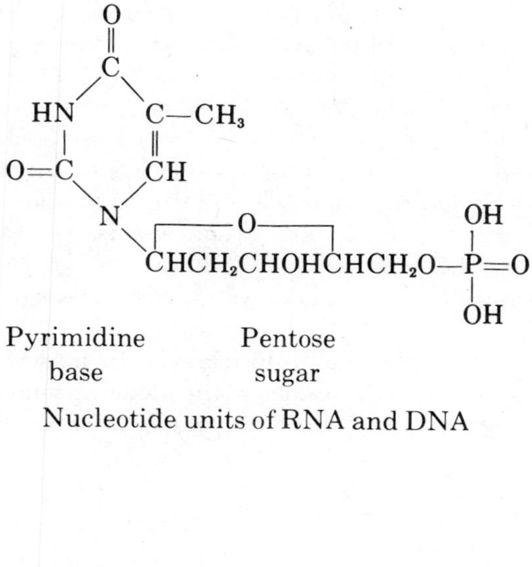

$$
\begin{array}{c}
NH_2 \\
| \\
C \\
N \diagup \ \diagdown \ N \\
\ | \qquad C \\
\ | \qquad \| \qquad CH \\
HC \qquad C \diagup \\
\diagdown N \diagdown \ N \\
\end{array}
$$

Purine base   Pentose sugar

$$
\begin{array}{c}
\overbrace{\qquad O \qquad} \\
CHCHOHCHCHCH_2OH \\
| \\
O \\
| \\
HO\!-\!P\!-\!OH \\
\| \\
O
\end{array}
$$

$$
\begin{array}{c}
O \\
\| \\
C \\
HN \diagup \ \diagdown C\!-\!CH_3 \\
\ | \qquad\quad \| \\
O\!=\!C \qquad CH \\
\diagdown N \diagup
\end{array}
$$

$$
\begin{array}{c}
\overbrace{\qquad O \qquad} \qquad OH \\
CHCH_2CHOHCHCH_2O\!-\!\overset{\|}{\underset{OH}{P}}\!=\!O
\end{array}
$$

Pyrimidine   Pentose
base   sugar

Nucleotide units of RNA and DNA

The bulk of the phosphorus in the bacterial cell is in RNA, this nucleic acid usually accounting for one-third to somewhat more than one-half of all the phosphorus. Large quantities are found in acid-soluble compounds, both organic and inorganic, a total commonly of 15 to 25 percent of the total cell phosphorus content. The acid-soluble fraction of bacterial protoplasm contains ortho- and metaphosphate, sugar phosphates, and many of the coenzymes and adenosine phosphates. The concentration of phospholipids in bacteria varies greatly, depending on species and age of the microorganism, but phospholipids usually represent less than 10 percent of the cell phosphorus. DNA contributes from 2 to 10 percent of the total. Although there is no evidence for the presence in microorganisms of phytin, inositol phosphates may occur. Inorganic polyphosphates may sometimes be quite abundant in certain fungi.

In soil, from 15 to 85 percent of the total phosphorus is organic. The absolute quantity of organic phosphorus generally declines with increasing depth. At the same time, the proportion of total phosphorus that is organic is greater in surface than in subsurface horizons. Of the inorganic forms, large quantities occur in minerals where the phosphate is part of the mineral structure, as insoluble calcium, iron, or aluminum phosphates. The calcium salts predominate in neutral or alkaline conditions, the iron and aluminum salts in acid surroundings.

The organic compounds making up the humus fraction are derived from surface vegetation, microbial protoplasm, or metabolic products of the micro-flora. Thus, the components of humus are directly related to the constituents making up the tissues of plants and cells of microorganisms or their derivatives. Of these substances, the inositol phosphates, nucleic acids, phospholipids, and molecules containing them are significant components of the soil organic fraction. The inositol phosphates are of several sorts. On the one hand, the molecules may have from one to six phosphates, and all of these exist in nature. The inositol hexaphosphate is dominant, the pentaphosphate is less abundant, and the tetra-, tri-, di-, and monophosphates exist in still smaller quantities (20). On the other hand, the inositol portion of the compound is present in various isomeric forms. The chief isomer is *myo*-inositol, but *scyllo*-, *neo*-, and DL-inositols are also present in lower concentration. These various inositol phosphates are often classified together as phytin and related substances. Such components are major constituents of the organic phosphorus fraction, frequently accounting for 10 to 80 percent of all the organic phosphorus. It is believed that several of the inositol phosphates are of microbial origin because they are not known to occur in higher plants. At least part of the inositol is not present in a simple form, but rather is bound into molecules with molecular weights considerably in excess of 1000 (19).

The evidence for the existence of nucleic acids or nucleotide derivatives in soil is indirect. Typical of such results is the demonstration of purine and pyrimidine bases—constituents of the RNA and DNA molecules—or the bases linked through a sugar to phosphate. In most localities, the nucleic acid-type compounds probably contribute less than 1 to a maximum of 10 percent of the total organic phosphorus. As the microbial cell is rich in nucleic acids, it is possible that a reasonably large part of the nucleic acid in soil, which is undoubtedly primarily of the RNA type, is bound within the cells of viable members of the microflora. Once the cells die, the nucleic acids would be readily mineralized.

The phospholipid content of humus is invariably small. Often as little as 0.1 percent or sometimes up to 5 percent or sometimes slightly more of the organic phosphorus is tied up in such compounds. A significant part of this fraction may be phosphatidylethanolamine and phosphatidylcholine (15), compounds also found in plants and microorganisms.

$$
\underset{\text{Phosphatidylethanolamine}}{\overset{\displaystyle \overset{O}{\underset{\displaystyle \underset{O^-}{|}}{\overset{\|}{\text{HOPOCH}_2\text{CH}_2\overset{+}{\text{N}}\text{H}_3}}}}{}
\qquad\qquad
\underset{\text{Phosphatidylcholine}}{\overset{\displaystyle \overset{O}{\underset{\displaystyle \underset{O^-}{|}}{\overset{\|}{\text{HOPOCH}_2\text{CH}_2\overset{+}{\text{N}}(\text{CH}_3)_3}}}}{}
$$

At present, the balance sheet of organic phosphorus remains incomplete. The known major components frequently do not account for the sum total of this element. Other as yet unidentified compounds may be formed by microbial action or, alternatively, the inositol phosphates, nucleic acids, or phospholipids may be modified to such an extent that existing analytical methods do not detect them.

Soils rich in organic matter contain abundant organic phosphorus. A good correlation, moreover, exists between the concentrations of organic phosphorus, organic carbon, and total nitrogen, almost all of the latter being organic (Table 20.1). Ratios of organic carbon to organic phosphorus of 100 to 300:1 are common for mineral soils. Similarly, the nitrogen:organic phosphorus ratio may range from 5 to 20 parts of nitrogen for each part of phosphorus; the ratio is commonly wider for virgin than for comparable cultivated land. The organic phosphorus level, therefore, is directly related to the concentration of other humus constituents, the phosphorus content being 0.3 to 1.0 and 5 to 20 percent of the carbon and nitrogen concentration, respectively. These results are not inconsistent with the hypothesis that the relative constancy of the ratios in humus is governed by the composition and activities of the microflora. On the other hand, although microorganisms are the probable cause of these ratios approaching certain equilibrium values, it is not valid to conclude that any one of these three constituents, carbon, nitrogen, or phosphorus, is present in soil largely or entirely in the cells of the microflora.

TABLE 20.1

Organic Phosphorus, Carbon, and Total Nitrogen in Several Iowa Soil Profiles (22)

Soil Type	Depth cm	Organic C %	Total N %	Organic P %	C:N:P
Carrington	0–15	2.40	0.20	0.0246	98:8.3:1
silt loam	15–23	2.28	0.20	0.0200	114:9.8:1
	31–38	1.43	—	0.0130	—
	46–53	0.81	0.076	0.0079	107:9.6:1
Grundy silt	0–15	2.84	0.21	0.0205	138:10.4:1
loam	15–25	2.36	0.16	0.0133	180:12.3:1
	35–46	1.64	0.12	0.0069	238:19.1:1
Garwin	0–13	5.11	0.41	0.0393	130:10.5:1
silty	13–35	1.74	0.12	0.0170	102:7.3:1
clay loam	35–56	0.68	0.06	0.0070	97:8.6:1

SOLUBILIZATION OF INORGANIC PHOSPHORUS

Insoluble inorganic compounds of phosphorus are largely unavailable to plants, but many microorganisms can bring the phosphate into solution. This attribute is apparently not rare since one-tenth to one-half of the bacterial isolates tested usually are capable of solubilizing calcium phosphates, and counts of bacteria solubilizing insoluble phosphates may range from 10^5 to 10^7 per gram. Such bacteria are often especially abundant on root surfaces (23). Species of *Pseudomonas, Mycobacterium, Micrococcus, Bacillus, Flavobacterium, Penicillium, Sclerotium, Fusarium, Aspergillus,* and others are active in the conversion. These bacteria and fungi grow in media with $Ca_3(PO_4)_2$, apatite, or similar insoluble materials as sole phosphate sources. Not only do the microorganisms assimilate the element but they also make a large portion soluble, releasing quantities in excess of their own nutritional demands. If the insoluble phosphate is suspended in an agar medium, the responsible strains are readily detected by the zone of clearing produced around the colony. The solubilization is not restricted to calcium salts for iron, aluminum, magnesium, manganese, and other phosphates are acted on also.

The major microbiological means by which insoluble phosphorus compounds are mobilized is by the production of organic acids. In the special case of the ammonium- and sulfur-oxidizing chemoautotrophs, nitric and sulfuric acids are responsible. The organic or inorganic acids convert $Ca_3(PO_4)_2$ to di- and monobasic phosphates with the net result of an enhanced availability of the element to plants. The amount brought into solution by heterotrophs varies with the carbohydrate oxidized, and the transformation generally proceeds only if the carbonaceous substrate is one converted to organic acids.

Nitric or sulfuric acids produced during the oxidation of nitrogenous materials or inorganic compounds of sulfur react with rock phosphate, thereby effecting an increase in soluble phosphate. The oxidation of elemental sulfur is a simple and effective means of providing utilizable phosphates. For example, a mixture may be prepared with soil or manure, elemental sulfur, and rock phosphate; as the sulfur is oxidized to sulfuric acid by *Thiobacillus*, there is a parallel increase in acidity and a net release of soluble phosphate. Nitrification of ammonium salts also leads to a slight but significant liberation of soluble phosphorus from rock phosphate composts. Biological sulfur or ammonium oxidation has never been adopted on a commercial scale because of the availability of cheaper and more efficient means of preparing fertilizers.

Although phosphate solubilization commonly requires acid production, other mechanisms may account for ferric phosphate mobilization. In flooded soil, the iron in insoluble ferric phosphates may be reduced, a process leading to the formation of soluble iron with a concomitant release of phosphate into solution (21). Such increases in the availability of phosphorus on flooding may explain why rice cultivated under water often has a lower requirement for fertilizer phosphorus than the same crop grown in dry-land agriculture. Phosphorus may also be made more available for plant uptake by certain bacteria that liberate hydrogen sulfide, a product that reacts with ferric phosphate to yield ferrous sulfide, liberating the phosphate (26).

The many phosphate-dissolving microorganisms in the vicinity of roots may appreciably enhance phosphate assimilation by higher plants. The data depicted in Figure 20.2 demonstrate that the yield of oats grown in sterile and nonsterile conditions with the addition of phosphorus as ferrophosphate, $CaHPO_4$, $Ca_3(PO_4)_2$, and bonemeal is consistently greater in infected than in noninfected pots, the microorganisms solubilizing the added phosphate. Similar studies have been reported for other plants and for other inorganic phosphorus sources.

MINERALIZATION OF ORGANIC PHOSPHORUS

The existence in soil of a large reservoir of organic phosphorus that cannot be utilized by plants emphasizes the role of microorganisms in converting the organic phosphorus to inorganic forms. By their actions, the bacteria, fungi, and actinomycetes make the bound element in remains of the vegetation and in soil organic matter available to succeeding generations of plants.

Mineralization is generally more rapid in virgin soils than in their cultivated counterparts. Not only is the total amount mobilized greater in virgin areas but the percentage of total organic phosphorus that is mineralized is higher in virgin than in cultivated land. The decomposition is also favored by warm

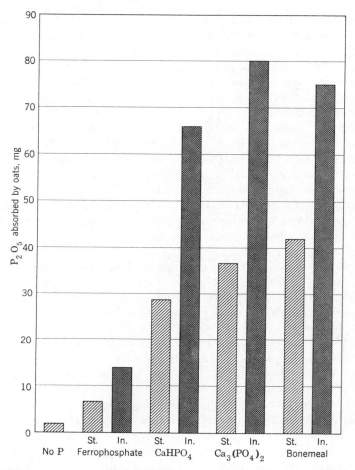

Figure 20.2. Phosphorus uptake by oats grown in sterile (St.) and in infected (In.) quartz sand. Inoculum: 1 percent nonsterile soil (7).

temperatures, with the thermophilic range being more favorable than the mesophilic range. The rate of mineralization also is enhanced by adjusting the pH to values conducive to general microbial metabolism, and a shift from acidity to neutrality increases phosphate release. Furthermore, the rate of mineralization is directly correlated with the quantity of substrate; hence, soils rich in organic phosphorus will be the most active. The degradation is not inhibited by inorganic phosphorus so that mineralization proceeds rapidly even at sites with adequate phosphate (6). As expected, phosphorus uptake by plants is correlated with the mineralization rate (25).

The mineralization and immobilization of this element are related to the analogous reactions of nitrogen. As a rule, phosphate release is most rapid under conditions favoring ammonification. Thus, a highly significant correlation is observed between the rates of nitrogen and phosphorus conversion to inorganic forms, the nitrogen mineralized being from 8 to 15 times the amount of phosphate made available. There is also a correlation between carbon (CO_2 release) and phosphorus mineralization, a ratio of ca. 100 to 300:1 (27). These results show that the ratio of C:N:P mineralized microbiologically at the equilibrium condition is similar to the ratios of these three elements in humus.

The enzymes that cleave phosphorus from the more frequently encountered organic substrates are collectively called *phosphatases*. These enzymes catalyze the following reaction.

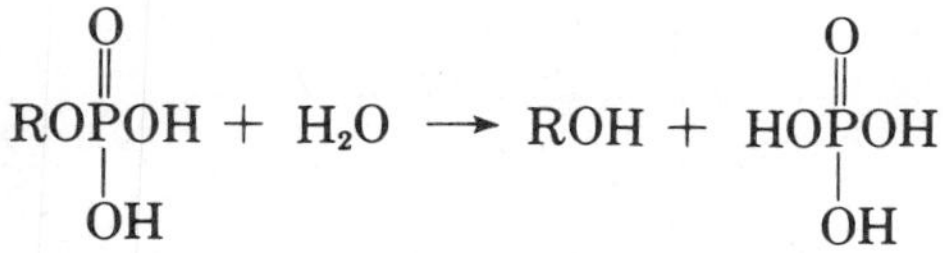

$$
\begin{array}{c}
\text{O} \\
\| \\
\text{ROPOH} + \text{H}_2\text{O} \rightarrow \text{ROH} + \text{HOPOH} \\
| \\
\text{OH}
\end{array}
\qquad\qquad
\begin{array}{c}
\text{O} \\
\| \\
\\
| \\
\text{OH}
\end{array}
\qquad (I)
$$

A single phosphatase, as a rule, may act on many different substrates; that is, R may have numerous structures. Thus, one enzyme may catalyze the cleavage of ethyl phosphate, glycerophosphate, and phenyl phosphate. On the other hand, molecules with two R groups (known as diesters, in contrast with the monoesters having only one) may require different enzymes for their breakdown.

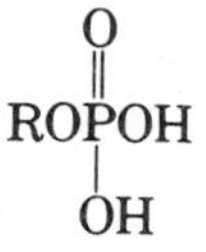

$$
\begin{array}{c}
\text{O} \\
\| \\
\text{ROPOH} \\
| \\
\text{OH} \\
\text{Phosphoric} \\
\text{monoester}
\end{array}
\qquad\qquad
\begin{array}{c}
\text{O} \\
\| \\
\text{ROPOR}' \\
| \\
\text{OH} \\
\text{Phosphoric} \\
\text{diester}
\end{array}
$$

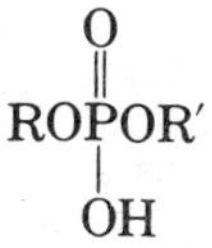

Phosphatases acting on phospholipids and hydrolyzing nucleic acids have diesters as their substrates. The enzymes catalyzing hydrolysis of the monoesters often have distinct optima at low or high pH ranges; because of the differences in pH for maximum activity, such enzymes are designated as acid or alkaline phosphatases. The abundance of organisms possessing these enzymes is shown in Table 20.2, the data being representative of certain soils.

The finding in soil of inositol phosphates with five or fewer phosphorus atoms per molecule suggests that a breakdown of the inositol hexaphosphate

TABLE 20.2

Relative Abundance of Microorganisms Able to Cleave Phosphate from Organic Compounds (8)

Substrate	Percent of Isolates Able to Cleave Substrate	
	In Soil	In Soil Near Roots
Glycerophosphate	22–26	18–36
Lecithin	11–20	12–38
Phytin	30	42
Phenolphthalein diphosphate	70–74	58–76

takes place. The enzyme phytase liberates phosphate from phytic acid or its calcium-magnesium salt, phytin, with the accumulation of inositol.

phytic acid   inositol

The enzyme acts on the hexaphosphate and removes the phosphates, presumably one at a time, to yield the penta-, tetra-, tri-, di-, and monophosphates and then finally free inositol. Some organisms cleave only the penta- or tetra- but not the hexaphosphate (5), and the enzymes that are formed differ also in that some species make intracellular phytase while others excrete an extracellular catalyst. Moreover, though some phytases are reasonably specific and act chiefly or solely on inositol phosphates, others are really nonspecific phosphatases removing phosphorus from dissimilar organic compounds.

Phytase activity is widespread, some 30 to 50 percent of the isolates from soil synthesizing the enzyme, and its activity in nature is enhanced by carbonaceous materials that increase the size of the community. Species of *Aspergillus, Penicillium, Rhizopus, Cunninghamella, Arthrobacter, Streptomyces, Pseudomonas,* and *Bacillus* can synthesize the enzyme. Yet, despite the great phytase potential, phytin is not readily metabolized in soil. The hydrolysis apparently is not limited by the phytase-producing capacity of microorganisms, which is appreciable, but by the small amount of phytic acid in the soil solution. The fact that phytate phosphorus added to soil is relatively unavailable to crops growing in acid conditions when compared with neutral environments is thus not a result of insufficient phytase synthesis but is a consequence of the small amount of phytic acid in the soluble phase of acid soils, where the substrate is bound into iron and aluminum complexes. There is little adsorption at neutral to alkaline pH values. Nevertheless, the degree of retention of inositol phosphates may not be the sole factor making such substrates unavailable to the microflora, and sorption or inactivation of phytase by clays may also lead to a low rate of breakdown (9). Not all of the phosphorus plants obtain from these substrates may be the result of actions of the subterranean community because roots, even under aseptic conditions, can use inositol hexaphosphate as a phosphorus source (17).

Pure nucleic acids added to soil are rapidly dephosphorylated (Figure 20.3). Indeed, a large number of different heterotrophs can develop in media containing nucleotides as the sole sources of carbon, nitrogen, and phosphorus. The mineralization is affected by pH, and the rate declines as the acidity increases. The transformation proceeds by an initial depolymerization of RNA by ribonuclease and DNA by deoxyribonuclease and subsequent cleavage of the phosphate from the products generated by the depolymerizing enzymes. The phosphorus in microbial cells, possibly because much is in the form of RNA and DNA, is liberated rapidly in some organisms although it is liberated slowly in others (18).

Bacteria, fungi, and actinomycetes are able to use phospholipids as a phosphorus source. Lecithin is the common substrate for the assessment of such actions. In the process, phosphate is cleaved from the organic compound, and it then is assimilated by the responsible populations. To the extent that phosphorus release exceeds the heterotrophic demand, some will be available for plant uptake too.

The phosphatase activity of many soils has been measured. For this purpose, a soil sample is incubated with a substrate such as glycerophosphate, β-naphthylphosphate, phenyl phosphate, or another organic compound, and the liberation of either inorganic phosphate or the organic portion of the substrate (e.g., β-naphthol or phenol) is determined. These assays show that the enzymatic activity is usually high and is affected by season, depth, and type of vegetation. The hydrolysis occurs at low and high pH, a possible indication of both acid and alkaline phosphatases.

Mycorrhizal fungi frequently have a dramatic and important effect on plants whose roots harbor these symbionts. For example, growth of plants in phosphorus-poor soils is markedly enhanced if the roots develop mycorrhizae as contrasted with those not bearing the fungus. The symbiotic association frequently allows for as extensive phosphate uptake in phosphorus-deficient environments as would occur on the addition of fertilizer phosphorus (11, 13).

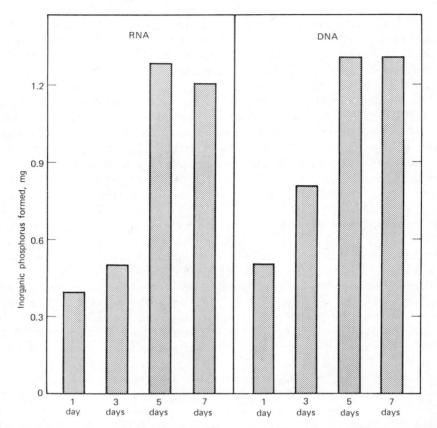

Figure 20.3. Formation of inorganic phosphate in soil amended with 20 mg of RNA and DNA (10).

Such an influence of the subterranean microsymbiont has been noted for corn, soybeans, onions, pine, and fir seedlings, to mention a few. The promotion of phosphate assimilation is so pronounced that inoculation with the fungus is a major consideration. Exactly how the mycorrhizae bring about the enhanced uptake and which soil constituents provide the element are not adequately understood. Much of the phosphorus may be derived from insoluble inorganic phosphates that the extensive hyphal network may be more able to exploit than the uninfected roots, but greater mineralization may also be involved.

IMMOBILIZATION

Microbial growth requires the presence of available forms of phosphorus. Because the element is essential for cell synthesis, the development of the microflora is governed by the quantity of utilizable phosphorus compounds in the habitat. In environments where phosphorus is limiting, its addition will therefore stimulate microbiological activities. Deficiencies are not frequently encountered although they may be induced artificially by the addition of carbohydrates. By and large, phosphate amendments have little effect on the microflora since the microscopic inhabitants appear to be highly efficient in mobilizing the large, natural reservoir of the element.

The assimilation of phosphorus into microbial nucleic acids, phospholipids, or other protoplasmic substances leads to the accumulation of nonutilizable forms of the element. Hence, during the decomposition of organic matter added to soil, the increase in microbial abundance puts a great demand on the phosphate supply. Consequently, should the carbonaceous residue be deficient in phosphorus, the microbial assimilation of available phosphate may depress crop yields; such decreases in yield can be prevented by the application of phosphatic fertilizers. As the decomposition of phosphorus-deficient substrates proceeds, the percentage of phosphorus in the decaying residue increases.

An indication of the microbial utilization of phosphate can be gained by comparing the growth and chemical composition of plants grown in the presence and absence of microorganisms. Thus, when barley is seeded into sterile and nonsterile samples of a soil with a low level of available phosphate, the yield and phosphorus content are less in the nonsterile samples (Table 20.3). This suggests that the subterranean community is competing with the cereal. The reduction in growth by the microflora in natural soil disappears if the phosphate supply is high (2).

Phosphorus, like nitrogen, is therefore both mineralized and immobilized. The process that predominates is governed by the percentage of phosphorus in the plant residues undergoing decay and the nutrient requirements of the responsible populations. Should the concentration exceed that required for

TABLE 20.3

Development of Barley in Sterile and Nonsterile Samples of a Phosphorus-Poor Soil (1)

Response Measured	Plant Part	Quantity per Plant (mg)	
		Sterile Soil	Nonsterile Soil
Dry matter	Shoot	370	240
	Root	240	120
P content	Shoot	0.35	0.23
	Root	0.14	0.06

microbial nutrition, the excess appears as inorganic phosphate; if inadequate for the microflora, the net effect is one of immobilization. Consequently, in the decomposition of substrates poor in phosphorus or which have a wide C:P ratio, a portion of the available nutrient supply is immobilized from the surroundings. As the ratio narrows with time because of CO_2 volatilization, phosphate will accumulate.

Many studies have been conducted of the mineral composition of bacterial and fungus cells. Commonly, phosphorus accounts for 0.5 to 1.0 percent of fungus mycelium and 1.0 to 3.0 percent of the dry weight of bacteria and probably of actinomycetes. However, microorganisms frequently exhibit a luxury consumption in culture media; that is, they assimilate excessive quantities, so that the values reported from cultural studies are undoubtedly too high for organisms growing in nature. Consider a hypothetical case of organic matter decomposition in which 100 parts of carbonaceous material containing 40 percent carbon are acted on by fungal, bacterial, and actinomycete populations. Using the phosphorus composition of microorganisms cited above and assuming further that 30 to 40, 5 to 10, and 15 to 30 percent of the substrate carbon is assimilated during decomposition by the fungi, bacteria, and actinomycetes, respectively, then the following calculations present a first approximation to the critical phosphorus content of organic materials:

$0.40 \times 100 = 40$ parts substrate C added
Fungi
  $(0.30$ to $0.40) \times 40 = 12$ to 16 parts cell C formed
  Fungi commonly contain 50 percent C
    hence $(12$ to $16)/0.50 = 24$ to 32 parts fungal mass
  $(0.005$ to $0.01) \times (24$ to $32) = 0.12$ to 0.32 parts
    of P required per 100 parts of organic matter

Similar calculations give values of 0.06 to 0.20 and 0.18 to 0.60 parts of phosphorus needed for the decomposition of 100 parts of substrate by bacteria and actinomycetes.

In actual trials in media containing glucose as the carbon source, *Aspergillus niger* assimilates 0.24 to 0.40 parts, *Streptomyces* sp. assimilates 0.27 to 0.63, while the figures for a mixed soil flora range from 0.16 to 0.36 parts of phosphorus for each 100 parts of glucose oxidized. With a carbohydrate like cellulose, from 0.35 to 0.45 parts of phosphorus are assimilated by the microflora for each 100 parts of cellulose (4, 14). Similarly, in a soil receiving a continuous supply of glucose, 0.37 parts of phosphorus are removed by the community for every 100 parts of the sugar (16). A value of 0.3 may be taken as an average figure for aerobic conditions provided the substrate is readily and completely metabolized. Anaerobically, microorganisms derive less energy from decomposition and, consequently, fewer cells are synthesized; hence, less phosphate need be immobilized for the same quantity of fermentable carbon.

Although an approximation of 0:3 percent may be generally valid for simple organic molecules, that is, phosphorus equivalent to 0.3 percent of the weight of the organic compound is required for the community to develop to its full extent, not all the carbon in natural products is readily available. In the decomposition of natural substrates, the quantity of phosphorus immobilized is diminished because less of the total carbon is degraded in a finite time interval. As a consequence, the amount of phosphorus immobilized during the decomposition of plant residues is closer to 0.2 percent of the dry weight of the organic matter. Therefore, the critical level of phosphorus in natural carbonaceous products that serves as a balance point between immobilization and mineralization is ca. 0.2 percent. If the substrate contains more phosphorus than the critical level, some is released. If the material has less than the critical level, less than needed by the microflora, phosphorus disappears from the environment as the net effect is one of immobilization. On the other hand, as the phosphorus-poor organic matter is decomposed and the phosphorus content of the residue increases, a point is eventually reached when the 0.2 percent figure is exceeded, and phosphate reappears.

As long as there is decay and microbial cell synthesis, both mineralization and immobilization are taking place. The phosphorus content governs not the absence of one or the other transformation but, instead, the greater rate of uptake or release of the nutrient. Nevertheless, few crop residues or animal manures bring about a net phosphorus immobilization. Only straw and certain similar materials are causes of biological phosphorus depletion. Because most plants contain 40 to 45 percent carbon on a dry weight basis, mineralization will be the net effect when the C:P ratio of crop remains is less than about 200:1 while immobilization predominates during the initial stages of decomposition when the C:P ratio of the added organic matter is greater than approximately 300:1.

The relatively constant organic C:organic P ratio in soil organic matter may be associated with the comparatively fixed phosphorus demand by the microflora for each unit of organic matter metabolized. Similarly, the N:P ratios for humus and for microbial cells are essentially identical at ca. 10:1. These observations on the composition of humus and of microorganisms are in conformity with the tenfold differences in critical levels in the immobilization-mineralization balance for nitrogen and for phosphorus, approximately 2.0 and 0.2 percent, respectively.

OXIDATION-REDUCTION REACTIONS

Phosphorus, like nitrogen, may exist in a number of oxidation states ranging from the -3 of phosphine, PH_3, to the oxidized state, $+5$, of orthophosphate. In contrast to nitrogen, however, little attention has been given to the inorganic transformations of phosphorus, but there is some evidence for biologically catalyzed changes in the oxidation state of this element too.

Biological oxidation of reduced phosphorus compounds is evident when phosphite is added to soil. The phosphite disappears with a corresponding increase in the concentration of phosphate.

$$HPO_3^= \rightarrow HPO_4^= \tag{III}$$

The conversion is brought about microbiologically since the reaction is eliminated on the addition of a biological inhibitor such as toluene. A number of heterotrophic bacteria, fungi, and actinomycetes utilize phosphite as sole phosphorus source in culture media and oxidize the phosphite within the cell to organic phosphate compounds. Bacteria utilize phosphate in preference to phosphite so that, in media containing both anions, the former disappears first. There is no evidence that the oxidation is capable of providing energy for the development of chemoautotrophic bacteria. Hypophosphite ($HPO_2^=$) can also be oxidized by heterotrophs in vitro to phosphate (12).

The possibility of the reverse process, a reductive pathway, has also received some attention. When certain soil samples are incubated anaerobically in a mannitol-$NH_4H_2PO_4$ medium, the phosphate disappears relatively rapidly. This decrease is not a result solely of assimilation, which can only account for a small proportion of the loss. Phosphate apparently is reduced to phosphite and hypophosphite.

$$H_3PO_4 \xrightarrow{2H} H_3PO_3 \xrightarrow{2H} H_3PO_2 \tag{IV}$$
$$+5 \qquad\qquad +3 \qquad\qquad +1 \text{ Oxidation state}$$

In the presence of nitrate or sulfate, phosphate reduction is retarded since the nitrate and sulfate seem to be more readily utilized as electron acceptors. Moreover, pure cultures of *Clostridium butyricum* and *Escherichia coli* form phosphite and hypophosphite from orthophosphate (24, 28). The process seems analogous biochemically to denitrification or to the bacterial conversion of sulfate to sulfide. However, by contrast with sulfate reduction, a process in which the final product is the most reduced state of sulfur (H_2S), no evidence exists that the ultimate product of phosphate reduction is phosphine (3). It is unlikely that the reduction takes place in well-aerated environments. This process may be only of limited practical significance even in waterlogged soils, but there is as yet too little information to provide definite answers.

REFERENCES

Reviews

Anderson, G. 1975. Other organic phosphorus compounds. In J. E. Gieseking, ed., *Soil components.* Springer-Verlag, New York, vol. 1, pp. 305–331.

Cosgrove, D. J. 1967. Metabolism of organic phosphates in soil. In A. D. McLaren and G. H. Peterson, eds., *Soil biochemistry.* Marcel Dekker, New York, vol. 1, pp. 216–228.

Halstead, R. L. and R. B. McKercher. 1975. Biochemistry and cycling of phosphorus. In E. A. Paul and A. D. McLaren, eds., *Soil biochemistry.* Marcel Dekker, New York, vol. 4, pp. 31–63.

Literature Cited

1. Barber, D. A. 1973. *Pestic. Sci.,* 4:367–373.
2. Benians, G. J. and D. A. Barber. 1974. *Soil Biol. Biochem.,* 6:195–200.
3. Burford, J. R. and J. M. Bremner. 1972. *Soil Biol. Biochem.,* 4:489–495.
4. Chang, S. C. 1940. *Soil Sci.,* 49:197–210.
5. Cosgrove, D. J., G. C. J. Irving, and S. M. Bromfield. 1970. *Aust. J. Biol. Sci.,* 23:339–343.
6. Daughtrey, Z. W., J. W. Gilliam, and E. J. Kamprath. 1973. *Soil Sci.,* 115:18–24.
7. Gerretsen, F. C. 1948. *Plant Soil,* 1:51–81.
8. Greaves, M. P. and D. M. Webley. 1965. *J. Appl. Bacteriol.,* 28:454–465.
9. Greaves, M. P. and D. M. Webley. 1969. *Soil Biol. Biochem.,* 1:37–43.
10. Greaves, M. P. and M. J. Wilson. 1970. *Soil Biol. Biochem.,* 2:257–268.
11. Hayman, D. S. and B. Mosse. 1971. *New Phytol.,* 70:19–27.
12. Heinen, W. and A. M. Lauwers. 1974. *Arch. Microbiol.,* 95:267–274.
13. Jackson, N. E., R. E. Franklin, and R. H. Miller. 1972. *Soil Sci. Soc. Amer. Proc.,* 36:64–67.
14. Kaila, A. 1949. *Soil Sci.,* 68:279–289.
15. Kowalenko, C. G. and R. B. McKercher. 1971. *Can. J. Soil Sci.,* 51:19–22.
16. Macura, J. and F. Kunc. 1965. *Folia Microbiol.,* 10:36–43.
17. Martin, J. K. 1973. *Soil Biol. Biochem.,* 5:473–483.
18. Mills, A. L. and M. Alexander. 1974. *J. Environ. Qual.,* 3:423–428.
19. Moyer, J. R. and R. L. Thomas. 1970. *Soil Sci. Soc. Amer. Proc.,* 34:80–83.
20. Omotoso, T. I. and A. Wild. 1970. *J. Soil Sci.,* 21:216–223.
21. Patrick, W. H., Jr., S. Gotoh, and B. G. Williams. 1973. *Science,* 179:564–565.
22. Pearson, R. W. and R. W. Simonson. 1939. *Soil Sci. Soc. Amer. Proc.,* 4:162–167.
23. Raghu, K. and I. C. MacRae. 1966. *J. Appl. Bacteriol.,* 29:582–586.
24. Rudakow, K. J. 1929. *Zent. Bakteriol.,* II, 79:229–245.
25. Sekhon, G. S. and C. A. Black. 1968. *Plant Soil,* 29:299–304.
26. Sperber, J. I. 1957. *Nature,* 180:994–995.
27. Thompson, L. M., C. A. Black, and J. A. Zoellner. 1954. *Soil Sci.,* 77:185–196.
28. Tsubota, G. 1959. *Soil Plant Food,* 5:10–15.

21
Microbial Transformations of Sulfur

Sulfur is an essential nutrient for members of the plant and animal kingdoms. Yet, despite its abundance in the earth's crust, sulfur is often present in soil in suboptimal quantities or in unavailable states so that responses to sulfur-containing fertilizers are not uncommon. The major reserve of the element in soil is the organic fraction, and the storehouse is only unlocked through biological decomposition. In the decomposition, there is a resemblance to the transformations affecting nitrogen availability, and the microscopic inhabitants are the sole agents converting the organic compounds of both elements into available, inorganic forms. The atmosphere also contains considerable sulfur from the burning of coal, the operation of factories, and even from microbiological action.

Sulfur in its various organic and inorganic forms is readily metabolized in soil. The dominance of one or another transformation is governed to a large extent by the environmental circumstances that affect the composition and activity of the microflora. Four distinct processes can be delineated: (*a*) decomposition of organic sulfur compounds, a process in which large molecules are cleaved to smaller units and the latter are then converted to inorganic compounds, (*b*) microbial assimilation or immobilization of simple compounds of sulfur and their incorporation into bacterial, fungal, or actinomycete cells, (*c*) oxidation of inorganic ions and compounds such as sulfides, thiosulfate, polythionates, and elemental sulfur, and (*d*) reduction of sulfate and other anions to sulfide.

The individual steps and generalized reactions of the biological sulfur cycle are presented diagramatically in Figure 21.1. The earth's vegetation gets the bulk of its sulfur from sulfate, but some may be obtained directly from the atmosphere. Animals, on the other hand, satisfy their demand for the element by feeding on plants or other animals. When incorporated into soil, the proteins

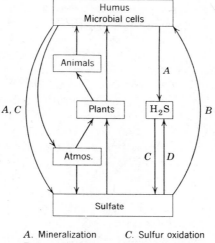

A. Mineralization   C. Sulfur oxidation
B. Immobilization   D. Sulfate reduction

Figure 21.1. The sulfur cycle.

of plant and animal tissues are hydrolyzed by the microflora to the amino acid stage. Sulfate and sulfide in turn accumulate following the microbiological attack on the amino acids and other sulfur-containing molecules. In aerated environments, the combined sulfur is ultimately metabolized to sulfate. Under waterlogged or other anaerobic circumstances, H_2S accumulates. Sulfide accumulation results in part from sulfate reduction and in part from the mineralization of organic sulfur. Between sulfide and sulfate in both the oxidative and reductive sequences are several intermediates, but these do not persist for extended periods, and their concentration in nature is usually low.

The transformations of sulfur resemble in many ways the microbial conversions of nitrogen. Because both elements are constituents of protoplasm, they must be assimilated during proliferation. Similarly, both nitrogen and sulfur are largely in organic combination in soil, and microbiological decomposition is therefore required to make the elements available. Inorganic sulfur compounds are oxidized in a fashion analogous to the nitrification of ammonium and nitrite, and the conditions necessary for the reduction of sulfate are quite similar to those for nitrate. The similarity is not limited to environmental factors but extends also to the physiology and biochemistry of the responsible organisms.

SOIL SULFUR

Sulfur enters the soil in the form of plant residues, animal wastes, chemical fertilizers, and rainwater. The element is also found as sulfide in several primary minerals. Occasionally, elemental sulfur is deliberately added for the control of certain plant pathogens or in the reclaiming of alkali soils. The quantity of volatile sulfur compounds returned to the earth's surface varies considerably. Figures for sulfur derived from the atmosphere range from 1 kg in parts of Africa, Australia, and New Zealand to more than 100 kg per hectare per year in industrialized sections of Europe and the United States.

A large part of the sulfur in the soil profile is in organic combination, usually half to three-fourths. The inorganic sulfate concentration is invariably low, typically accounting for less than one-tenth of the total sulfur present. A common characteristic of virgin as well as cultivated land is the ratio between organic sulfur, carbon, and nitrogen. Considering sulfur as unity, the C:S ratio of the organic fraction is approximately 100:1, and the total N:organic S ratio is approximately 10:1 in different soils and in the several horizons at a single site. Experimentally observed values, however, may range up to twofold greater or less for the C:organic S or the N:organic S ratios.

The soil organic fraction contains two groups of characterized components, ester sulfates and amino acids. The former are organic sulfates bearing C—O—S linkages, such molecules accounting for 20 to 65 percent of the total sulfur. Compounds of this class include choline sulfate, aromatic compounds such as tyrosine-O-sulfate, and sulfate-containing polysaccharides, although the actual substances in humus are unknown.

$$^-O_3SCH_2CH_2\overset{+}{N}(CH_3)_3$$

Choline
sulfate

$$^-O_3SO\!-\!\!\langle\ \rangle\!-\!CH_2\overset{\overset{\displaystyle NH_2}{|}}{C}HCOOH$$

Tyrosine-O-sulfate

The amino acids, which are largely bound into proteins or other polymers, are represented by cystine (or cysteine) and methionine. From about 5 to 35 percent of the sulfur is in this form. The identity of a fourth to half of the organic sulfur in humus remains unknown. Sulfate dominates the inorganic fraction providing that aeration is adequate, but sulfide, elemental sulfur, thiosulfate, and tetrathionate have also been observed in small amounts.

MINERALIZATION

Sulfur is taken up by the plant root system largely as the sulfate ion although several amino acids may be assimilated without prior degradation. Atmospheric sulfur dioxide supplies some of the element as well. Within plant tissues, however, the sulfate is reduced to the sulfhydryl (—SH) form. Since agricultural crops and other vegetation require for growth the sulfate found in their rooting medium, the mineralization of organic sulfur plays an important part in the microbiological reactions required for higher life.

A diverse group of organic compounds containing sulfur is presented as substrates to the microflora. The element occurs in plant, animal, and microbial proteins in the amino acids, cystine and methionine, and in the B vitamins, thiamine, biotin, and lipoic acid. It is also found in the tissues and excretory products of animals as free sulfate, taurine and, to some extent, as thiosulfate and thiocyanate.

Cystine	$HOOCCHNH_2CH_2SSCH_2CHNH_2COOH$
Cysteine	$HSCH_2CHNH_2COOH$
Methionine	$H_3CSCH_2CH_2CHNH_2COOH$
Taurine	$H_2NCH_2CH_2SO_2OH$
Thiosulfate	$^-SSO_3^-$
Thiocyanate	SCN^-

Upon the addition to soil of plant or animal remains, the sulfur contained therein is mineralized. A portion of the inorganic products is utilized by the microflora for cell synthesis, and the remainder is released into the environment. Aerobically, the terminal, inorganic product is sulfate. In the absence of atmospheric O_2, particularly during the putrefaction of proteinaceous matter, H_2S and the odoriferous mercaptans accumulate. The ability to form H_2S from partially degraded proteins is a property common to many genera of bacteria. Hence, it is likely that sulfides are among the major inorganic substances released during the decomposition of proteinaceous substrates under anaerobiosis.

The mineralization of sulfur in humus is often quite slow, and the rate is frequently not sufficiently rapid to satisfy the entire demand of growing plants. Several components of the soil organic fraction are attacked at similar rates,

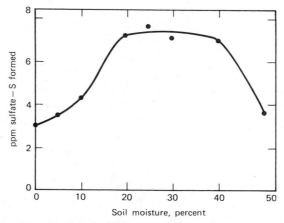

Figure 21.2. Role of moisture status in the mineralization of sulfur in soil (34).

moreover, so that no especially labile components appear to exist (15, 31). The mineralization of humus sulfur tends to be faster in the presence than in the absence of O_2, and the process is favored by increasing temperature in the mesophilic range and by the addition of lime to acid sites. As shown in Figure 21.2, moisture has a marked influence on the transformation. The lack of accumulation of significant amounts of free amino acids or other simple organic sulfur compounds may be taken as evidence that these products are degraded almost as rapidly as they are generated.

The sulfur in cystine and cysteine is recovered quantitatively as sulfate when either of these amino acids is applied to well-aerated soils. The conversion is rapid because many microorganisms attack the two compounds. The decomposition may proceed by any one of several known mechanisms. In soil, cystine can be formed by a chemical oxidation of added cysteine. The sulfur of the molecule in turn is oxidized to sulfate with possibly cysteine sulfinic acid as an intermediate, a reaction sequence not involving H_2S (14).

$$(I)$$

The cysteine sulfinic acid and cysteic acid proposed above as intermediates in many soils are oxidized to sulfite ($SO_3^=$) and sulfate by microorganisms in culture (29). If sulfite is released, it can be oxidized readily to sulfate, a reaction that proceeds even in the absence of microbial activity. The capacity to oxidize the sulfur in cystine to sulfate is not rare, and several fungi active in this conversion have been described. The sulfur in other compounds with the structure R-SH can similarly be converted to sulfate by heterotrophs. On the other hand, many bacteria in pure culture bring about a desulfhydration of cysteine by means of the enzyme cysteine desulfhydrase, which liberates equimolar quantities of pyruvic acid, H_2S, and NH_3.

$$HSCH_2CHNH_2COOH + H_2O \rightarrow CH_3COCOOH + H_2S + NH_3 \qquad (II)$$

The transformation of methionine frequently proceeds in an entirely different manner. Sulfate is sometimes produced when this amino acid is applied to soil, but often none is detected since the decomposition proceeds largely by way of volatile compounds. The added sulfur is lost through volatilization of methane thiol (CH_3SH) and dimethyl disulfide (CH_3SSCH_3). A similar conversion has been shown in culture media with several fungi and bacteria. The reaction appears to proceed as follows (28):

$$
\overset{\displaystyle \overset{NH_2}{|}}{H_3CSCH_2CH_2CHCOOH} \xrightarrow{-NH_3} CH_3SCH_2CH_2\overset{\displaystyle \overset{O}{\|}}{C}COOH \tag{III}
$$

$$
\tfrac{1}{2}CH_3SSCH_3 \longleftarrow CH_3SH \qquad CH_3CH_2\overset{\displaystyle \overset{O}{\|}}{C}COOH
$$

Other fungi can form sulfate from methionine. Although bacteria of numerous genera produce H_2S readily from peptone, cystine, and cysteine in culture, methionine is characteristically more resistant to attack.

As in the ammonification of organic nitrogen, the extent of mineral sulfur formation is influenced by the sulfur content and the C:S ratio of the decomposing substrate. Sulfate accumulates only when the sulfur level in the organic matter exceeds the microbial needs. In lieu of precise data, it can be proposed that the percentage of sulfur mineralized per annum is similar to the figure for nitrogen, that is, some 1 to 3 percent of the total supply in soils of the humid-temperate zone. It is also likely that environmental factors that govern microbial growth in general would affect the rate of sulfur mineralization.

MICROBIAL ASSIMILATION

Many compounds serve as sulfur sources for microbial growth, although any single strain may be limited to a few substances. Among the materials that may supply this element are sulfate, hyposulfite, sulfoxylate, thiosulfate, persulfate, sulfide, elemental sulfur, sulfite, tetrathionate, and thiocyanate of the inorganic substances and cysteine, cystine, methionine, taurine, and undecomposed proteins of the organic group. Sulfate is commonly included in culture media, but this anion is not known to be produced in environments devoid of O_2 so that anaerobes in soil probably assimilate reduced sulfur compounds. Indeed, various heterotrophs may be unable to make use of sulfate or sometimes any inorganic form of sulfur, and these are commonly cultivated in media with sulfur-containing amino acids. The sulfur content of most microorganisms lies between 0.1 and 1.0 percent of the dry weight, and the most conspicuous cellular constituents containing the element are the amino acids, cystine and methionine.

The microbiological immobilization of nitrogen is frequently of concern in field practice, but deficiencies of sulfur arising in the same way are uncommon. However, a deficiency can be induced in soil by treatment with carbohydrates. For example, the addition of cellulose may bring about a significant decline in sulfate levels as the populations growing on the polysaccharide assimilate the inorganic sulfur (Figure 21.3). Plants growing in such amended soils would generally suffer as a consequence of the immobilization, but this detrimental effect can be prevented if sulfate is applied. As long as there is less sulfur in the organic matter than required for microbial proliferation, immobilization will be dominant; when the element is in excess, mineral sulfur will be liberated as a waste product. The critical C:S ratio in carbonaceous materials above which immobilization is dominant to mineralization is reported to be in the range of 200:1 to 400:1 (4, 30), equivalent to about 0.1 to 0.2 percent sulfur. Mineralization will predominate with organic materials having lower C:S ratios or higher sulfur percentages, but immobilization will be more prominent with substances with wider C:S ratios or lower sulfur contents.

INORGANIC SULFUR OXIDATION

The inorganic compounds of sulfur that are transformed biologically represent various oxidation states from -2 of sulfide to $+6$ of sulfate. Not all of the

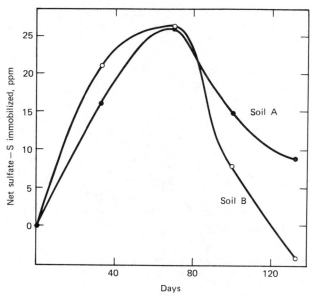

Figure 21.3. Changes in sulfate levels in two soils amended with cellulose (23).

reactions in soil are enzymatic, and many individual steps are nonbiological. In soil, sulfides, elemental sulfur, and thiosulfate can be oxidized slowly by chemical means, but the microbiological oxidation is far more rapid when conditions are favorable. At near-optimum moisture and temperature, chemical changes are insignificant in comparison to microbiological conversions.

The soil inhabitants capable of oxidizing inorganic sulfur compounds may either be autotrophs or heterotrophs. The bacteria using such molecules for energy are chiefly members of the genus *Thiobacillus*. This genus contains eight species. Of these, five have been the subjects of considerable investigation. *T. thiooxidans* is a strict chemoautotroph which oxidizes elemental sulfur and is capable of active growth at pH 3.0 or below. *T. thioparus*, on the other hand, is an acid-sensitive, obligate chemoautotroph. *T. novellus* cannot use elemental sulfur but will oxidize organic compounds as well as inorganic sulfur salts. The common species able to develop in the absence of O_2, *T. denitrificans*, uses nitrate as the electron acceptor in anaerobic conditions. The distinguishing trait of *T. ferrooxidans*, the fifth species, is its ability to use the oxidation of either ferrous or sulfur salts for energy. Elemental sulfur, sulfide, thiosulfate, tetrathionate, and thiocyanate serve as energy sources for one or more members of the genus, and CO_2 or bicarbonate supplies the carbon for chemoautotrophic growth. With the exception of *T. novellus*, the aforementioned species are all apparently obligate autotrophs, deriving no energy from the oxidation of organic carbon. The pH optima also serve to distinguish the five species. The optimum for *T. thiooxidans* and *T. ferrooxidans* is often in the vicinity of pH 2.0 to 3.5; the remaining three, *T. denitrificans*, *T. thioparus*, and *T. novellus*, prefer near-neutral or even slightly alkaline conditions.

Representatives of the group are generally obligate aerobes, the chief exception being *T. denitrificans*, which can utilize nitrate as the terminal electron acceptor. When grown thus anaerobically, the bacterium converts nitrate to gaseous nitrogen compounds and, at the same time, oxidizes thiosulfate or some other sulfur compound. In addition to the classical strains, thermophilic and halophilic variants of autotrophic thiobacilli are isolated with ease, and one species, *T. perometabolis*, is a heterotroph whose growth is enhanced as it oxidizes thiosulfate to sulfate (21).

The following equations typify the transformations catalyzed by the thiobacilli. The reactions shown are also carried out by other species of *Thiobacillus* than those listed.

T. thiooxidans and *T. novellus*

$$Na_2S_2O_3 + 2O_2 + H_2O \rightarrow 2NaHSO_4 \tag{IV}$$

T. thioparus

$$5Na_2S_2O_3 + 4O_2 + H_2O \rightarrow 5Na_2SO_4 + H_2SO_4 + 4S \tag{V}$$
$$Na_2S_4O_6 + Na_2CO_3 + \tfrac{1}{2}O_2 \rightarrow 2Na_2SO_4 + 2S + CO_2 \tag{VI}$$

T. thiooxidans

$$S + 1\tfrac{1}{2} O_2 + H_2O \rightarrow H_2SO_4 \tag{VII}$$

T. denitrificans

$$5S + 6KNO_3 + 2H_2O \rightarrow K_2SO_4 + 4KHSO_4 + 3N_2 \tag{VIII}$$

The abundance of chemoautotrophic sulfur-oxidizing bacteria can be measured by inoculating soil dilutions into mineral salts media containing inorganic sulfur compounds and observing the change in acidity. By such procedures, it has been shown that mineral soils usually have less than 100 or 200 thiobacilli per gram, but counts as high as 10,000 are occasionally encountered. In peats, the population is typically less than 500, and often 50 per gram is a maximum figure. The population, therefore, is never dense unless sulfur compounds are added deliberately.

Sulfur oxidation is not restricted to the genus *Thiobacillus*, for a number of other organisms carry out the same transformation. One of these is also an autotroph, but it is restricted to soils that are quite warm because they are in regions where the ground is subject to geothermal heating, as in Yellowstone National Park. The organism, placed in the genus *Sulfolobus*, oxidizes the elemental sulfur naturally present in these hot acidic environments and produces sulfuric acid at temperatures up to 85°C (25).

Heterotrophic bacteria, actinomycetes, and fungi also oxidize inorganic sulfur compounds. It is assumed that no energy is made available to the organism by such oxidations and that the transformations are incidental to the main metabolic pathways. For example, species of *Arthrobacter, Bacillus, Flavobacterium,* and *Pseudomonas* convert elemental sulfur or thiosulfate to sulfate (33), and species of *Streptomyces* are able to generate thiosulfate from elemental sulfur (36). Filamentous fungi and yeasts oxidize powdered sulfur, and several heterotrophic bacteria convert thiosulfate to tetrathionate in the presence of organic nutrients. As a rule, these reactions are slower than the corresponding thiobacillus step. Filamentous fungi produce sulfate from organic substrates such as cystine, thiourea, methionine, and taurine; the active genera are represented by *Aspergillus, Penicillium,* and *Microsporum,* but further investigation will undoubtedly disclose additional groups. The high rate of sulfate formation by these ubiquitous fungi, the many heterotrophic bacteria able to oxidize inorganic sulfur, and the few thiobacilli in many areas suggest that heterotrophs may be more important than chemoautotrophs in the production of sulfate from organic matter.

In culture, a wide variety of inorganic sulfur compounds are metabolized by one or another microorganism. The same is true in vivo; for example, soils treated with colloidal sulfur, thiosulfate, trithionate, or tetrathionate form sulfate after an initial lag period lasting for several days. These reactions are biological as they are abolished by microbial inhibitors. Sulfides are readily oxidized in soil although this process may not be entirely microbiological because sulfides can be converted to elemental sulfur by chemical means. It is interesting that tetrathionate appears prior to sulfate in soils perfused with thiosulfate, an observation suggesting that tetrathionate is a naturally occurring product of thiosulfate oxidation.

The oxidation of powdered sulfur produces considerable sulfuric acid, as shown by equation VII. The addition to soil of elemental sulfur is essentially equivalent to sulfuric acid application, such is the activity of the thiobacilli (Figure 21.4). At high rates of application, the pH of a neutral soil may fall to as low as pH 3 or sometimes even 2 after several months. *T. thiooxidans* is usually the major organism responsible, but *T. thioparus* and *T. denitrificans* metabolize free sulfur as well. The transformation is enhanced by decreasing the size of the sulfur particle and increasing the temperature through the mesophilic range, and it is affected by pH and water content. Such amendments are sometimes proposed to overcome the sulfur deficiency of crops, the slow oxidation to the available, sulfate form being advantageous because large quantities of sulfate added in one increment are easily lost by leaching.

Considerable effort has been devoted to establish how the thiobacilli make

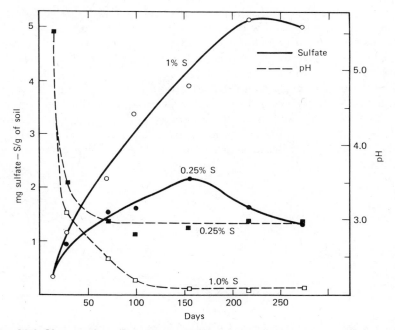

Figure 21.4. Changes in sulfate level and pH in soil receiving 0.25 and 1.0 percent elemental sulfur (1).

sulfate, yet controversy still exists about the pathway of oxidation. It is possible that dissimilar pathways function in the various species, and it is certain that the products accumulating in culture depend on the conditions of incubation. Compounding the difficulties in elucidating the reaction sequence is the likelihood that some of the presumed intermediates arise not directly in the oxidation but instead from reactions involving the actual intermediates and products outside the cell. Moreover, some of the thiobacilli do but others do not cause the accumulation of sulfur compounds during the oxidation of sulfide and elemental sulfur. Among the products that thus accumulate are thiosulfate and the polythionates. The polythionates may be represented by

$$^-O_3SS_nSO_3{}^-$$

where n is usually 1 (trithionate), 2 (tetrathionate), or 3 (pentathionate). Frequently, it is the tetra- but sometimes it is the tri- or pentathionate that accumulates in culture. Nevertheless, those of the compounds that appear in vitro are usually oxidized further to sulfate.

Some of the ways proposed for sulfate formation are given in Figure 21.5. One sequence is the conversion of elemental sulfur to sulfite, which is then oxidized to sulfate. A second assumes that some of the sulfite reacts with residual sulfur, giving rise to thiosulfate. In the third, the thiosulfate may be either cleaved to sulfite and sulfur or converted to tetrathionate; the latter can then be metabolized to sulfur or sulfite, which are then oxidized to sulfate. In the oxidation of sulfide, elemental sulfur may be generated in microbial cultures. These reactions are not the only ones that have been postulated, and others may even be more frequent or more important.

The thiobacilli can have a profound agricultural significance in several ways in addition to their possible role in the formation of the sulfate needed for plant nutrition. One such way is in the alteration of soil acidity to reduce the incidence or severity of potato scab or of the rot of sweet potatoes. These diseases are caused by acid-sensitive actinomycetes, *Streptomyces scabies* and

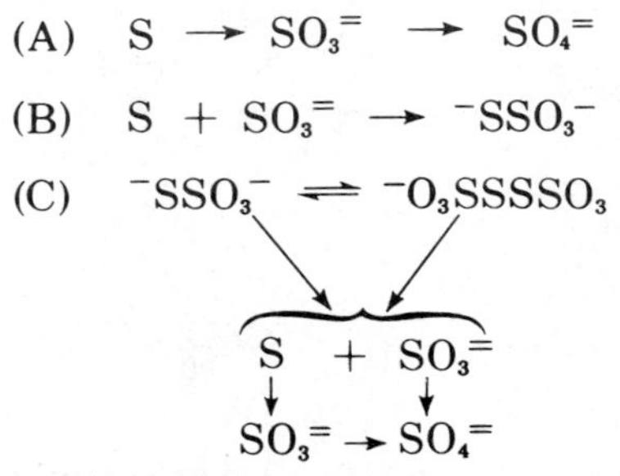

Figure 21.5. Proposed pathways for oxidation of inorganic sulfur by thiobacilli.

Streptomyces ipomoeae for the scab and rot, respectively. The diseases are not severe at reactions below ca. pH 5.0; therefore, the pathogens are often controlled by the addition of sulfur in quantities sufficient to bring the reaction to a point below the limiting level. With these streptomycete diseases, control is associated with the sulfuric acid formed by the thiobacilli. Similar treatments have been adapted to the reclamation of alkali land. If free sulfur is added to these soils, providing that thiobacilli are present, the sulfuric acid generated will neutralize the alkalinity and bring the soils into potential productivity. The corrosion of concrete may also result from the activities of *Thiobacillus*. Atmospheric H_2S is frequently the source of the element for the deterioration of concrete.

The oxidation of elemental sulfur causes a solubilization of soil minerals. The sulfuric acid formed reacts with minerals and other insoluble materials, leading to nutrient mobilization. Thus, the oxidation increases the quantity of soluble phosphate, potassium, calcium, manganese, aluminum, and magnesium. Indeed, manganese deficiency can be corrected by the application of sulfur or thiosulfate, treatments which increase the concentration of the divalent manganous ion. As discussed in Chapter 20, the composting together of colloidal sulfur, soil, and rock phosphate is one means of solubilizing phosphate, and composts of this type were at one time recommended for the preparation on the farm of available phosphate.

Thiobacilli are involved in the formation of acid sulfate soils. The occurrence of soils of these sorts is linked with the drainage of marine sediments, certain mangrove swamps, and other waterlogged lands. The sites are initially rich in sulfides, typically pyrite (FeS_2), but often FeS may be prominent as well. Upon drainage and exposure to O_2, the sulfides are oxidized to H_2SO_4 and the pH falls abruptly, usually to below pH 4.0. The sulfate concentration may be exceedingly high when the oxidation has run its course (6). The chemoautotrophs are often the chief agents of this change, but the process may also be largely chemical. Individual species of *Thiobacillus*—usually *T. ferrooxidans* or *T. thiooxidans*—also may participate in the oxidation of copper, zinc, lead, antimony, nickel, cobalt, or cadmium sulfides, commonly with the release of the soluble cation and sulfate and an increase in acidity. As with the iron sulfides, the transformations of the other sulfides may involve bacterial or nonbiological reactions or both, but the microorganisms characteristically accelerate greatly the slow chemical reactions. Pyrite and FeS have been proposed as sulfur fertilizers in deficient soils, the element being slowly released in a form that plants easily assimilate (Figure 21.6).

Soils in the vicinity of sulfur mines at times are polluted by the materials that are transported from the mines. In regions in which the natural deposits are exploited, the pH of the soil may fall to pH 4.0 or sometimes to less than 2.0, and the locality becomes largely free of higher plants. The microbial community is upset, and few bacteria and actinomycetes remain. The fungi,

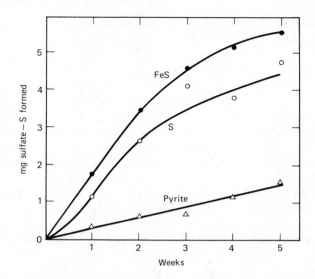

Figure 21.6. Formation of sulfate in soil amended with iron sulfides and elemental sulfur (5).

however, may flourish (19). At sites where sulfide-containing ores are mined and the residue is left behind with access to the air, the sulfides are converted to sulfate and bring about an acid condition. Water passing through such materials transports heavy metals downward, contaminating the groundwater. At the low pH prevailing in the remains of the mining operation, moreover, the acidity will solubilize aluminum, manganese, and other cations, and these too may be highly toxic and prevent the development of vegetation.

REDUCTION OF INORGANIC SULFUR COMPOUNDS

In soils that become deficient in O_2 as by flooding, the sulfide level increases to relatively high concentrations, often in excess of 150 ppm. At the same time, the sulfate concentration falls, and not infrequently a distinct zone of ferrous sulfide deposition appears in the profile. As these processes take place, there is a concomitant increase in the number of sulfate-reducing bacteria. The population of the sulfate reducers is commonly less than 10^4 and frequently less than 10^3 per gram, but the numbers may become greater than several million per gram after about two weeks in flooded soils. Much of the sulfide that accumulates originates by sulfate reduction, but the mineralization of organic sulfur compounds leads to the same product.

The formation of sulfide by sulfate reduction in nature is enhanced by increasing water levels, additions of organic materials, and rising temperatures. The process requires low oxidation-reduction potentials, which are characteristic of anaerobic habitats, and is usually limited to a range of pH values of 6.0 or above (9). This transformation is retarded by aeration and nitrate amendments, both of which allow for the maintenance of higher oxidation-reduction potentials than are associated with normally flooded land. Ferric or manganic salts also promote oxidizing conditions and hence delay sulfide production (10). Sulfide accumulation may be particularly pronounced in sulfate-rich saline areas, in which plant excretions may promote appreciable sulfate reduction.

The predominant microorganisms concerned with the reduction of sulfate are bacteria of the genus *Desulfovibrio*. These organisms are non-spore-forming, obligate anaerobes that produce H_2S from sulfate at a rapid rate. Although several species have been described, *Desulfovibrio desulfuricans* seems to be the most ubiquitous species in nature. Its pH range is narrow, and no growth occurs in media more acid than pH 5.5. This fact has a direct bearing on the lack of appreciable sulfide formation in many acid soils. A second but apparently less common group of sulfate reducers is made up of species of *Desulfotomaculum*. Members of this genus form spores, and individual strains are either mesophilic or thermophilic. They are similar to *Desulfovibrio* species in reducing sulfate to sulfide. Although the thermophilic habit is common to representatives of *Desulfotomaculum*, isolates of *Desulfovibrio* also may proliferate at elevated temperatures.

Desulfovibrio and *Desulfotomaculum* use sulfate and other forms of inorganic sulfur but not atmospheric oxygen or organic sulfur compounds as electron acceptors for growth. The electron donors or energy sources for the reaction include a number of carbohydrates, organic acids, and alcohols.

$$2CH_3CHOHCOONa + MgSO_4 \rightarrow H_2S$$
$$+ 2CH_3COONa + CO_2 + MgCO_3 + H_2O \quad (IX)$$

Some isolates of *D. desulfuricans* also utilize molecular hydrogen for the reduction of sulfate, sulfite, and thiosulfate with the consumption of four, three, and four moles of H_2 per mole of electron acceptor, respectively.

$$SO_4^= + 4H_2 \rightarrow S^= + 4H_2O \qquad (X)$$
$$SO_3^= + 3H_2 \rightarrow S^= + 3H_2O \qquad (XI)$$
$$S_2O_3^= + 4H_2 \rightarrow 2SH^- + 3H_2O \qquad (XII)$$

However, the bacteria are not H_2 autotrophs because they need organic molecules as sources of carbon. In addition to the anaerobic *Desulfovibrio* and *Desulfotomaculum*, there is evidence that *Bacillus, Pseudomonas,* and *Saccharomyces* strains liberate H_2S from sulfate, but the significance of the latter three organisms in soil has not been ascertained.

In spite of the few microorganisms adapted to reduce the most oxidized natural form of sulfur, sulfate, partially reduced inorganic sulfur compounds such as thiosulfate, tetrathionate, and sulfite are readily converted to sulfide by many bacteria, fungi, and actinomycetes. The active microorganisms are in no way unique, and numerous genera of aerobes and anaerobes are implicated. In each instance, an organic electron donor is required for the reduction of the inorganic sulfur compound.

The mechanism of H_2S formation from sulfate is still not fully resolved. The first step in the reduction is known to involve the conversion of sulfate to sulfite, a reaction that requires ATP. In the process, the ATP within the cell is metabolized to release two of its phosphorus atoms. The sulfite is then further reduced to sulfide, but disagreement remains in regard to whether thiosulfate or trithionate are intermediates (2, 8, 20). Some investigators feel they are in the pathway from sulfate, others believe they are not intermediates but may be products of side reactions. That thiosulfate and trithionate are often produced from sulfate, however, is beyond dispute. Three hypothetical pathways for the metabolism of sulfite are depicted in Figure 21.7: (a) a direct reduction to yield sulfide with no free sulfur products being formed, (b) an initial formation of thiosulfate, which is then cleaved to yield sulfide and regenerate some sulfite, and (c) an initial production of trithionate, which is subsequently converted to a mixture of thiosulfate and sulfite.

Microorganisms that reduce the availability of sulfate have a profound influence on soil fertility because they diminish the supply of the major sulfur source for agricultural crops. Beyond this fact, however, the sulfate-reducing bacteria can have a great economic influence. Low concentrations of the product of their metabolism are quite toxic to rice, citrus, and undoubtedly other crops and trees of practical importance, and under certain conditions linked with flooding or poor drainage, phytotoxicity may cause significant economic losses (12, 32). It is apparently the free H_2S that does injury to roots, and ferrous iron that precipitates the sulfide as FeS reduces or prevents the toxicity. The H_2S that *Desulfovibrio* makes may also be the cause of death of nematodes and fungi present in waterlogged regions (24, 27). These anaerobes may also assume prominence in saline soils of the arid zone. When these soils become flooded, as they occasionally do, the sulfate that is reduced by the bacteria results in the production of equivalent quantities of carbonates; the carbonates in turn precipitate calcium as $CaCO_3$, thereby reducing soil salinity (26).

Sulfate-reducing bacteria also apparently participate in the formation of

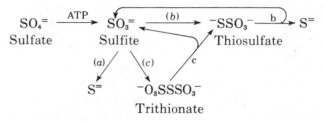

Figure 21.7. Possible pathways of sulfate reduction by *Desulfovibrio*.

sulfur deposits. It is hypothesized that *Desulfovibrio* species form sulfide from sulfate, and then the elemental sulfur is precipitated in the presence of O_2 either by chemical autooxidation or by biological means (18). Sulfides may also be implicated in the corrosion of stone and concrete, particularly in anaerobic tanks used for sewage treatment.

VOLATILE SULFUR COMPOUNDS

Attempts to characterize the global cycling and behavior of sulfur in the atmosphere have pointed to a hitherto unsuspected activity of the soil micro-flora. Models of the movement of sulfur on a global scale are designed to quantify the transport of compounds from the land mass and oceans to the atmosphere and back again and also to identify the substances involved. Most models that have been prepared suggest that substantially more sulfur is volatilized as a result of microbiological activity than comes from all pollution sources combined; that is, more than from the sum of that liberated by coal and petroleum combustion, petroleum refining, and smelting. The pollution sources give rise to SO_2, whereas it is assumed that the microbial product released to the atmosphere is H_2S. The latter product is postulated to arise from decomposing vegetation and organic materials undergoing decay in swamps, bogs, flooded soils, and those muds that exist near the water surface. Any H_2S that would be discharged to the atmosphere would be oxidized readily in the air to SO_2 so that microorganisms may indirectly make more SO_2 than that created by all industrial, mining, and combustion processes combined. However, the actual magnitude of H_2S discharge has not been evaluated experimentally, instead it is calculated by differences noted when attempting to balance the atmospheric sulfur cycle (16).

It has been argued, however, that the amounts of H_2S needed to balance the global atmospheric cycle are not present in the air. Moreover, the surface waters—from which much of the H_2S might be discharged to the atmosphere—are too oxidizing and would bring about the oxidation of H_2S before it is evolved. Interest has therefore focused on the possibility that dimethyl sulfide (CH_3SCH_3) may be a major product of microbial metabolism in nature (22). Gas chromatographic analysis has indeed shown that dimethyl sulfide is released from soil and decaying plant remains, but so too are a number of other volatile organic sulfur compounds. These include methane thiol (CH_3SH), dimethyl disulfide (CH_3SSCH_3), methyl thioacetate (CH_3SCOCH_3), ethane thiol (CH_3CH_2SH), diethyl disulfide ($CH_3CH_2SSCH_2CH_3$), carbon disulfide (CS_2), and carbonyl sulfide (COS) (3, 13). Some of the same compounds emanate in the decomposition of cattle manure and may account in part for the foul odor. At the present time, it is not yet certain which of the several compounds, H_2S or one or more of the organics, are major products evolved from soil, although the likeliest candidates remain dimethyl sulfide and H_2S. Nevertheless, it is clear that microorganisms are a previously unrecognized source of appreciable atmospheric sulfur.

The same metabolites that have been characterized as coming from soil have also been recognized as products in cultures of individual heterotrophs. In addition, compounds not yet detected in soils have also been found in vitro. Among the products thus observed are: (a) compounds with a type structure RSH where R is CH_3, CH_3CH_2, $CH_3CH_2CH_2$, and $CH_3CH_2CH_2CH_2$; (b) chemicals of type structure RSR where R is CH_3 or CH_3CH_2; (c) those of type structure $CH_3S_nCH_3$ where n is 1, 2, or 3; and (d) others, including SO_2 and COS. The organisms active in such reactions are species of *Clostridium* and *Pseudomonas* among the bacteria, *Aspergillus* and *Schizophyllum* among the fungi, and yeasts of the genus *Saccharomyces*.

Soils may also serve to remove sulfur compounds from the atmosphere. H_2S likely reacts in soil by a chemical process and hence disappears from the gas phase. Methane thiol can also be removed from air by soil, and it probably is a substrate for microbial populations. The absorption of SO_2 is especially rapid, but its removal from the air does not involve components of the microbiota (17). The ecological or environmental significance of soil as a means of ridding the air of volatile sulfur compounds has not been assessed as yet.

The volatile compounds may have an effect on individual microbial populations. Principal interest has focused on SO_2 or the bisulfite (HSO_3^-) formed from SO_2 in water. Lichens are remarkably sensitive to SO_2, often being inhibited by levels in the air of less than 0.1 ppm, and they have even been used as biological indicators of SO_2 pollution (11). Blue-green algae are more sensitive than other algae, and they are inhibited by far lower levels of bisulfite than are heterotrophic microorganisms (35). Nitrification may be retarded by methane thiol, dimethyl sulfide, and H_2S (7).

REFERENCES

Reviews

Anderson, G. 1975. Sulfur in soil organic substances. In J. E. Gieseking, ed., *Soil components.* Springer-Verlag, New York, vol. 1, pp. 333–341.

Freney, J. R. and R. J. Swaby. 1975. Sulphur transformations in soils. In K. D. McLachlan, ed., *Sulphur in Australasian agriculture.* Sydney Univ. Press, Sydney, pp. 31–39.

LeGall, J. and J. R. Postgate. 1973. The physiology of sulphate-reducing bacteria. *Advan. Microbial Physiol.*, 10:81–133.

Roy, A. B. and P. A. Trudinger. 1970. *The biochemistry of inorganic compounds of sulphur.* Cambridge Univ. Press, New York.

Literature Cited

1. Adamczyk-Winiarska, A., M. Krol, and J. Kobus. 1975. *Plant Soil,* 43:95–100.
2. Akagi, J. M., M. Chan, and V. Adams. 1974. *J. Bacteriol.,* 120:240–244.

3. Banwart, W. L. and J. M. Bremner. 1975. *Soil Biol. Biochem.,* 7:359–364.
4. Barrow, N. J. 1960. *Aust. J. Agr. Res.,* 11:960–969.
5. Barrow, N. J. 1971. *Aust. J. Exptl. Agr. Anim. Husb.,* 11:217–22.
6. Bloomfield, C. and J. K. Coulter. 1973. *Advan. Agron.,* 25:265–326.
7. Bremner, J. M. and L. G. Bundy. 1974. *Soil Biol. Biochem.,* 6:161–165.
8. Chambers, L. A. and P. A. Trudinger. 1975. *J. Bacteriol.,* 123:36–40.
9. Connell, W. E. and W. H. Patrick, Jr. 1968. *Science,* 159:86–87.
10. Engler, R. M. and W. H. Patrick, Jr. 1973. *Soil Sci. Soc. Amer. Proc.,* 37:685–688.
11. Ferry, B. W., M. S. Baddeley, and D. L. Hawksworth, eds., 1973. *Air pollution and lichens.* Athlone Press, Univ. of London, London.
12. Ford, H. W. 1973. *J. Amer. Soc. Hort. Sci.,* 98:66–68.
13. Francis, A. J., J. M. Duxbury, and M. Alexander. 1975. *Soil Biol. Biochem.,* 7:51–56.
14. Freney, J. R. 1967. In A. D. McLaren and G. H. Peterson, eds., *Soil biochemistry,* Marcel Dekker, New York, vol. 1, pp. 229–259.
15. Freney, J. R., G. E. Melville, and C. H. Williams. 1975. *Soil Biol. Biochem.,* 7:217–221.
16. Friend, J. P. 1973. In S. I. Rasool, ed., *Chemistry of the lower atmosphere.* Plenum Press, New York, pp. 177–201.
17. Ghiorse, W. C. and M. Alexander. 1976. *J. Environ. Qual.,* 5:227–230.
18. Jones, G. E. and R. L. Starkey. 1957. *Appl. Microbiol.,* 5:111–118.
19. Krol, M., W. Maliszewska, and J. Siuta. 1972. *Polish J. Soil Sci.,* 5:25–33.
20. Lee, J.-P., J. LeGall, and H. D. Peck, Jr. 1973. *J. Bacteriol.,* 115:529–542.
21. London, J. and S. C. Rittenberg. 1967. *Arch. Mikrobiol.,* 59:218–225.
22. Lovelock, J. E., R. J. Maggs, and R. A. Rasmussen. 1972. *Nature,* 237:452–453.
23. Massoumi, A. and A. H. Cornfield. 1965. *J. Sci. Food Agr.,* 16:565–568.
24. Mitchell, R. and M. Alexander. 1962. *Soil Sci.,* 93:413–419.
25. Mosser, J. L., A. G. Mosser, and T. D. Brock. 1973. *Science,* 179:1323–1324.
26. Ogata, G. and C. A. Bower. 1965. *Soil Sci. Soc. Amer. Proc.,* 29:23–25.
27. Rodriguez-Kabana, R., J. W. Jordan, and J. P. Hollis. 1965. *Science,* 148:524–526.
28. Ruiz-Herrera, J. and R. L. Starkey. 1970. *J. Bacteriol.,* 104:1286–1293.
29. Stapley, E. O. and R. L. Starkey. 1970. *J. Gen. Microbiol.,* 64:77–84.
30. Stewart, B. A., L. K. Porter, and F. G. Viets, Jr. 1966. *Soil Sci. Soc. Amer. Proc.,* 30:355–358.
31. Tabatabai, M. A. and J. M. Bremner. 1972. *Agron. J.,* 64:40–44.
32. Takai, Y. and T. Kamura. 1966. *Folia Microbiol.,* 11:304–313.
33. Vitolins, M. I. and R. J. Swaby. 1969. *Aust. J. Soil Res.,* 7:171–183.
34. Williams, C. H. 1967. *Plant Soil,* 26:205–223.
35. Wodzinski, R. S., D. P. Labeda, and M. Alexander. Unpublished observations.
36. Yagi, S., S. Kitai, and T. Kimura. 1971. *Appl. Microbiol.,* 22:157–159.

22

Microbial Transformations of Iron

Despite the fact that it is only a minor nutrient for the growth of most of the microscopic life of the soil, iron is an element that readily undergoes transformation through the activity of the microflora. Iron is always abundant in terrestrial habitats, and it is one of the major constituents of the earth's crust. Yet, the element is often in a form unavailable for plant utilization, and serious deficiencies are occasionally encountered.

Microorganisms are implicated in the transformations of iron in a number of distinctly different ways, and the form of the element may be affected through a variety of biological means. (*a*) Certain bacteria are able to oxidize ferrous iron to the ferric state, the latter precipitating as ferric hydroxide. (*b*) Many heterotrophic species attack soluble organic iron salts, and the iron, now in an inorganic and only slightly soluble form, is precipitated from solution. (*c*) Microorganisms alter the oxidation-reduction potential of their surroundings. Decreases in the oxidation-reduction potential resulting from microbial growth lead to the formation of the more soluble ferrous from the highly insoluble ferric ion. (*d*) Innumerable bacteria and fungi produce acidic products such as carbonic, nitric, sulfuric, and organic acids. Increases in acidity bring iron into solution. (*e*) Under anaerobiosis, the sulfide formed from sulfate and organic sulfur compounds may remove iron from solution as ferrous sulfide. (*f*) The liberation by microorganisms of certain organic acids and other carbonaceous products of metabolism often results in the formation of soluble organic iron complexes. This process is the reverse of *b*.

Iron may thus be precipitated in nature by iron-oxidizing bacteria, the action of heterotrophs in decomposing the organic moiety of salts of the metal, the liberation of O_2 by algae, and the creation of an alkaline reaction. Conversely, solubilization may occur through acid formation, the synthesis of certain organic products, or by the creation of reducing conditions. Chemically, the ferrous ion predominates in solution below pH 5 while the ferric ion is favored above pH 6. Even in culture media containing soluble ferrous salts, the formation of alkaline products causes an oxidation and, therefore, a precipitation of the metal as ferric iron. On the other hand, if the oxidation-reduction potential falls below 0.2 volts, most of the iron will be found in the ferrous state. At potentials greater than 0.3 volts, the ferric ion is the major form. Hence, an increase in reducing intensity leads to the accumulation of the soluble ferrous iron; the reverse occurs when the environment becomes more oxidized. The effects of O_2, oxidation-reduction potential, and pH on growth and on the ferrous-ferric equilibrium make it often difficult to determine whether the process of iron precipitation is enzymatic or merely chemical. In the latter case, the surface of the cell may serve as a selective accumulating structure for the ferric iron that is generated by nonbiological reactions.

FERROUS OXIDATION

One way of investigating the microbiological oxidation of ferrous iron is to add an appropriate salt to sterile media or quartz sand and measure the rate of ferrous oxidation when soil is introduced. As the chemical oxidation is appreciable at near-neutral pH, the nonbiological changes are estimated by means of controls containing soil inocula subjected to sterilization. The results of such studies suggest that normal soil effects a greater change than the sterile sample, that is, an apparent microbial ferrous oxidation. The data are, even in the best of circumstances, difficult to evaluate because of the dominant nonbiological action.

An unequivocal demonstration of the biological oxidation of ferrous salts in soil has been provided by Gleen (9), who used the fact that ferrous iron is stable at pH 3.0. An oxidation at the acid reaction, therefore, must be catalyzed by living agents. By perfusing soil with $FeSO_4$ solution at pH 3.0, Gleen showed that the soluble ferrous iron was oxidized to the ferric form, the latter precipitating on the soil column. An initial lag was observed in the rate of oxidation during the $FeSO_4$ perfusion, but no such lag period was noted on subsequent treatment with a fresh solution (Figure 22.1). The lag in the first treatment and its elimination on the addition of fresh substrate are indications of a microbiological conversion, the initial lag reflecting the time necessary for the populations to become sufficiently large to produce chemically detectable changes and the subsequent absence of the lag indicating that the soil has been enriched with an active population. That the oxidation is biological was further demonstrated by adding low concentrations of a poison like sodium azide, which eliminated the transformation.

In bituminous coal deposits, an interesting process takes place through the metabolism of iron-oxidizing bacteria. In such deposits, vast quantities of sulfuric acid are produced microbiologically, the acid entering the mine waters and creating a highly corrosive drainage effluent. The acid mine effluent

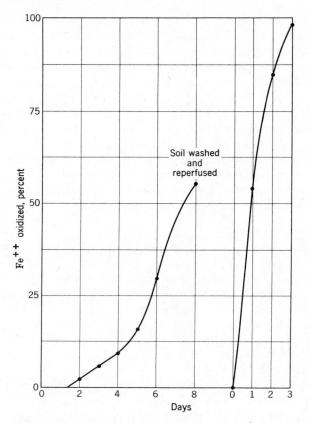

Figure 22.1. Ferrous oxidation in soil maintained at pH 3.0. After the initial oxidation was completed, the soil column was washed and fresh FeSO₄ added (9).

frequently completely destroys the aquatic life and makes the vegetation along adjacent stream banks scarce. In addition to sulfuric acid, a considerable amount of ferrous iron is present in the mine waters, both the iron and the sulfate being derived from iron disulfides in the coal veins. It has been estimated that the bituminous mines of western Pennsylvania alone add about a million tons of sulfuric acid each year to the Ohio River drainage area (13).

Pyrite, a typical iron disulfide, is slowly oxidized by nonbiological means, yet the addition of a small amount of acid mine water brings about a rapid liberation of iron from the sulfide ore. This activity is eliminated by sterilization and is negligible at 0°C. Paralleling the release of iron into solution is an increase in acidity and the appearance of sulfate (8). The chief organism responsible for the biological release of iron from sulfide ores is *Thiobacillus ferrooxidans*. The observation that this bacterium oxidizes ferrous salts at pH 3.5

in the absence of organic materials (Figure 22.2) is clear evidence of the autotrophic nutrition of the organism, a fact difficult to establish for the many microorganisms that contain deposits of ferric salts on their cells when they are growing in solutions at neutral pH. At pH values near neutrality, the ferrous iron is readily oxidized nonbiologically and may precipitate on microbial cells. Ferrous oxidation by the chemoautotrophic *T. ferrooxidans* proceeds at about pH 2.0 to 4.5 with an optimum often in the vicinity of 2.5 to 3.5, a pH range that indicates the organism will only be active in acid soils. Different strains have different pH ranges and optima, but all are restricted to acid conditions.

The reaction that yields energy to support the proliferation of *T. ferrooxidans* is expressed in several ways. The energy is obtained by an aerobic process in which ferrous iron is oxidized to the ferric form.

$$4Fe^{++} + O_2 + 4H^+ \rightarrow 4Fe^{+3} + 2H_2O \tag{I}$$

Inasmuch as ferric sulfate may be the main product, the energy-yielding process is sometimes written to give that molecule.

$$4FeSO_4 + O_2 + 2H_2SO_4 \rightarrow 2Fe_2(SO_4)_3 + 2H_2O \tag{II}$$

The organism is often coated with ferric hydroxide, and this possibly may arise by a nonbiological reaction.

$$Fe_2(SO_4)_3 + 6H_2O \rightarrow 2Fe(OH)_3 + 3H_2SO_4 \tag{III}$$

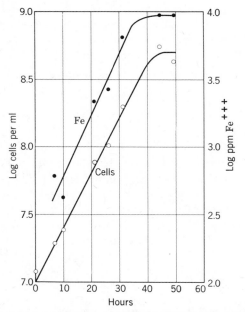

Figure 22.2. Growth and ferrous oxidation by *Thiobacillus ferrooxidans* (18).

The bacterium is able to grow and derive energy by oxidizing ferrous iron but, as its genus name indicates, also by the oxidation of inorganic sulfur. Sulfide, sulfur, and thiosulfate can be used by many strains (24). Indeed, in the oxidation of ores containing both iron and sulfide, as with pyrite and chalcopyrite ($CuFeS_2$), the autotroph is often capable of oxidizing both ferrous and sulfide ions (8). Because the energy yield in oxidizing iron is small, about 10 kcal per gram atom of iron (55.8 g), appreciable substrate turnover is necessary for cell synthesis. As transformation of the ores progresses, not only are the iron and sulfur contained therein made soluble but so too may be the copper, arsenic, nickel, cobalt, and other elements present in the particulate material.

Considerable debate exists on the precise role of *T. ferrooxidans* in the release of sulfate and iron and in the formation of acid from pyrite. Resolution of the problem is made difficult because FeS_2 is also oxidized solely by chemical means. Without question, these thiobacilli markedly promote the process, an enhancement that could result from their increasing either the rate of the overall conversion or of one step in a reaction sequence having several steps. A series of reactions that may account for the changes noted in nature may involve an initial oxidation of pyrite or other metal sulfide, a process that might be microbial or nonbiological.

$$2FeS_2 + 7O_2 + 2H_2O \rightarrow 2FeSO_4 + 2H_2SO_4 \qquad \text{(IV)}$$

The next step might then be the enzymatic oxidation shown in equation II, and that transformation would then be followed by the chemical reaction given in equation III. Alternatively, the ferric iron generated in equation II might spontaneously oxidize the metal sulfide and in turn be reduced to the ferrous form, the ion that the bacteria once again oxidize.

$$14Fe^{+3} + FeS_2 + 8H_2O \rightarrow 15Fe^{+2} + 2SO_4^= + 16H^+ \qquad \text{(V)}$$

Heterotrophs also are associated with ferric iron precipitation, but often it is unclear whether the reaction leading to the appearance of ferric iron in such instances is enzymatic or not. Little or none of the energy released in the oxidation is utilized by the heterotrophs. *Metallogenium*, for example, deposits ferric hydroxide on its flexible filaments; strains of this heterotrophic bacterial genus are able to catalyze iron oxidation from pH 3.5 to 5.0, thereby appreciably enhancing the very slow abiotic oxidation (26). However, many organisms may merely present surfaces onto which are adsorbed the ferric precipitates appearing spontaneously in nonacidic areas containing ferrous iron (14). The iron then encrusts the cells or filaments. Such a process sometimes is conspicuous in drains placed in fields to allow for better water relations in the soil. Thus, ferric hydroxide may appear in copious amounts in tile drainage systems installed in poorly drained land. The ferrous iron in the originally O_2-deficient sites is oxidized nonbiologically, and the ferric oxide precipitating on the heterotrophic bacteria—possibly together with occluded organic matter—

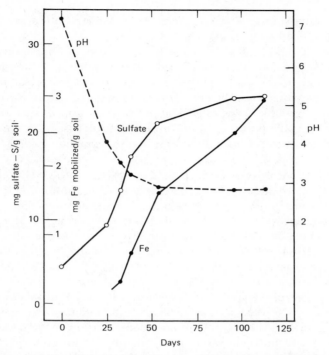

Figure 22.3. Changes in acidity and products of the oxidation by *Thiobacillus* in a pyrite-containing soil (23). By permission of the Oxford University Press.

creates a mass so large that the tile drain fails to operate properly (20). By contrast, in drains installed in acid sulfate soils containing pyrite undergoing oxidation, the ferric iron-containing deposits impeding water movement may be a consequence of the action of *T. ferrooxidans* (23). The changes in pH, ferrous iron, and sulfate levels in pyrite-rich soil exposed to O_2 are depicted in Figure 22.3.

DECOMPOSITION AND FORMATION OF ORGANIC IRON COMPOUNDS

Cationic iron forms complexes with both simple and highly complex molecules, and substantial quantities of the element are bound as iron-organic complexes in soil. The identities of the compounds or complexes are uncertain, but model studies in the laboratory have shown that sugars, simple organic acids, and highly polymerized humus constituents can combine with cationic iron to yield a variety of complexes. Organically bound iron is available for microbial attack, however, because many heterotrophic species can degrade the complexes and release the bound iron. Research has focused on the simple iron-containing compounds not because of their greater importance but rather because the chemistry of the substrate is known.

The precipitation of the iron found in certain water-soluble organic compounds is a major means of altering the availability of the element. The organic portion of the molecule provides energy for microbial proliferation, and as the carbonaceous moiety is decomposed, the iron is released and precipitates as insoluble ferric salts. The precipitation thus results from a direct action on the organic portion of the compound rather than on the iron.

When a sample of soil is added to a solution containing ferric ammonium citrate, the decomposition of the citrate portion of the molecule results in a rapid accumulation of ferric hydroxide. The reaction proceeds in the presence or absence of air. In addition to its release from ferric ammonium citrate, iron is precipitated from solutions of the citrate, lactate, acetate, malate, malonate, oxalate, and gallate salts. The responsible microorganisms, chiefly bacteria, are found abundantly in soil, and strains of diverse genera can in this manner remove iron from solution by attacking the organic portion of the salts. Representatives of the bacterial genera *Pseudomonas, Bacillus, Serratia, Acinetobacter, Klebsiella, Mycobacterium,* and *Corynebacterium,* several types of filamentous fungi, and species of *Nocardia* and *Streptomyces* are active in the conversion. As demonstrated by the data in Table 22.1, the abundance of organisms varies with the soil and with the identity of the organic portion of the molecule. Plants growing in sterile nutrient solutions containing ferric complexes may fare poorly because the iron is only slowly available, but the microbial cleavage of these compounds releases the element and thereby favors development of the plant (2).

Iron-organic compounds characteristic of humus may be attacked by a novel group of bacteria. These organisms are members of the genera *Pedomicrobium, Metallogenium,* and *Seliberia,* and their action results in the deposition of

TABLE 22.1

Abundance of Microorganisms in Three Soils Capable of Degrading Iron-Organic Complexes (6)

	Substrate	
Soil	Ferric Citrate	Ferric Oxalate
	Number per gram of soil, $\times\ 10^3$	
Calcareous brown soil	4400	14.6
Forest mollisol	950	350
Spodosol	4.4	24

ferric hydroxides that coat the bacteria (1, 25). *Pedomicrobium* reproduces by buds that appear at the ends of extensions of the organism. *Metallogenium* cells are coccoidal, and these cells produce flexible filaments with tapered ends. In *Seliberia,* by contrast, the rod-shaped cells are twisted in a spiral, and the bacteria are characterized by the star-shaped clusters they produce.

The iron in microbial cells may also be complexed with organic molecules, but these compounds undoubtedly are mineralized with the liberation of the element. The element may also combine with microbial or plant polysaccharides, a combination that sometimes renders the polysaccharide resistant to degradation (15).

Organic iron complexes appear to have a role in iron movement downward through certain kinds of profiles. The microflora produces an array of simple organic acids, amino acids, and other metabolites able to bind and solubilize insoluble forms of iron. If the iron that is solubilized originated in the surface horizon, then the water-soluble complexes could easily move downward with the passage of water. In the lower horizons, the soluble organic iron complex might be degraded by the many heterotrophs possessing this sort of biochemical activity, and the iron so released would precipitate from solution.

Iron in rocks and minerals may also be solubilized by a variety of populations. Lichens, for example, will weather rocks and bring about the slow but still significant destruction of the rock's structure. In the process, which is in part linked with the solubilizing action of unique products of the lichen, a portion of the iron is brought into solution (27). Many fungi can solubilize a considerable part of the iron present in minerals when these minerals are included in culture media, and often this dissolution results merely from the increase in acidity associated with growth (19). Bacteria also carry out analogous conversions. However, sometimes release of the element may be a consequence of particular metabolites rather than being merely a reflection of changes in acidity; thus, bacteria excreting 2-ketogluconic acid solubilize iron from a variety of minerals provided to them in laboratory media (7). Nevertheless, the significance of these transformations in nature has yet to be established.

IRON REDUCTION

In soils that are well-drained, most of the iron occurs in the higher oxidation state, and only small amounts of the ferrous ion are found. Should the soil become waterlogged or otherwise subjected to anaerobiosis, its ferrous iron content rapidly rises. This process is almost entirely the result of biological agencies as little or no change occurs in sterile waterlogged soil (3) or if the soil is treated with a potent metabolic inhibitor such as KCN. The processes suggested by the results in Figure 22.4 are typical of the transformations in waterlogged habitats—the drop in oxidation-reduction potential (E_h) and the rapid reduction of iron in the period following flooding. Manganous ions characteristically appear before the ferrous ions. Treatment with organic

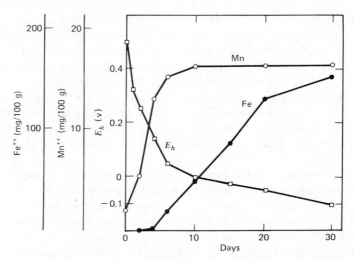

Figure 22.4. Changes in concentration of ferrous and manganous ions and oxidation-reduction potential in a soil after flooding (22).

matter enhances the reduction, and the quantity of ferrous iron appearing in the soil solution is directly related to the amount of fermentable substrate added. In terms of the oxidation-reduction potential, ferrous becomes prominent at E_h values below ca. 0.2 volt during periods of intense microbiological action. In acidic areas, not as low an E_h value needs to be achieved in order for the water-soluble ferrous form to be prominent (10). When the water status of poorly drained soils is improved, there is a reversion to the ferric state.

Several mechanisms may account for the microbiological ferric reduction and the stimulatory effect of fermentable substrates. For example, an increase in acidity accompanying fermentation favors iron mobilization. Furthermore, the depletion of O_2 as a consequence of microbial metabolism will tend to lower the E_h and lead to ferric reduction. Another possible mechanism for the transformation is the direct reaction of fermentation products with ferric hydroxides and oxides. Alternatively, reduction may be the result of electron transport, the iron functioning as an electron acceptor in cell respiration in a manner analogous to the reduction of nitrate by denitrifying bacteria.

Many bacteria, when growing in organic media at suboptimal O_2 levels, bring a portion of the added $Fe(OH)_3$ or ferric oxide into solution. Transformations resulting in ferrous production are not peculiar to any single genus but are attributes of a variety of organisms, and as many as 10^4 to 10^5 or sometimes 10^6 bacteria per gram of soil have the capacity to reduce iron actively. On occasion, as many as 10 percent of the bacterial colonies appearing on agar plates prepared from soil dilutions are associated with the reduction. Among the genera containing species able to convert ferric to ferrous iron are *Bacillus,* *Clostridium, Klebsiella, Pseudomonas,* and *Serratia*. Some of the isolates produce divalent iron only in media that become distinctly acid so that, for these at least, pH is probably responsible for the transformation. With others, no increase in acidity occurs so that another mechanism of solubilization is operative. With some aerobes, the presence of ferric salts may permit slight growth in the absence of O_2, as in the instance of *Fusarium oxysporum* (11).

The conversion of tri- to divalent iron by at least certain heterotrophs seems to be enzymatic, the ferric ion probably serving in respiration as an electron acceptor. The trivalent cation may thus substitute for O_2 in cellular metabolism when O_2 is no longer available. In culture, nitrate suppresses the iron-reducing capacity of many of the active isolates. In addition, many but not all species that reduce ferric ions also can convert nitrate to nitrite. On the basis of these findings, it has been postulated that the enzymatic reduction of iron proceeds by one of two mechanisms: (*a*) the reaction is catalyzed in some heterotrophs by the same enzyme concerned with the production of nitrite from nitrate—nitrate reductase—or (*b*) the transformation may involve an enzyme not functioning in nitrate metabolism (12, 17).

A phenomenon possibly associated with the microbial metabolism of iron is known as *gleying*. Sites in the soil profile that have undergone gleying are sticky and exhibit a gray or light greenish-blue coloration. Gleys are common where the water table is high, and such horizons are characteristically associated with waterlogged areas. The color of the gleyed zone is attributed to the ferrous sulfide produced under anaerobiosis by the reaction of the end-products of the microbial reduction of sulfate and iron. In a model system designed to simulate gleying, a clay soil is incubated with a sugar solution under partial or complete anaerobiosis; as a result of bacterial action, the clay becomes bleached, and iron is observed in the fermentation liquid. The color of the bleached clay varies with the soil; sometimes it is white, sometimes gray, and occasionally brown. In flooded soils treated with glucose, the rate of disappearance of the sugar and the formation of ferrous iron give the sigmoid curves characteristic of bacterial growth, and the most rapid rates of both processes occur at the same time. This suggests that bacteria are the responsible agents, and gleyed clays may indeed contain as many as 10^7 iron-reducing bacteria per gram. In the gleyed sites that have been investigated, the predominant aerobic and facultatively anaerobic iron reducers are *Bacillus* and *Pseudomonas* (4, 5, 16). There is therefore evidence that, where fermentable organic matter is available, gleying may at least in part be a result of bacterial action.

In the absence of O_2, another biological process of importance to the transformations of iron may take place. This is the production of sulfide either through organic sulfur mineralization or by the reduction of sulfate. Microorganisms that thus form H_2S cause the precipitation of iron as ferrous sulfide by a reaction of the H_2S with iron salts. The responsible organisms can be easily recognized on agar media containing iron lactate and $(NH_4)_2SO_4$ because the colonies become surrounded by a dark halo of FeS. A precipitate will also be formed in iron-rich media by the release of H_2S during the decomposition of proteins or other sulfur-containing molecules.

Iron and steel materials buried underground are subject to corrosion in conditions of O_2 deficiency. Corrosion of this type may be so severe that iron pipes become useless after a few years, and the economic loss through deterioration of buried pipes is in the hundreds of millions of dollars per annum. At least part of the effect is brought about by microorganisms. Corrosion is particularly severe in poorly drained soils which remain wet for long periods. There is, moreover, a direct correlation between the oxidation-reduction potential and the occurrence and severity of anaerobic deterioration of iron pipes. The corrosion does not appear in soils whose potential is greater than 400 mv, it is usually slight at potentials from 200 to 400 mv, moderate at E_h of 100 to 200 mv, and almost always severe in environments with potentials below 100 mv.

The optimum conditions for the destruction of iron pipes in soil are moderate temperatures, pH values greater than 5.5, low concentrations of free O_2, and the presence of sulfate. The population concerned can be deduced from the stimulatory role of sulfate, and there is no question that it is the sulfide formed by sulfate-reducing *Desulfovibrio* populations that effects a change in the iron by precipitation of ferrous sulfide. As the bacteria are strict anaerobes utilizing sulfate as electron acceptor for growth, the need for sulfate, low E_h, and anaerobiosis can be understood. The pH range for these bacteria parallels that of the iron corrosion, i. e., ca. pH 5.5 and above. The net reaction is best represented by the equation

$$4Fe + SO_4^= + 4H_2O \rightarrow FeS + 3Fe(OH)_2 + 2OH^- \qquad \text{(VI)}$$

with ferrous sulfide and ferrous hydroxide as products. The individual steps have yet to be characterized (21).

REFERENCES

Review

Aristovskaya, T. V. and G. A. Zavarzin. 1971. Biochemistry of iron in soil. In A. D. McLaren and J. Skujins, eds., *Soil biochemistry*. Marcel Dekker, New York, vol. 2, pp. 385–408.

Literature Cited

1. Aristovskaya, T. V. 1963. *Soviet Soil Sci.*, pp. 20–29.
2. Barber, D. A. and R. B. Lee. 1974. *New Phytol.*, 73:97–106.
3. Berthelin, J. and A. Kogblevi. 1974. *Rev. D'Ecol. Biol. Sol.*, 11:499–509.
4. Bloomfield, C. 1949. *J. Soil Sci.*, 1:205–211.

5. Bromfield, S. M. 1954. *J. Soil Sci.,* 5:129–139.
6. Dommergues, Y. and G. Beck. 1966. *Soil Biol.,* No. 6, pp. 32–34.
7. Duff, R. B., D. M. Webley, and R. O. Scott. 1963. *Soil Sci.,* 95:105–114.
8. Duncan, D. W., J. Landesman, and C. C. Walden. 1967. *Can. J. Microbiol.,* 13:397–403.
9. Gleen, H. 1950. *Nature,* 166:871–872.
10. Gotoh, S. and W. H. Patrick, Jr. 1974. *Soil Sci. Soc. Amer. Proc.,* 38:66–71.
11. Gunner, H. B. and M. Alexander. 1964. *J. Bacteriol.,* 87:1309–1316.
12. Hammann, R. and J. C. G. Ottow. 1974. *Z. Pflanzenernaehr. Bodenk.,* 137:108–115.
13. Leathen, W. W., N. A. Kinsel, and S. A. Braley, Sr. 1956. *J. Bacteriol.,* 72:700–704.
14. MacRae, I. C. and J. F. Edwards. 1972. *Appl. Microbiol.,* 24:819–823.
15. Martin, J. P., J. O. Ervin, and R. A. Shepherd. 1966. *Soil Sci. Soc. Amer. Proc.,* 30:196–200.
16. Ottow, J. C. G. and H. Glathe. 1971. *Soil Biol. Biochem.,* 3:43–55.
17. Ottow, J. C. G. and A. von Klopotek. 1969. *Appl. Microbiol.,* 18:41–43.
18. Silverman, M. P. and D. G. Lundgren. 1959. *J. Bacteriol.,* 77:642–647.
19. Silverman, M. P. and E. F. Munoz. 1970. *Science,* 169:985–987.
20. Spencer, W. F., R. Patrick, and H. W. Ford. 1963. *Soil Sci. Soc. Amer. Proc.,* 27:134–137.
21. Starkey, R. L. 1958. *Producers Monthly,* 22(9):12–30.
22. Takai, Y. and T. Kamura. 1966. *Folia Microbiol.,* 11:304–313.
23. Trafford, B. D., C. Bloomfield, W. I. Kelso, and G. Pruden. 1973. *J. Soil Sci.,* 24:453–460.
24. Tuovinen, O. H. and D. P. Kelly. 1974. *Arch. Microbiol.,* 98:351–364.
25. Ten, K.-M. 1967. *Mikrobiologiya,* 36:337–344.
26. Walsh, F. and R. Mitchell. 1972. *Environ. Sci. Technol.,* 6:809–812.
27. Williams, M. E. and E. D. Rudolph. 1974. *Mycologia,* 66:648–660.

23
Transformations of Other Elements

Microorganisms metabolize the elements in soil in numerous ways. The principles set forth with regard to the metabolism of nitrogen frequently can be extended to the transformations of several elements, and consideration of the various steps of the nitrogen cycle will lead to some understanding of the microbiologically induced changes of other elements. Some of the reactions these elements undergo have no known counterparts in the nitrogen cycle, however.

Compounds or ions of the various elements may undergo one or more of the following types of reactions.

a. The release of inorganic ions during the decomposition of organic materials. As in ammonification and phosphorus mineralization, the accumulation of the inorganic ion is dependent on the concentration of the element in excess of the microbiological demand.

b. Removal of inorganic ions from solution and the disappearance of the available form of the element to satisfy the nutrient demands of the microflora. Immobilization into microbial protoplasm for the purposes of cell synthesis is in continuous competition with mineralization, one process supplying the element for plant use, the other removing it.

c. Oxidation of inorganic ions and compounds. The process is often catalyzed by chemoautotrophic bacteria, as in the oxidation of ammonium, nitrite, inorganic sulfur compounds, and ferrous iron by species of *Nitrosomonas, Nitrobacter,* and *Thiobacillus.* However, heterotrophs may also oxidize inorganic ions, but the organisms derive little or no energy from the process; thus, the capacity for sulfate and nitrate biosynthesis is not restricted to the few species of autotrophs present in nature.

d. Reduction of an oxidized state of the elements. Among the substances serving as electron acceptors in the absence of adequate O_2 are nitrate,

380

sulfate, carbonate, and ferric ions. Nevertheless, the oxidized forms of certain elements that are reduced microbiologically do not serve as electron acceptors to sustain growth in anaerobic environments.

e. Indirect transformations resulting from the activities or the products of microorganisms; for example, changes in acidity and alkalinity or alterations in the partial pressure of O_2 by respiration modify the oxidation state of several elements.

f. Changing the total quantity of an element in the soil. Nitrogen fixation and chemoautotrophic CO_2 assimilation increase the amount of nitrogen and carbon while denitrification and H_2S formation diminish the nitrogen and sulfur levels. In some instances, volatilization of an element may be attributable to a *methylation* reaction. In such transformations, an organic compound (RCH_3) serves as a donor of one, two, three, or possibly four methyl groups ($-CH_3$) to an anionic or cationic form of the element; if the element is designated X, the product may then be CH_3X, $(CH_3)_2X$, $(CH_3)_3X$, or $(CH_3)_4X$.

The microflora is usually active in interconverting various of the oxidation states of elements that exist in more than one valence state in soil provided that toxicity is not a hindrance. As one or more of the ions may be unavailable or even toxic to agronomic or horticultural crops, inorganic transformations can exert important effects on plant development. Furthermore, various of the conversions may yield products more or less mobile than their precursors so that the movement of an element through soil may be enhanced or diminished. In addition, the process may generate substances that have greater or lesser human toxicity than the molecules or ions that were acted on. The problem of toxicity has become more acute in recent years as the levels in soil of certain elements have risen markedly because they are present in pesticides, commercial fertilizers, lime, or municipal sludge applied to soil. Copper, lead, and arsenic compounds were once widely used as pesticides, and arsenicals and mercury-containing fungicides are still occasionally employed in agriculture. Some of the modern organic pesticides also contain manganese and zinc, and the use of fertilizers and lime may enrich soil with copper, nickel, lead, or uranium. An array of metals is introduced when municipal sludge is spread on the land, and the wastes of cities containing certain kinds of industries are notorious as sources of potential toxicants of these sorts.

Many elements undergo microbiologically induced transformations. Several of the more conspicuous changes have been discussed in previous chapters. In addition to the elements already cited, there is evidence for direct or indirect biological alterations in the availability, solubility, or oxidation state of potassium, manganese, selenium, tellurium, arsenic, zinc, copper, calcium, magnesium, aluminum, molybdenum, mercury, silicon, nickel, uranium, lead, antimony, vanadium, and various halogens. Some typical curves illustrating the

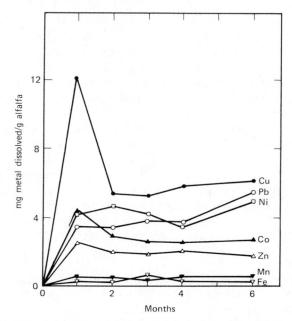

Figure 23.1. Solubilization of several elements when their oxides are incubated with decomposing alfalfa residues (8).

solubilization of several elements when their oxides are incubated with decomposing plant remains are shown in Figure 23.1.

POTASSIUM MOBILIZATION

A major cation that plants must obtain from the soil is potassium. Because the quantity in soil is often inadequate, the element is one of the macronutrients supplied in chemical fertilizers. Despite the potential agricultural importance of microbial transformations of this element, little is known of the conversions of potassium that can be effected by the microscopic inhabitants.

Potassium is readily retained by soil constituents, but the various reactions are inadequately delineated. A portion of the soil's reserve of the element is soluble while a large part is bound in the structure of various minerals where it is nonexchangeable. The external sources of potassium are agricultural fertilizers and the tissues of plants and animals. In crop residues, the element is not strongly bound in organic combination so that microbial action is not as critical to the release of potassium during organic matter breakdown as it is in the mineralization of bound nitrogen or sulfur. Moreover, the element exists in only the monovalent state in biological systems, and there are thus none of the

inorganic oxidations and reductions that typify the microbiological transformations of nitrogen, sulfur, and iron.

The microflora does have an influence on the level of available potassium,
however. The cation is solubilized through the liberation of organic or inorganic
acids that react with potassium-containing minerals. Alternatively, the element
disappears through the assimilation necessary for the formation of new
microbial cells. Some may be released during the decomposition of plant
residues, but approximately two-thirds of the potassium of plants is not strongly
bound and is immediately soluble in water so that only about one-third of the
total amount requires microbial intervention for its release. Only that portion of
the potassium that occurs in organic complexes needs to be liberated biologically.

Certain bacteria are capable of decomposing aluminosilicate minerals and
releasing a portion of the potassium contained therein. These organisms grow
in potassium-deficient media to which is added the insoluble aluminosilicate.
Typical of this group of microorganisms are species of *Bacillus* and *Pseudomonas*
among the bacteria and *Aspergillus*, *Mucor*, and *Penicillium* among the fungi;
however, many other groups of bacteria, fungi, and actinomycetes release
potassium in culture in this way. The amount liberated depends to a great
extent on the organism (25). Potassium may thus be solubilized from biotite,
muscovite, microcline, nephelite, leucite, orthoclase, and undoubtedly a number
of other silicates.

Acid production is the major mechanism for solubilization of the insoluble
potassium in minerals. The important acids in the solubilization are carbonic,
nitric, sulfuric, and several organic acids. Carbonic acid is formed from the CO_2
produced by the vast number of heterotrophic populations, and many cultures
that produce no organic acids mobilize potassium through the release of CO_2.
Microorganisms such as *Clostridium pasteurianum* and *Aspergillus niger* are active
because of the organic acids they synthesize. The production of nitric and
sulfuric acids in autotrophic metabolism has the same effect in releasing
potassium. This is illustrated by the results of Table 23.1 where the sulfuric and
nitric acids formed from sulfur and ammonium are the active principles. The
autotrophic oxidation has been used to release potassium from greensand,
which, if composted with sulfur, has its potassium made soluble. The potassium
released can then bring about an increase in crop yield.

Some potassium may be released from clay minerals by a shift of the
equilibrium between soluble and insoluble forms as microorganisms remove the
cation from solution.

$$\text{K in protoplasm} \leftarrow \text{soluble-K} \rightleftharpoons \text{mineral-K} \qquad (I)$$

Where this mechanism is operative, the potassium obtained from clay minerals
to satisfy the demands of microbial nutrition ultimately becomes available,

TABLE 23.1

Effect of Nitrification and Sulfur Oxidation on Water-Soluble Cations in a Silt Loam (3)[a]

	Concentration of Water-Soluble Cations, ppm				
Addition	K	Ca	Mg	Al	Mn
None	18	111	30	0	0
CaCO₃, 0.4%	7	270	13	0	0
(NH₄)₂SO₄, 0.8%	30	355	80	0	38
(NH₄)₂SO₄ + CaCO₃	26	1067	85	0	18
Sulfur, 0.4%	18	370	103	660	468
Sulfur + CaCO₃	53	785	64	413	471

[a] Incubated for 19 weeks at 30°C.

because the soluble cation is liberated at the time of decomposition of the microbial cells.

POTASSIUM IMMOBILIZATION

As a rule, microorganisms require the same inorganic ions as higher plants, and they may therefore be expected to compete with the macroorganisms in environments in which the nutrient supply is suboptimal. It is simple to demonstrate a competition for nitrogen between the microscopic soil inhabitants and field crops; potassium competition, on the other hand, is quite difficult to verify.

The quantity of water-insoluble, nonexchangeable potassium in soil fluctuates even when the physical and chemical environment is maintained relatively constant. Because of these fluctuations, a hypothesis has been advanced that part of the nonexchangeable potassium fraction is microbial in origin; that is, the potassium is immobilized into protoplasmic constituents. Therefore, it would seem that the microflora is participating in reducing the concentration of available potassium. In the same vein, a stimulation of microbial transformations was proposed to be the cause of the decrease in exchangeable potassium noted on liming.

The validity of these hypotheses is questionable because of chemical reactions that tend to obscure biological changes. Thus, the concentration of available potassium varies as the moisture level fluctuates, both in sterile and in nonsterile soil. Such changes cannot be biological. Furthermore, the alterations in exchangeable potassium resulting from liming can be duplicated in sterile samples; therefore, this phenomenon also cannot be attributed to the micro-flora. On theoretical grounds, the detection of microbial potassium immobiliza-tion should be difficult in the absence of added organic matter since even with nitrogen, a nutrient required in much greater amounts, immobilization in the absence of supplemental carbon is never appreciable.

Potassium is essential for the growth of all microorganisms. During their proliferation in soil, they must assimilate the ion, even if they must themselves render it soluble. In order to predict the extent of immobilization, the amount of decomposable organic matter, the efficiency of converting substrate carbon to microbial carbon, and the potassium content of the microbial cells must be known. In microorganisms, usually 0.5 to 3.0 percent of the dry weight is potassium, but higher and lower values have been recorded. Consider an example in which 10,000 kg of readily oxidizable organic matter containing 40 percent carbon is added to soil. A total of 4000 kg of carbon is present for oxidation. Assume that an aerobic community containing 50 percent carbon in its total mass assimilates 30 percent of the substrate carbon so that 1200 kg of microbial carbon is formed or a total of about 2400 kg of cell mass. If this population as a whole has a potassium content of 1.0 to 2.0 percent, then some 24 to 48 kg of potassium is assimilated.

The figures cited above are hypothetical, but they do serve as a first approximation, in the absence of definitive experimental results, of the order of change to be expected. Precise data on the potassium content of the soil flora would permit greater accuracy in the estimates. Nevertheless, even when immobilization is proceeding rapidly, it is unlikely that there would be serious competition for potassium between microorganisms and plants.

OXIDATION OF MANGANESE

Manganese is an essential micronutrient for the growth and development of higher plants. Because the element exists in several oxidation states of dissimilar availability to plants, the ability of the microflora to transform manganese has considerable importance. Manganese occurs in soil in the tetravalent form and as the divalent manganous ion. Plants are known to assimilate the divalent manganous form, while the tetravalent manganic ion presumably is not utilized. The exchangeable cation, Mn^{++}, is water-soluble while Mn^{+4} is essentially insoluble, the latter typically occurring as the manganic oxides, represented as MnO_2. The trivalent form also occurs in nature, and a significant part of the element may also be bound in organic complexes.

Of the two major ionic species, the ion that predominates is a function of the pH. At reactions more acid than pH 5.5, manganese is present largely as Mn^{++}. At reactions more alkaline than about pH 8.0, Mn^{++} is unstable and is oxidized to manganic oxides. Because manganic oxides are not assimilated appreciably by plants, alkaline conditions frequently are associated with defi-ciencies of the element. Below pH 8.0, there is little chemical oxidation of divalent manganese, the process of Mn^{++} autooxidation being characteristic of low hydrogen ion concentrations.

$$MnO_2 + 4H^+ + 2e^- \underset{\text{alkaline}}{\overset{\text{acid}}{\rightleftharpoons}} Mn^{++} + 2H_2O \tag{II}$$

In the intermediary ranges, between pH 5.5 and 8.0, the prominence of microbiological phenomena becomes evident.

When a solution containing water-soluble manganous ions is percolated through a neutral soil in the perfusion apparatus, there is an initial adjustment between the concentration of the cation in solution and that in the soil phase. This initial stage is followed by a disappearance of the manganous ion from the liquid (Figure 23.2). The loss of divalent manganese from solution is accompanied by an increase in the soil of the quantity of insoluble, oxidized manganese compounds. The process is biological because (*a*) the oxidation exhibits the logarithmic transformation typical of the bacterial growth curve, (*b*) the disappearance is most rapid at relatively low concentrations of added manganous

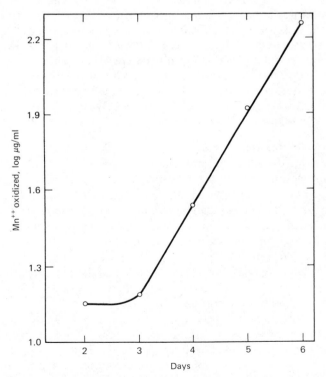

Figure 23.2. Oxidation of manganous ions in a soil perfused with a solution containing a manganese salt (36).

ions whereas chemical reactions are typically faster at high levels of reactants, and (c) manganous oxidation is eliminated by antimicrobial inhibitors. Metabolic inhibitors are convenient tools for the differentiation of biological and nonbiological reactions provided that the inhibitors have no effect on the chemical transformations.

The oxidation process can be demonstrated readily in laboratory cultures by the addition of soil crumbs to an agar medium containing $MnCO_3$; the products of microbiological action are seen as brown spots developing in the agar. It is characteristic of the active organisms to form dark brown specks in media containing $MnSO_4$ or $MnCO_3$, a result of manganic oxide accumulation. The active organisms include strains of the bacterial genera *Arthrobacter, Bacillus, Corynebacterium, Klebsiella, Metallogenium, Pedomicrobium,* and *Pseudomonas* and of the fungal genera *Cladosporium, Curvularia, Fusarium,* and *Cephalosporium.* However, species of additional genera of fungi as well as actinomycetes exhibit the same capacity in culture (30). Occasionally, two organisms are required for the oxidation to the insoluble MnO_2 (36).

There is little conclusive evidence that microorganisms obtain energy for growth by the oxidation of the manganous ion, and proliferation of the active species usually and possibly always requires organic carbon. Therefore, although the oxidation does release energy, the existence of chemoautotrophic manganese bacteria is subject to doubt. Autotrophy associated with manganese metabolism has been claimed for the aquatic bacterium, *Sphaerotilus,* but such reports require verification (2, 32).

The number of manganese-oxidizers varies considerably between soil types, but they often account for some 5 to 15 percent of the total viable microflora. The abundance is affected by proximity to plant roots as the number of active organisms is increased in the zone under the influence of the root system (Table 23.2). Manganous oxidation is brought about by microorganisms in conditions as acid as pH 5.5 and as alkaline as pH 8.9. The biological transformation is not too sensitive to acidity, but it is generally most rapid at pH 6.0 to 7.5. Individual species will be more or less sensitive to the hydrogen ion concentration, and several isolates carry out the oxidation in slightly acid circumstances.

The precise chemical structure of the manganic oxides is not understood despite their usual designation as MnO_2. Manganese dioxide or similar compounds do exist in soil, but there is also some evidence for the presence of Mn_2O_3 and Mn_3O_4. The three oxides Mn_3O_4, Mn_2O_3, and MnO_2 have progressively increasing oxidation states in addition to different physical and physiological properties. The oxides produced microbiologically seem to be intermediary between the trivalent and tetravalent forms, Mn_2O_3 and MnO_2. While it is usually assumed that only the divalent ion is available to plants, the higher oxidation states presumably not being assimilated, certain of the manganic oxides may indeed serve as nutrient sources.

Manganese oxides may become deposited in tile drains in fields in certain

TABLE 23.2

Microorganisms in Rhizosphere[a] of Oats Resistant and Susceptible to Manganese Deficiency (29)

Oat Variety	Soil Treatment	Bacteria/g × 10^3	Actinomy-cetes/g × 10^3	Mn Oxidizers × 10^3	Mn Oxidizers % of Total
		Rhizosphere Soil			
Resistant	None	266,000	2,500	41,800	15.7
	Manure, MnSO$_4$	509,000	2,100	42,200	8.3
Suscep-tible	None	564,000	2,300	225,000	42.2
	Manure, MnSO$_4$	638,000	1,200	181,000	28.2
		Nonrhizosphere Soil			
	None	129,000	1,500	8,000	6.2
	Manure, MnSO$_4$	224,000	600	9,400	4.2

[a] Soil associated with the root system.

areas, and the extent of precipitation may sometimes be so great as to plug the tile lines. Serious clogging requires mechanical cleaning of the lines otherwise the crop will suffer. The problem, insofar as it is now recognized, only is evident above pH 6.0, and it appears to result from microbial processes (21).

Several mechanisms for manganese oxidation have been proposed. One involves the oxidation of the manganous ion by salts of hydroxy acids. This reaction occurs in alkaline solution containing citrate, tartrate, lactate, malate, or gluconate, all of which are produced during the decomposition of carbohydrates. This hypothesis is supported by observations that pure cultures of a number of organisms transform soluble compounds of manganese to insoluble, brown oxides when grown on agar media containing salts of various hydroxy acids; the same reaction is noted when a sterile sodium carbonate solution is added to the surface of the agar in order to increase the alkalinity. From this, it would seem that microorganisms contribute to manganese oxidation by the production of hydroxy acids or by decreasing the hydrogen ion concentration, or both. In either instance, the initial step in forming the organic acid or in altering the pH is biological, but the subsequent oxidation of the cation is chemical. Thus, in the presence of a hydroxy acid, a strain that creates an alkaline environment will bring about a nonbiological oxidation; alternatively, the production of hydroxy organic acids in alkaline systems effects the same change.

A second mechanism may account for the reaction in other microorganisms. Certain heterotrophs possess enzymes that catalyze the oxidation of manganous ions (12). However, manganese-oxidizing enzymes of heterotrophs have yet to be adequately characterized.

MANGANESE DEFICIENCY

Deficiencies of micronutrients are often a significant factor in reducing crop yields. Manganese is no exception since it is essential for growth. Deficiencies of manganese occur commonly in soils rich in organic matter and at pH values of 6.5 to 8.0. Heavy applications of $MnSO_4$ may alleviate the condition, yet such additions frequently have no effect or, at best, a transitory effect as the element becomes oxidized in the soil. Alternatively, dilute $MnSO_4$ solutions are sometimes sprayed on the foliage to reverse the deficiency. The availability of manganese can also be increased by the use of sulfur or $(NH_4)_2SO_4$, which, by the autotrophic formation of sulfuric or nitric acids, alters the pH to favor the available state of the element. Flooding the soil also increases the amount of assimilable manganese.

A deficiency of manganese is responsible for the gray speck disease of oats and for marsh spot in peas, but similar nutrient imbalances have been reported in other cereals, various grasses, a wide variety of vegetable crops, and for several fruit trees. A biological influence on these deficiencies is readily demonstrated. Thus, when oats are grown in solution culture, gray speck symptoms do not appear, but the symptoms develop quickly if the culture solution is inoculated with a small amount of the "diseased" soil. Should the diseased soil be sterilized with formaldehyde, a treatment that has a minimum of effect on the manganese status, the oats developing therein are healthy. When the sterile soil is subsequently reinfected with a small quantity of infested soil and then planted to oats, the gray speck reappears (17). Thus, biological agencies are superimposed on the simple mineral deficiency and intensify the condition.

Ample evidence can be cited to support the contention that the manganese-oxidizing microflora is associated with the gray speck of oats. The data of Table 23.2, for example, show a correlation between the severity of the condition and the numbers of Mn^{++} oxidizers in the oat rhizosphere. The microorganisms presumably reduce the available manganous level in soil. Fumigants such as chloropicrin reduce or eliminate the symptoms and also markedly suppress the responsible populations; a straw mulch, on the other hand, intensifies the condition and produces a large number of manganous oxidizers (29). Gray speck disease serves as an excellent example of the saprophytic flora causing a plant disorder by virtue of its peculiar biochemical properties rather than through the more common means of inducing pathological responses.

MANGANESE REDUCTION

In the cyclic sequence of manganese interconversions, the divalent ion may be regenerated through acid production or by bacterial reduction. Hence, a decrease in pH, lowering of the oxidation-reduction potential, or removal of O_2 as a result of microbial metabolism will increase the level of exchangeable manganese. Soil organic matter itself, with no direct biological intervention, also reduces the higher oxides, but the extent of the reduction is considerably less than that resulting from microbial action. In practice, when soils are water-logged and the partial pressure of O_2 in them declines, MnO_2 is reduced to Mn^{++}, and the content of manganese in the liquid phase rises (Figure 23.3). The appearance of the soluble divalent cation is stimulated by the addition of available carbohydrates or plant residues, and the amount of the element in solution is related to the quantity of organic matter added and the length of the flooding period. Not only in poorly drained areas does this occur, for the addition of glucose to a well-drained soil results in a decrease in manganic oxides; once the sugar has been entirely metabolized, however, the microflora or chemical autooxidation reforms the insoluble manganic compounds.

In pure culture, many bacteria reduce MnO_2 in the presence of an

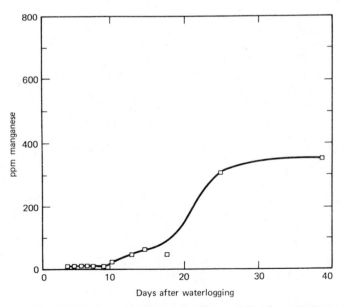

Figure 23.3. Changes in form of manganese in a soil that is waterlogged (23)

oxidizable organic nutrient. MnO_2 may serve here as an electron acceptor for respiratory enzymes, replacing O_2 in this regard.

$$RH_2 + MnO_2 \rightarrow Mn(OH)_2 + R \tag{III}$$

Cell respiration can in this way be linked with the manganic-manganous system, and enzyme preparations catalyzing MnO_2 reduction have been isolated from bacteria (5). Carbonaceous nutrients in soil may thus function metabolically in the same manner for manganic oxide reduction, that is, by providing a need for electron acceptors. On the other hand, products of microbial metabolism may also contribute to the conversion because some of them bring about a nonenzymatic transformation of tetravalent manganese to the divalent state (35). A high percentage of the bacteria, actinomycetes, and fungi isolated from soil is able to effect this reaction in culture (5), species of *Bacillus, Clostridium, Micrococcus,* and *Pseudomonas* of the bacteria as well as other groups all having confirmed activity.

As pointed out above, microbiologically synthesized acids increase manganese availability because of the effect of the hydrogen ion on the manganous-manganic equilibrium. It is not surprising, therefore, that the divalent manganese content of soil rises following sulfur or thiosulfate application, the biogenesis of sulfuric acid making more of the element available to crops and frequently relieving symptoms of manganese deficiency. An anomaly is noted, however, during the oxidation of sulfur by *Thiobacillus thiooxidans* grown in media containing MnO_2; more soluble manganese is released than in the uninoculated control medium to which is added sulfuric acid to bring the pH to the level produced during growth of the bacterium. A similar phenomenon occurs in soil. Consequently, only part of the MnO_2 solubilization can be accounted for in terms of acid formation by thiobacilli, and pH is not the sole cause of Mn^{++} release under these conditions. The additional effect may arise from the coupling of manganic reduction with sulfur oxidation, that is, the Mn^{+4} serves as an alternate electron acceptor for the bacteria (15, 33).

The high concentrations of divalent manganese present in acid soils are at times responsible for poor plant growth because excessive levels of the ion are phytotoxic. The biogenesis of this form of the element by microbial agencies may thus lead to plant injury in poorly drained or flooded fields or in areas where drainage is inadequate during wet periods. Such effects have been noted in both orchard trees and agronomic crops.

It is clear from the foregoing discussion that there is a manganese cycle in soil, a cycle involving divalent, tetravalent, and probably other oxidation states of the element. The form which is favored depends on the acidity, the community, the presence of O_2, and the availability and abundance of organic matter. In soils of pH below 5.5, Mn^{++} predominates because of the chemical equilibrium. Increasing pH brings biological forces into play, and the microbiological production of MnO_2 and other oxides becomes apparent. In the same general pH range, biological reduction regenerates divalent manganese. At reactions higher than pH 8.0, chemical autooxidation favors the oxidized states of the element.

$$Mn^{++} \underset{\substack{\text{biological} \\ H+}}{\overset{\substack{OH- \\ \text{biological} \\ \text{autooxidation}}}{\rightleftharpoons}} \underset{(MnO_2)}{Mn^{+4}} \tag{IV}$$

In the microbiological transformations of manganese, immobilization is of no consequence since microbial cells rarely contain more than 0.05 percent of the element. Consequently, the net amount assimilated would not greatly affect the more general processes of oxidation and reduction.

METABOLISM OF MERCURY

A series of events led to a dramatic upsurge in interest in mercury, an element long known for its toxicity. A number of people in Japan in recent years were poisoned because of consumption of mercury-containing fish and shellfish they ate, and several of those affected died. Hundreds of individuals in developing countries also died owing to the consumption of cereals treated with mercurials. Populations of certain birds in Sweden diminished rapidly, and the decline was traced to the eating of seed treated with fungicides containing mercury. Because of the attention then given to the element, environmental monitoring programs were initiated in many regions, and these revealed that fish often contained dangerously high levels.

Mercury is naturally present in soil, but the concentration is typically less than 1.0 ppm. Some soils are richer than others in the metal, but these differences are often of geological origin. Fungicides represent a common but highly localized source, such pesticides being designed to control fungi causing plant diseases or to prevent seeds from becoming moldy during storage. Appreciable amounts of these pesticides have been employed in agricultural practice, but many are now being replaced with less objectionable seed protectants or pesticides. The mercury present from natural or man-made sources is invariably toxic. Not only is the mercuric ion poisonous but so too are the organic forms, and frequently organic mercury—including the microbial product, methylmercury—is more detrimental to animals and microorganisms than are inorganic ions.

Metallic mercury ($Hg°$) is volatilized from soil treated with phenylmercuric acetate, ethylmercuric acetate, and mercuric and mercurous ions, and the volatilization is largely a microbial process—at least for the two organic chemicals—because the reaction is almost abolished in autoclaved samples (18, 19). The evidence for a microbial role is supported by observations that metallic mercury is liberated from cultures of several microorganisms incubated in media containing trace amounts of phenylmercuric acetate, ethylmercuric chloride, or $HgCl_2$. The reactions seem to be reductions.

$$RHgCl + 2H \rightarrow RH + Hg + HCl \tag{V}$$
$$HgCl_2 + 2H \rightarrow Hg + 2HCl \tag{VI}$$

Monomethylmercury (CH_3Hg^+) or dimethylmercury (CH_3HgCH_3) is also formed in soil from mercuric ions or mercury-containing fungicides (4, 6). Methylmercury is both volatile and quite poisonous. Its production represents a methylation reaction, with some organic donor of methyl groups carrying out a reaction of this sort.

$$Hg^{++} \xrightarrow{\ RCH_3\ } CH_3Hg^+ \xrightarrow{\ RCH_3\ } CH_3HgCH_3 \tag{VII}$$

Such methylations have also been noted in vitro in the presence and absence of O_2, with mercuric ions usually being the substrate of choice. Laboratory-grown cultures of *Bacillus, Clostridium, Mycobacterium,* and *Pseudomonas* among the bacteria, *Aspergillus, Neurospora,* and *Scopulariopsis* of the fungi, and also yeasts bring about such methylations, but the chief organisms functioning in nature remain unknown. Methylmercury, in turn, is destroyed by bacteria in vitro, and the demethylation takes place by a process analogous to the one described in equation V (28).

$$CH_3Hg^+ + 2H \rightarrow Hg + CH_4 + H^+ \tag{VIII}$$

Mercuric or mercurous ions may react with the H_2S produced in the bacterial reduction of sulfate. The mercury sulfide thus generated is insoluble and is also considerably less toxic than the cationic mercury precursor. The metal sulfide has been found in soil as well as in liquid media supplemented with sulfate.

SELENIUM

Selenium is of considerable practical importance for two reasons: its essentiality and its toxicity. Warm-blooded animals and possibly humans require selenium, and feed supplies deficient in this element may be the cause of serious difficulties in livestock production. The absence of sufficient amounts in the diet of animals has led to programs concerned with fertilization with suitable forms of the metal. At the same time, concentrations not too much higher than those needed to sustain normal development are detrimental to animals, and a variety of growth effects, organ damage, changes in susceptibility to tooth decay, paralysis, and even death of animals and occasionally people are attributable to excess quantities. Cattle, sheep, hogs, and horses have all been reported to suffer one form of damage or another by consuming selenium-rich forage. Because of the small margin of safety between the level that is required and that which is harmful, the microbial transformations of various forms of the element assume considerable prominence.

Plants that assimilate selenium provide a number of organic substrates for the microflora. Some plants contain small amounts of these compounds, but others accumulate such high levels of the toxic molecules that the vegetation constitutes a hazard to grazing herbivores. The element is commonly present in proteins in the form of seleno-amino acids, these amino acids characteristically containing selenium in place of sulfur. The microflora thus may, on death of the plant, be provided with selenomethionine or Se-methylselenocysteine, the analogs of the sulfur-amino acids, methionine and S-methylcysteine.

Selenomethionine $CH_3SeCH_2CH_2CH(NH_2)COOH$
Se-Methylselenocysteine $CH_3SeCH_2CH(NH_2)COOH$

Selenocystathionine and dimethylselenide (CH_3SeCH_3) are also plant constituents. Molecules of these sorts are mineralized by microorganisms in culture (10). The mineralization must occur in nature, too, inasmuch as a substantial part of the selenium in soil appears to exist as selenite and selenate, and a mechanism must operate in nature to convert the organic complexes derived from the vegetation back to the oxidized anions revealed by chemical analysis.

The oxidation of elemental selenium in soil has been verified. Part of the oxidation may be by nonbiological means, but studies involving the use of inhibitors suggest that part results from microbial metabolism (16). How the microbial oxidation occurs and what organisms are responsible remain totally obscure.

Selenite added to soils is converted to elemental selenium. A number of bacteria, actinomycetes, and fungi bring about the reduction of selenium salts. In filamentous microorganisms and in bacteria, the end product of the reaction in media commonly is elemental selenium. Aerobic, facultatively anaerobic, and obligately anaerobic bacteria are represented in the group that reduces selenite and selenate salts to the metallic state. The reduction typically leads to a color change, the colonies taking the brick-red coloration of the selenium metal. Active species are simple to detect by the red coloration associated with their growth.

The ability to form elemental selenium from selenite and selenate is not an uncommon attribute among members of the subterranean community (5). Species of *Candida, Clostridium, Corynebacterium, Micrococcus, Rhizobium,* and others effect the transformation. The formation of elemental selenium can be detected microscopically because the metal is deposited within the cell as distinct, red granules. In contrast to sulfate reduction in which the final product is H_2S, elemental selenium rather than hydrogen selenide, H_2Se, accumulates. Bacteria do not seem to be capable of using the reaction to permit anaerobic growth, that is, to use selenate or selenite as an electron acceptor for proliferation as the denitrifying bacteria use nitrate when O_2 is unavailable. Considerations of comparative biochemistry would suggest that such organisms do exist, but experimental verification is lacking. Selenium appears to be required for the growth of certain bacteria, and it has been detected in proteins of microorganisms. It is likely, therefore, that selenium immobilization into cells takes place. This assimilation would reduce the quantity available for uptake by roots, but the trace amount needed for microbial proliferation suggests that the extent of immobilization would be small.

Selenium, like mercury, is subject to methylation. The product in this instance is dimethylselenide. The reaction may be brought about by isolates of *Aspergillus, Candida, Cephalosporium, Corynebacterium, Fusarium, Penicillium,* and undoubtedly other fungi and bacteria. Dimethylselenide is generated in soil and released to the overlying atmosphere (Figure 23.4). The volatilization of the methylated compound may be pronounced when the microflora is provided with readily available carbon sources, and selenate, selenite, and seleno-amino acids are all capable of being metabolized to yield dimethylselenide (10). Selenium has been found in the atmosphere of several areas of the world, and it is possible that this atmospheric constituent is the dimethylselenide made by components of the community.

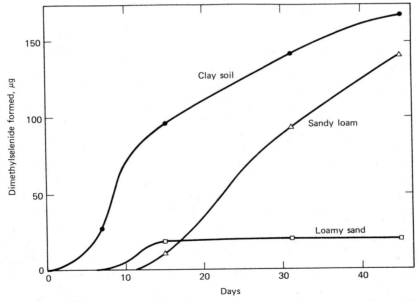

Figure 23.4. Dimethylselenide evolution from soils amended with selenite and glucose (14).

A series of selenium transformations thus is evident, but it is premature to suggest the details of any possible selenium cycle. The conversions involve mineralization of organic molecules derived from plants and probably humus, assimilation into cellular proteins, oxidation of inorganic forms, reduction of selenate and selenite to the elemental state, and methylation of the two anions.

Tellurium resembles selenium in many of its properties, and although little is known about tellurium metabolism, the scant information currently available suggests that the element is subject to many, if not all, of the conversions undergone by selenium. Thus, the capacity to reduce tellurate and tellurite to black elemental tellurium is widespread among bacteria, fungi, and actinomycetes obtained from soil (5). Colonies growing on agar media containing anionic tellurium are often dark gray to black as a consequence of the accumulation of elemental tellurium. As with selenium, metallic tellurium is deposited within the bacterial cell or in the hyphae of filamentous microorganisms, the accumulation of the metal resulting in the black colony. The reactions may be written as follows:

$$TeO_4^= + 6H \rightarrow Te + 2H_2O + 2OH^- \qquad (IX)$$
$$TeO_3^= + 4H \rightarrow Te + H_2O + 2OH^- \qquad (X)$$

As with selenium, fungi are capable of methylating inorganic compounds of tellurium, but in this instance the product is dimethyltelluride (CH_3TeCH_3) (13).

ARSENIC

Arsenic first attracted the interest of microbiologists because of instances of human poisoning attributed to volatile compounds associated with the use of wallpapers containing arsenical pigments. When moisture is adequate, the wallpaper itself supports a fungal growth that liberates the volatile arsenic compounds. The fungi participating in the conversion include species of *Aspergillus, Mucor, Scopulariopsis, Fusarium,* and *Paecilomyces*. Soil microorganisms likewise produce gaseous arsenic-containing compounds, and such substances are released from soils treated with salts of arsenic.

Arsenic is a minor constituent of soil, but fields or crops are sometimes treated with arsenical pesticides. At one time, lead arsenate was widely applied to fruit trees, and although sprays containing this chemical are no longer in use, soils in these old orchards retain large quantities of the element. Calcium arsenate was also employed for insect control. Arsenate and arsenite were at one time commonly included in sprays for weed control or to destroy vegetation along railroad tracks. Organic arsenicals such as methylarsonic acid (or its sodium salts) and dimethylarsinic acid (synonymous with cacodylic acid) still find use as agricultural herbicides and defoliants.

$$\underset{\overset{\displaystyle |}{OH}}{H_3C\overset{\overset{\displaystyle O}{\|}}{A}sOH}$$

$$\underset{\overset{\displaystyle |}{CH_3}}{H_3C\overset{\overset{\displaystyle O}{\|}}{A}sOH}$$

Methylarsonic   Dimethylarsinic
acid   acid

Soils near some smelters are often polluted with large amounts of the element.

Volatilization and probably methylation of dimethylarsinic acid are evident in soils treated with the pesticide. However, some arsenate is also generated (34). Although the product evolved from soil has not yet been identified, analogy with pure culture studies indicates it may be either tri- or dimethylarsine. Several fungi are capable of making trimethylarsine, $(CH_3)_3As$, from methylarsonate and dimethylarsinate, and *Candida humicola* synthesizes the same volatile metabolite from arsenite and arsenate as well (9). By contrast, a strain of *Methanobacterium* produces dimethylarsine, $(CH_3)_2AsH$, from arsenate (20). The methylation may proceed by a pathway such as that given in Figure 23.5.

Arsenite oxidation has been the subject of some scrutiny. If applied at low rates to soil, arsenite disappears and arsenate is produced; the reaction is biological as it is abolished by enzyme poisons such as sodium azide. Moreover, the transformation follows the logarithmic rate typical of the growth of bacterial cultures. A soil enriched with an arsenite-metabolizing flora rapidly acts on the substrate and, at the same time, consumes O_2. The amount of O_2 utilized is in agreement with that predicted by equation XI (24).

$$2NaAsO_2 + O_2 + 2H_2O \rightarrow 2NaH_2AsO_4$$

arsenite   arsenate   (XI)

Repeated perfusion of soil with arsenite leads to an ultimate diminution in the rate of oxidation. This unexpected decline in rate is suggestive of a heterotrophic process that requires a reserve of available organic matter although

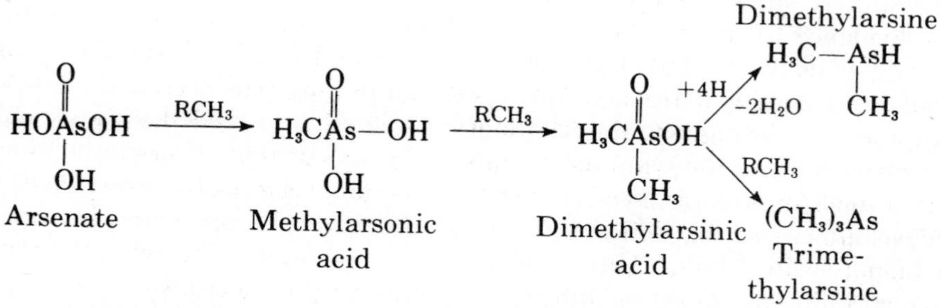

Figure 23.5. Possible pathway for the formation of methylarsines.

arsenite oxidation might seem to be an autotrophic parallel to the *Nitrobacter* metabolism of nitrite.

Attempts to isolate chemoautotrophs capable of deriving energy from the conversion of arsenite to arsenate have failed, the reaction in enrichments being abolished when organic carbon is absent. However, heterotrophic bacteria that convert arsenite to the more oxidized state have been isolated. In this process, therefore, the inorganic transformation is clearly the result of heterotrophic metabolism. Many heterotrophs, moreover, can reduce arsenate to arsenite.

TRANSFORMATION OF OTHER MINERALS

Zinc is of biological importance because of its role in plant nutrition, but the introduction of significant quantities of the element together with sewage sludge applied to the land has raised the question of its potential phytotoxicity. Zinc is required or is stimulatory to the growth of a number of fungi, yeasts, and bacteria, and analysis of their cells also shows its presence; however, because so little is required in liquid media (less than 1.0 ppm) and the cells contain such small quantities (usually 100 to 400 ppm), microbial assimilation probably has an inconsequential impact on plant growth.

The microflora may increase the solubility of zinc in several ways. (*a*) Organic acids may make soluble the cation in zinc silicates. For example, soluble zinc is released as the pH falls and organic acids are generated by microorganisms when a soil sample is inoculated into a medium containing Zn_2SiO_4. Subsequently, the soluble zinc level diminishes, probably a result of the fall in acidity in the liquid as the organisms utilize the organic acids (Figure 23.6). (*b*) The fall in pH arising from the oxidation of ammonium salts by the nitrifiers will make zinc available. Increasing acidity favors zinc uptake by plants, increasing alkalinity diminishes it. (*c*) The decomposition of plant remains leads to a release of the soluble cation (8). The mechanism may be associated with organic acid production. (*d*) The oxidation of the sulfide in ZnS by *Thiobacillus* will release the metal in a water-soluble form, and significant amounts may remain in solution if the pH stays low. A process of this sort may be pronounced when such ores are mined and exposed to the air.

Evidence for a microbiological effect on zinc availability also comes from investigations of the "little-leaf" condition of several fruit trees, an abnormality resulting from zinc deficiency. When soils supporting fruit trees with little-leaf symptoms are brought to the laboratory, sterilized, and seeded with a suitable test plant, none of the symptoms appears. The inoculation of the sterile sample with a small amount of diseased soil or with certain bacteria, however, leads to a redevelopment of the deficiency. The condition can be prevented by supplementation with $ZnSO_4$. Thus, under certain conditions, the microflora may have an influence, direct or indirect, on the availability to crops of zinc.

The copper level also may be affected by the metabolism of the microflora. For example, the concentration of soluble copper decreases during the decom-

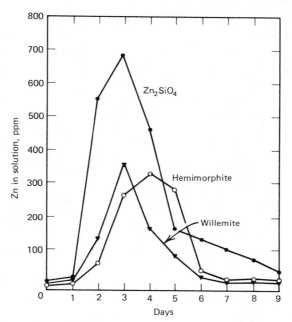

Figure 23.6. Zinc solubilization when soil is inoculated into a medium containing synthetic Zn₂SiO₄ and two zinc silicate minerals (1).

position of certain crop residues. The effect on copper is indirect and probably is a consequence of a chemical reaction involving products released during decay of the vegetation (8). Copper is found in a number of sulfide ores, and the exposure of these subterranean materials to O_2 is frequently a prelude to the appearance of considerable soluble copper in the resulting acidic circumstances (31). Two schools of thought exist on how the copper is thus liberated. One holds that the oxidation of the sulfide or ferrous ions in the ores by *Thiobacillus* yields sulfuric acid or ferric ions that then react with materials like chalcopyrite ($CuFeS_2$) to solubilize the copper by purely nonenzymatic means. Proponents of a second view of copper release from sulfides argue that *Thiobacillus ferrooxidans* is capable of bringing about an enzymatic oxidation of cuprous to cupric ions, a process that is believed to provide energy for growth of the autotroph when it is supplied with Cu_2S (22). Soluble copper may also be precipitated and thus made less mobile when microorganisms are producing H_2S, as they do when soils with sulfate become O_2-deficient.

The community may alter the chemistry, mobility, oxidation state, or solubility of many other elements. The evidence for such reactions is sometimes derived from field observation or tests with soil or rocks undergoing weathering, but often the information is solely from pure culture studies. The biological production of sulfuric and nitric acids from sulfur and ammonium salts can cause the solubilization of calcium and aluminum, an effect that is readily observable under natural conditions. Similarly, organic acids generated by heterotrophs, in culture at least but probably in nature too, will solubilize silicon, aluminum, magnesium, and calcium (7, 27), a process that may be implicated in the biological weathering of rocks and minerals and in soil formation. Bacteria and fungi also synthesize a variety of chelating agents, and these compounds are known to liberate silicon, calcium, magnesium, aluminum, sodium, and other elements from minerals or insoluble salts (11). Chelates of microbial origin have been postulated to play a role not only in solubilizing silicate minerals but also in the weathering of rocks and the transport of aluminum and iron downward through the soil profile. Many elements exist in the form of anions, and studies of individual heterotrophs have demonstrated that the oxidized states may be reduced; some of the elements acted on have been discussed already, but others—such as vanadium as vanadate and molybdenum as molybdate—are subject to the same sort of change (5). Decomposing plant materials, moreover, will bring insoluble oxides of cobalt, nickel, and lead into the aqueous phase by unknown mechanisms (8). *Thiobacillus* acting on sulfides or iron ores may be responsible, directly or indirectly, for the release of a host of elements in soluble form, and tests in the laboratory point to the possibility of many such reactions. Methylation is now confirmed for sulfur, selenium, tellurium, mercury, and arsenic, but it is quite likely that other elements may be similarly metabolized microbiologically. Lead appears to be one likely candidate for methylation in natural environments (26).

REFERENCES

Literature Cited

1. Agbim, N. N. and K. G. Doxtader. 1975. *Soil Biol. Biochem.*, 7:275–280.
2. Ali, S. H. and J. L. Stokes. 1971. *Antonie van Leeuwenhoek J. Microbiol. Serol.*, 37:519–528.
3. Ames, J. W. and G. E. Boltz. 1919. *Soil Sci.*, 7:183–195.
4. Bache, C. A., W. H. Gutenmann, L. E. St. John, R. D. Sweet, H. H. Hatfield, and D. J. Lisk. 1972. *J. Agr. Food Chem.*, 21:607–613.
5. Bautista, E. M. and M. Alexander. 1972. *Soil Sci. Soc. Amer. Proc.*, 36:918–920.
6. Berkert, W. F., A. A. Moghissi, F. H. F. Au, E. W. Bretthauer, and J. C. McFarlane. 1974. *Nature*, 249:674–675.
7. Berthelin, J., Y. Dommergues, and D. Boymond. 1972. *Rev. D'Ecol. Biol. Sol*, 9:397–406.
8. Bloomfield, C., W. I. Kelso, and M. Piotrowska. 1971. *Chem. Ind.*, pp. 59–61.
9. Cox, D. P. and M. Alexander. 1973. *Bull. Environ. Contam. Toxicol.*, 9:84–88.
10. Doran, J. W. and M. Alexander. Unpublished data.
11. Duff, R. B., D. M. Webley, and R. O. Scott. 1963. *Soil Sci.*, 95:105–114.
12. Ehrlich, H. L. 1968. *Appl. Microbiol.*, 16:197–202.
13. Fleming, R. W. and M. Alexander. 1972. *Appl. Microbiol.*, 24:424–429.

14. Francis, A. J., J. M. Duxbury, and M. Alexander. 1974. *Appl. Microbiol.,* 28:248–250.
15. Garey, C. L. and S. A. Barber. 1952. *Soil Sci. Soc. Amer. Proc.,* 16:173–175.
16. Geering, H. R., E. E. Cary, L. H. P. Jones, and W. H. Allaway. 1968. *Soil Sci. Soc. Amer. Proc.,* 32:35–40.
17. Gerretsen, F. C. 1937. *Ann. Bot.,* 1:207–230.
18. Gilmour, J. T. and M. S. Miller. 1973. *J. Environ. Qual.,* 2:145–148.
19. Kimura, Y. and V. L. Miller. 1964. *J. Agr. Food Chem.,* 12:253–257.
20. McBride, B. C. and R. S. Wolfe, 1971. *Biochemistry,* 10:4312–4317.
21. Meek, B. D., A. L. Page, and J. P. Martin. 1973. *Soil Sci. Soc. Amer. Proc.,* 37:542–548.
22. Nielsen, A. M. and J. V. Beck. 1972. *Science,* 175:1124–1126.
23. Patrick, W. H. and F. T. Turner. 1968. *Nature,* 220:476–478.
24. Quastel, J. H. and P. G. Scholefield. 1953. *Soil Sci.,* 75:279–285.
25. Savostin, P. 1972. *Z. Pflanzenernaehr. Bodenk.,* 132:37–45.
26. Schmidt, U. and F. Huber. 1976. *Nature,* 259:157–158.
27. Silverman, M. P. and E. F. Munoz. 1970. *Science,* 169:985–987.
28. Spangler, W. J., J. L. Spigarelli, J. M. Rose, and H. M. Miller. 1973. *Science,* 180:192–193.
29. Timonin, M. I. 1946. *Soil Sci. Soc. Amer. Proc.,* 11:284–292.
30. Timonin, M. I., W. I. Illman, and T. Hartgerink. 1972. *Can. J. Microbiol.,* 18:793–799.
31. Torma, A. E. and G. Legault. 1973. *Ann. Microbiol., (Paris),* 124A:111–121.
32. van Veen, W. L. 1972. *Antonie van Leeuwenhoek J. Microbiol. Serol.,* 38:623–626.
33. Vavra, J. P. and L. R. Frederick. 1952. *Soil Sci. Soc. Amer. Proc.,* 16:141–144.
34. Woolson, E. A. and P. C. Kearney. 1973. *Environ. Sci. Technol.,* 7:47–50.
35. Yoshida, K. and T. Kamura. 1972. *Nippon Dojo-Hiryogaku Zasshi,* 43:451–455.
36. Zavarzin, G. A. 1962. *Mikrobiologiya,* 31:586–588.

ECOLOGICAL INTERRELATIONSHIPS

24
Interactions Among Species

In natural environments, a number of relationships exist between individual microbial species and between individual cells. The interrelations and interactions of the various microbial groups making up the soil community, however, are in a continual state of change, and this dynamic state is maintained at a level characteristic of the flora. The composition of the microflora of any habitat is governed by the biological equilibrium created by the associations and interactions of all individuals found in the community. Environmental changes temporarily upset the equilibrium, but it is reestablished, possibly in a modified form, as the community shifts to become acclimated to the new circumstances.

In soil, many microorganisms live in close proximity, and they interact in a unique way that is in marked contrast to the behavior of pure cultures studied by the microbiologist in the laboratory. Members of the microflora rely on one another for certain growth substances, but at the same time they exert detrimental influences so that both beneficial and harmful effects are evident. The sum total of all of the individual interactions establishes the *climax community*, the native flora typifying a given habitat.

A number of possible interactions may occur between two species: (*a*) *neutralism*, in which the two microorganisms behave entirely independently, (*b*) *symbiosis*, the two symbionts relying on one another and both benefiting by the relationship, (*c*) *protocooperation*, an association of mutual benefit to the two species but without the cooperation being obligatory for their existence or for their performance of some reaction, (*d*) *commensalism*, in which only one species derives benefit while the other is unaffected, (*e*) *competition*, a condition in which there is a suppression of one organism as the two species struggle for limiting quantities of nutrients, O_2, or other common requirements, (*f*) *amensalism*, in which one species is suppressed while the second is not affected, typically the result of toxin production, and (g) *parasitism* and *predation,* the direct attack of one organism on another.

Because of these interrelationships, the introduction of an alien organism into soil rarely leads to its establishment. The fact that the species introduced is scarce or absent indicates of itself that the habitat is unfavorable for the microorganism's development. Bacteria and fungi not indigenous to a soil type rapidly die out when added, and changes resulting from the introduction of foreign organisms are always transient. The ecological axiom that *the community reflects the habitat* is an excellent rule in microbial ecology.

The decline of alien populations has long been known in plant pathology because susceptible plants grown one or two years after the plowing under of crop residues teeming with certain pathogens often show no symptoms of disease. In some soils, for example, in which more than 10^9 cells of the pathogen *Corynebacterium insidiosum* are added per gram, the bacterium can no longer be found after seven days (32). Similarly, the disappearance of the large numbers of enteric pathogens entering soil with animal or human excrement is evident in the rarity of diseases resulting from the eating of crops grown in the treated fields, provided adequate time has elapsed for the parasites to be destroyed by indigenous populations. The current interest in disposing of sewage sludge and manure from feedlots or large herds on the land has refocused attention on the human pathogenic bacteria, but the carefully conducted, modern studies confirm the rapid dying out of most of the cells that are deliberately added (Figure 24.1).

BENEFICIAL ASSOCIATIONS

The three beneficial relationships cited above—symbiosis, protocooperation, and commensalism—are found to operate among the soil inhabitants. Because of the high cell density in a restricted ecological zone, microorganisms develop certain relations in time that are beneficial and others that are detrimental. Sometimes the benefit is mutual, but commensal relationships are quite frequent.

One of the more important beneficial associations is that involving two species, one of which can attack a substrate not available to the second organism, but the decomposition results in the formation of products utilized by the second. This type of commensalism is not infrequent in nature, and it is the way many polysaccharides are transformed to nutrients supporting nonspecialized microorganisms; for example, cellulolytic fungi produce from cellulose a number of organic acids that serve as carbon sources for noncellulolytic bacteria and fungi. A second type of commensal association arises from the need of many microorganisms for growth factors. These compounds are synthesized by certain microorganisms, and their excretion permits the proliferation of nutritionally fastidious soil inhabitants. Thus, a surprisingly high percentage of isolates will not grow in the absence of individual amino acids, B vitamins, and

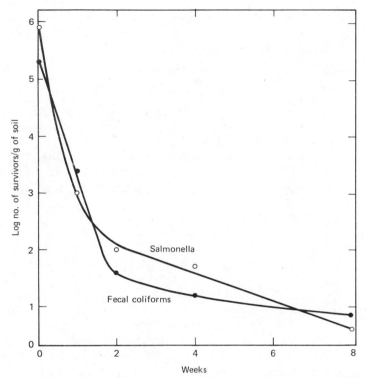

Figure 24.1. The decline of fecal coliforms and *Salmonella enteritidis* introduced into soil (10).

sometimes purines or pyrimidines, but the abundance of fastidious organisms is matched by heterotrophs with the capacity of excreting the needed metabolites, the latter apparently sustaining growth of the former in nature. The microbial decomposition of biologically produced inhibitors that prevent the proliferation of sensitive species is another instance of a beneficial relationship. Aerobes may permit the growth of obligate anaerobes by consuming the O_2 in the environment. Commensalism is also evident in the relationship between acid-sensitive organisms that grow in soils of low pH only in the immediate vicinity of populations that decrease the acidity of the microenvironment inhabited by the two populations (29). Similarly, commensalism operates between selected fungi that persist in soil in a resistant form and associates that stimulate the fungus to make the resistant structure (41).

The decomposition of a number of natural products and pesticides is occasionally more rapid in mixed than in pure culture. The explanation for the phenomenon is obscure. The greater rate may be the result of a removal of products that deter the primary organism, or it may arise from the production by the secondary population of growth factors required by the primary flora. The relationship may be commensalism if only one of the associates benefits, as when the secondary population synthesizes metabolites without which the primary heterotrophs will not multiply. On the other hand, protocooperation may be involved—as when the secondary organisms favor their associates by removing toxic wastes but simultaneously get carbon in the form of products made by their associates. This type of interaction is not uncommon, and it probably is important in many habitats.

Nutritional protocooperation has been demonstrated frequently in culture. For example, in a medium deficient in nicotinic acid and biotin, neither *Proteus vulgaris* nor *Bacillus polymyxa* will multiply as the former bacterium requires nicotinic acid and the latter, biotin. In mixed culture in the same medium, however, both grow since the partner bacterium synthesizes the missing vitamin (40). Similar nutrient interactions between bacteria and fungi have been reported for various vitamins, amino acids, and purines. In each instance, the partner is capable of synthesizing the appropriate growth factor. Many hitherto anomalous effects can now be explained on the basis of the excretion of specific nutrients by one of the two microbial partners.

In soil, mutual feeding assumes great prominence. A large proportion of the indigenous bacteria requires or is stimulated by water-soluble B vitamins and amino acids. They will not grow in simple laboratory media unless supplemented with the appropriate substances. The occurrence of nutritionally exacting species in nature is rather surprising because the organisms must be supplied continuously with the deficient nutrients in order for them to compete effectively for the limited amount of carbonaceous and inorganic nutrients. Many of the bacteria in soil require one or more vitamins for growth and many excrete these compounds. Thiamine is the most frequently required vitamin, but biotin and vitamin B_{12} also are essential for a large number of bacteria. Similarly, many isolates will not grow in the absence of amino acids, and many excrete the very same growth factors (34).

The presence of fastidious microorganisms would appear inexplicable except for the fact that vitamins and amino acids occur in soil. It is likely that these substances are produced by the microflora because the compounds in plant residues are destroyed during decomposition. The synthesis of vitamins and amino acids in vitro does not signify the same occurs in the field, yet it does suggest that potentiality. Thus, the prevalence of organisms requiring growth factors probably results from the synthesis and release by other microorganisms of these substances. Undoubtedly, the effect is nonspecific—that is, the organisms synthesizing the compounds do not have fixed partners; instead, the association is fortuitous. Because of the importance of such substances in nutrition, the interactions arising from the excretion of and need for growth factors probably are among the major biological determinants of the composition of the community.

Symbiotic associations are evident in soil among several groups of organisms: algae and fungi in lichens, bacteria residing within protozoan cells, bacteria and roots in the *Rhizobium*-legume symbiosis, fungi and roots in mycorrhizae, protozoa in underground termites, and—above rather than beneath the surface—fungi and ants. In lichens, the algae and fungi are in such an intimate physical and physiological relationship that the lichens they make up are classified as distinct organisms. The alga benefits in part because of the protection afforded to it by the hyphae that envelop and protect it from environmental stresses, while the fungus gains by making use of the CO_2 fixed by its photosynthetic partner. Where blue-green algae are participants in the lichen association, the heterotroph benefits from the fixed nitrogen acquired from the possessor of nitrogenase. That the symbiosis is ecologically successful is clear from the ubiquity of lichens on rocks, arid and semiarid sites, and other locations where neither of the free-living organisms is detected.

Bacteria reside in the cells of numerous protozoa, sometimes in the hundreds or more in each animal. The bacterial inhabitants gain from this association by obtaining nutrients and a protected environment from the protozoan, and indeed many of the bacteria will not grow outside of the host's cytoplasm. Nevertheless, a benefit accruing to the protozoan has rarely been demonstrated and hence the interaction may not always be symbiotic. However, in at least certain ciliates, the bacteria are essential in order to permit continued multiplication of the animal (19).

A few of the symbiotic relationships between micro- and macroorganisms have been thoroughly explored. The mutual benefits gained by legumes and rhizobia include nitrogen delivered to the plant by the N_2-metabolizing bacteria and organic carbon transferred to the rhizobia by the CO_2-metabolizing host. However, which compounds are actually transported, the basis for the specificity, and processes leading to nodulation are hardly known. The mycorrhizal, fungus-root association is a striking instance of symbiosis: the fungus gets essential organic nutrients and other (as yet undefined) benefits allowing it to multiply in conjunction with roots but frequently not in soil, and the plant has an enhanced rate of uptake of phosphorus, nitrogen, and other inorganic nutrients. Protozoa are also components of a symbiosis with termites, the microorganism being essential for the insect by converting cellulose in the wood consumed by the larger animal to a usable form; the microsymbionts obviously acquire much from their associates because they are unknown in land free of termites. A novel relationship operates between attine ants and specialized fungi, neither of which is detectable in nature in the absence of the other. The ants macerate leaf tissues, flowers, or other plant debris and introduce the specific fungi into this mass of prepared organic matter. The fungi flourish there and then provide the insects with their chief and possibly sole source of nutrients (30).

MICROBIAL COMPETITION

Microorganisms inoculated into sterile soil develop rapidly and attain large population sizes; similar inoculations into nonsterile soil lead to poor growth, and often the introduced species is eliminated in a period of days or weeks. The difference is entirely the result of biological interrelationships of an injurious nature. Detrimental effects of one species on its neighbors are quite common in soil, and they are detected by the decrease in abundance or metabolic activities of the more susceptible organism. The compounding of simple interactions found in two-culture systems into the multiorganism complex of the soil results in a diversity of harmful associations. Consequently, there is a permanent struggle for existence in the habitat, and only those species most fitted for the specific environment survive.

The categories of deleterious interactions are summarized by the terms competition, amensalism, parasitism, and predation, that is, (a) the rivalry for limiting nutrients or other common needs, (b) the release by one species of products toxic to its neighbors, and (c) the direct feeding of one organism on a second. Because the supply of nutrients in soil is perennially inadequate, competition for carbon, inorganic nutrients, or O_2 is quite common. Alteration of the environment to the detriment of certain microbial groups may occur through the synthesis of metabolic products that inhibit or kill microbial cells, by the utilization of O_2, which leads to the suppression of obligate aerobes, or by the autotrophic formation of nitric and sulfuric acids, which affect the proliferation of acid-sensitive microorganisms. Predatory and parasitic activities similarly are not rare. Predation and parasitism are observed in the feeding on bacteria by protozoa, the attack on nematodes by predacious fungi, the digestion of fungal hyphae by bacteria, and the lysis of bacteria and actinomycetes by bacteriophages.

It is not difficult to demonstrate competition in liquid media between populations of two unicellular organisms. Counts are made of the two species when they are developing separately and when growing together. The better competitor—typically the organism with the shorter generation time under the test conditions—multiplies at the same rate in pure or in mixed culture, but its final cell number is somewhat less in the mixture. The poorer competitor, by contrast, may initially grow as readily alone as in a two-membered mixture, but its multiplication rate falls markedly once the first species has metabolized nearly all of the limiting nutrient. The abundance of the second organism is, at the end of the experiment, markedly less than it would have been were it alone (Figure 24.2). By making one or another nutrient limiting, competition can be demonstrated for sources of energy, carbon, inorganic nutrients, or growth

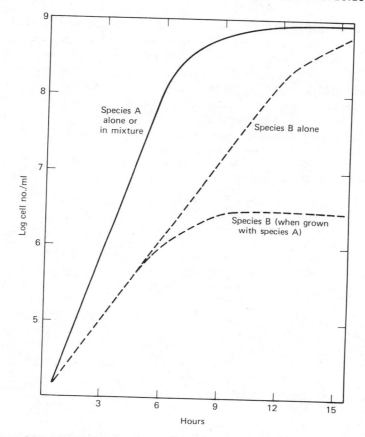

Figure 24.2. Idealized competition between two bacteria in liquid media.

factors. Similarly, if sterilized soil rather than culture solutions is inoculated with two bacteria, the outcome will shift with changing temperature or water content as one or the other of the interactants is benefited by the variable of particular interest (35). With filamentous species, as among the fungi, the organism with a greater rate of hyphal development and extension will colonize unexploited sites more readily than will those of its neighbors not able to bring about as rapid hyphal development (27). Competition itself does not lead to the demise of the less vigorous species, but populations deprived of nutrients are likely to die readily because of the stresses associated with starvation.

Inasmuch as competition is a rivalry for limiting resources, attention has been given to defining which factors in soil are in inadequate supply. It has sometimes been argued that space may be limiting and hence the basis for interspecific rivalries; however, microscopic studies of soil, decaying organic materials, and root surfaces reveal many sparsely or unpopulated sites so space probably is rarely a limitation. On the other hand, space may restrict the number or mass of cells or hyphae directly on a microsite on decomposing carbonaceous materials where the organisms are actively proliferating or in small pores among the soil particles.

The inadequate quantity of readily available carbon compounds is a more likely basis for competition, especially inasmuch as the chief limiting factor for community metabolism usually is the paucity of easily utilizable organic nutrients. Thus, at low levels of available carbon, fast growers will often hold slow growers in check when both are added to sterilized soil, but there is no such check on the less active heterotroph when the carbon supply is adequate. Under these circumstances, competitiveness is found to be directly correlated with growth rate (13). This exploitation of simple organic molecules by bacteria with short generation times may explain why the spores of many fungi fail to germinate in nature: the nutrients they need for germination are metabolized before the spores can utilize them to get energy for outgrowth from the dormant stage. Moreover, heterotrophs with high metabolic rates may die out in nature because they maintain their high rates of metabolism at the expense of cell constituents; they are unable to multiply using resistant substrates that are transformed and yield energy to the active populations only slowly (23).

In the presence of abundant and readily available organic materials—such as individual carbohydrates—or nitrogen-poor plant remains, a competition for nitrogen can be demonstrated. The rivalry is evident in the suppression of the poor competitor when the carbonaceous substances are added to soil and the reversal of the apparent inhibition when nitrogen is added. Such an interaction may be the reason why a lowering of the available nitrogen content by promoting immobilization or by cropping practices decreases the severity of certain plant diseases and nitrogen supplementation mitigates the suppression.

As a first approximation, the ability of an organism to compete is probably governed by its capacity to utilize the carbonaceous substrates found in soil, its growth rate, and its nutritional complexity. A simple nutrition could be advantageous, but the presence in soil of growth factors suggests that effective competitors need not be nutritionally independent, for they can develop at the expense of growth factors obtained from the environment.

Of considerable practical significance is the competition between strains of *Rhizobium* derived from soil and those applied with legume seeds at the time of sowing. The better competitor invades the root hairs more frequently and is responsible for a high percentage of the nodules. Should the resulting nodules be ineffective in N_2 fixation, the legume will not benefit from the root-nodule symbiosis (4).

AMENSALISM

When a diluted soil suspension is plated on a rich agar medium, many of the individual bacterial, actinomycete, and fungal colonies appearing on the petri dishes are found to be in close proximity. Nevertheless, one or more of the colonies occasionally is surrounded by a clear zone in which no other organism appears. This halo devoid of growth is good presumptive evidence that the colony surrounded by the zone of clearing is producing an *antibiotic*. An antibiotic is a substance formed by one organism that, in low concentrations, inhibits the growth of another organism. The capacity of an individual colony on a dilution plate to produce an antibiotic is confirmed by streaking the culture on fresh agar, and, after two to three days, crossing the line of growth with perpendicular streaks of one or more test species. Following a suitable period of incubation, antibiosis is observed as a suppression of the test organism. Antibiotic synthesis can also be demonstrated by ascertaining the toxicity of culture solutions following the development of the suspected microorganisms.

Many soil inhabitants produce inhibitory substances in laboratory media, and it is not difficult to isolate strains that, when tested in pure culture, suppress numerous microorganisms. The frequent isolation of antibiotic-producers demonstrates their wide distribution in soil. A variety of actinomycetes, bacteria, and fungi are able to synthesize antibiotics. Actinomycetes are particularly active in this regard, and streptomycin, chloramphenicol, cycloheximide, and chlortetracycline are but a few of the important chemotherapeutic substances synthesized by them. Most industrially prominent actinomycetes originally were obtained from soil. Antibiosis is especially common among *Streptomyces* isolates, but numerous strains of *Nocardia* and *Micromonospora* are also active. The most frequently encountered bacteria synthesizing antibiotics are species of *Bacillus* and strains of *Pseudomonas* that liberate pyocyanin and related compounds. Species of *Penicillium*, *Trichoderma*, *Aspergillus*, *Fusarium*, and other fungi also excrete antibiotic substances.

Antibiotics are effective in inhibiting or killing susceptible fungi, bacteria, and actinomycetes. Representative data are presented in Table 24.1. The relative abundance of organisms suppressing any individual test species varies markedly with the locality from which the soil sample was obtained. Furthermore, estimates of the numbers of antagonistic isolates will be governed by the species used for sensitivity determination because some organisms are inhibited by a large number of isolates while others are relatively insensitive to antibiotics. The degree of suppression therefore depends on the soil sampled, the producing strain, and the test species. In addition, many microorganisms produce more than one toxic metabolite in culture media, and each may act on a different group of organisms.

Despite the high proportion of soil inhabitants producing antibiotics in

TABLE 24.1

Percent of Actinomycetes From 15 Soils Synthesizing Antibiotics Toxic to Test Heterotrophs (21)

Test Organism	Percent of Actinomycetes Forming Antibiotics	
	October Samples	April Samples
Rhizoctonia solani	19	23
Candida albicans	10	10
Bacillus subtilis	17	24
Arthrobacter simplex	10	10
Escherichia coli	1.9	3.1

culture, the role of these organisms in the community and their significance in determining the composition of the soil microflora are unknown. Thus, though most chemotherapeutic antibiotics originate from soil-borne saprophytes, the significance of the toxic compounds in the natural environment of the active species remains a point of considerable controversy.

One of the strongest arguments for antibiosis as a natural phenomenon arises from the assumption that products synthesized by so many organisms must have some benefit to the cells producing them. Because of the abundance and ubiquity of the responsible microorganisms, it is necessary either to accept the importance of antibiosis or to postulate that the toxic products have no ecological value, being metabolic errors maintained through innumerable generations.

The sheer abundance of these microorganisms in soil may have led to an overemphasis of their ecological significance, however. The arguments opposing the view of antibiosis as a major factor regulating the composition of the community may be summarized by the following five points. (a) There is no evidence that the ability to produce antibiotics favors survival of the active cultures. Toxin synthesizers, despite their apparent competitive advantage, are not particularly more common than the innocuous organisms. (b) No relationship has been demonstrated between the predominant species in soil and their sensitivity or resistance to antibiotics as might be expected. Indeed, the predominant bacteria are generally quite sensitive to antibiotics. (c) The rapid disappearance in soil of alien microorganisms is generally not associated with the buildup of toxins effective against the invaders. Clearly, the mechanism of elimination is by means other than antibiosis. (d) Following the addition of antibiotic-forming organisms to natural soil, the active principle presumably synthesized usually cannot be detected. Often, the population in the inoculum dies off. Certain antibiotics are synthesized when the appropriate microorganism is added to sterile soil, but organic amendments are frequently required even in sterile samples. At this point, however, the habitat is no longer soil but only a laboratory medium containing sand, silt, clay, and organic matter. (*e*) Antibiotics introduced into or formed in soil may be inactivated through adsorption, by chemical reaction, or by biological decomposition.

Nevertheless, none of the arguments cited eliminates the possible microecological significance of antibiosis. Antibiotics may be a powerful force in small locales immediately surrounding the active organisms. The release of toxic products may indeed take place at those sites where conditions are favorable and the quantity of substrate is adequate. In spite of the rapid biological and chemical inactivation of many of these compounds, the zone immediately surrounding the active species may conceivably contain a concentration sufficient to exert a marked local effect yet be too small for detection by present techniques. According to this hypothesis, the influence of antibiotics is expressed only in the vicinity of the antagonist, and the phenomenon of antibiosis can be considered not as a general environmental characteristic but as a restricted although important microbial interaction. The ability of a species to colonize a microscopic locale, moreover, could well be conditioned by its ability to suppress its neighbors at that individual site. In this sense, therefore, antibiotic production is one of the several weapons in the struggle for existence in microenvironments, and it can be classified together with rapid growth, nutritional complexity, and physiological adaptability as mechanisms favoring colonization and survival in mixed populations.

Toxic substances have been observed in a wide variety of soils, however. The molecules are sometimes inorganic or are active only at reasonably high levels, and chemicals of these sorts are not generally classified as antibiotics. The chief evidence for the existence of antimicrobial substances in soil comes from studies of *fungistasis*, a topic discussed in Chapter 4. A fungistatic agent is one that inhibits but does not kill a fungus. Fungistasis affects conidia, hyphae, sclerotia, and ascospores of many but not all fungi, some species being reasonably resistant to the toxicants. The degree of inhibition is different among dissimilar soils and usually declines with depth. The number and identities of the compounds remain unknown, but volatile, nonvolatile, and heat inactivated toxins have all been noted. The fungistatic principles appear to be of microbial origin. This is suggested by experiments in which the thermolabile molecules are first destroyed by heat sterilizing the soil, and then the sterile sample is inoculated with a small quantity of untreated soil: the inoculated sample regains its power of suppressing test species. Presumably because of fungistasis, spores of sensitive organisms remain dormant for some time. The toxicity is relieved to some extent when soluble nutrients become available, and the spores then germinate. Spores of some organisms may not germinate owing to their need for carbon sources rather than their being inhibited, but the spores of those species that germinate even in distilled water fail to give rise to hyphae because of some deleterious factor rather than mere nutrient shortage. Fungistasis may even be advantageous; thus, a pathogen that emerges from a resistant stage at a distance from plant roots is susceptible to attack and elimination, but if it cannot germinate (because of the soil toxins) until the root grows nearby and releases exudations that overcome the inhibition, it is protected from destruction until the environment favors the outgrowth and subsequent penetration into host tissues.

Soils also contain one or more factors detrimental to growth of bacteria, and species of many genera are inhibited by such substances. The presence of harmful compounds can be shown simply by placing agar disks on soil for a period of time to allow for the chemicals to diffuse into the agar. The agar is then inoculated with a test bacterium, and the organism's development is evaluated after a suitable incubation period (6). Extracts of soil also show the presence of factors deleterious to bacterial replication. The antibacterial compounds have yet to be characterized.

Although the fungistatic and antibacterial compounds may—or may not —be antibiotics, some evidence can be cited suggesting that a few antibiotics may indeed be generated in nature. For example, buried straw colonized by *Cephalosporium gramineum* (the cause of a disease of winter wheat) contains the same antibiotic that the fungus makes in culture (7). Furthermore, an inhibitor produced in media inoculated with *Bacillus subtilis* is also found in soil into which the bacterium is introduced (38). Nevertheless, these and similar studies represent restricted or artificial conditions so that arguments for the ecological significance of antibiosis in soil remain tenuous. By contrast, a few fungi entering into mycorrhizal symbioses synthesize the same inhibitor in the root association as they do in culture media, and this activity may explain the well known protection afforded by diverse mycorrhizae against root infection (25, 33).

Identified microbial products generated in soil are also known to be harmful to the activities of native populations. To date, attention has been focused almost solely on simple metabolites: CO_2, NH_3, nitrite, ethylene, and sulfur compounds. The level of CO_2 necessary to inhibit conidial germination, sporulation, and mycelium development of various fungi is less than 2.0 and often lower than 1.0 percent, and CO_2 concentrations of these levels are common in the field. Moreover, the concentration would be higher in the vicinity of decomposing plant remains so that a suppression of sensitive populations by community respiration is highly likely. Many heterotrophs, nevertheless, are not influenced by CO_2 levels comparable to those found below ground. Ammonia is an effective inhibitor of *Nitrobacter*, but NH_3 produced during the decomposition of nitrogen-rich plant residues appears also to suppress some fungi (14). Moreover, one of the volatile fungal inhibitors of alkaline soils appears to be NH_3 (24), which enters the gas phase at pH values above neutrality. The nitrite that sometimes accumulates in soils is also deleterious to fungi. Nitrite is only known to build up when nitrite-oxidizers are suppressed as a result of the application of urea, anhydrous NH_3, ammonium fertilizers, or similar substances to alkaline soils or sites that become alkaline as NH_3 is formed from the urea or organic materials. A group of fungi may also be held in check because of their sensitivity to ethylene, a product of heterotrophs growing nearby in the profile (37). In addition, H_2S, methane thiol, dimethyl sulfide, and other volatile sulfur compounds inhibit selected populations and activities (5, 31).

PREDATION AND PARASITISM

Predation is one of the more dramatic interrelationships among microorganisms in nature. Of the many microscopic inhabitants of soil, the bacteria stand out as particularly prone to the attack of predators. The most numerous predators on bacteria are the protozoa, which, by feeding on the millions of bacteria, undoubtedly affect their populations. The prey, however, is never overwhelmed as the protozoa themselves are governed by the biological equilibrium. In the predator-prey relationship between protozoa and bacteria, a change in either group will bring about a qualitative and quantitative change in the other. The presence of a nutrient supply in the form of bacteria is essential for the development of soil protozoa, and large numbers of bacteria must be ingested for one protozoan cell division. The marked decline in numbers of a prey bacterium associated with the rise in protozoan numbers and the modest fall in abundance when the indigenous protozoa are suppressed are depicted in Figure 24.3. Such results suggest that protozoa are a key factor in limiting the size of bacterial populations, probably reducing the abundance of edible cells and serving to maintain a diverse community. The interactions between these predators and the organisms on which they prey have already been considered in Chapter 6, Protozoa, but it is worth emphasizing that unicellular animals seem to be (*a*) a major means for preventing the establishment of at least certain kinds of bacteria introduced into soil (9) and (*b*) significant consumers of the biomass of indigenous populations of prey cells (18).

Myxobacteria and cellular slime molds also affect bacteria by feeding directly on them. Both of these predaceous groups are common in arable land. Prior to its digestion, the bacterial cell is usually destroyed by extracellular enzymes produced by the myxobacteria, but the slime molds may consume intact prey cells. Of the two predatory groups, the myxobacteria are numerically preponderant so that their significance is probably greater. Bacteria edible by one micropredator type are often edible by others, but not a few bacteria are relatively resistant to predation. The information on predation is too scant, however, to help account for the dominance of certain bacterial genera and the scarcity of others. Myxobacteria feed not only on bacteria, moreover, for algae, fungi, and yeasts are frequently attacked (12).

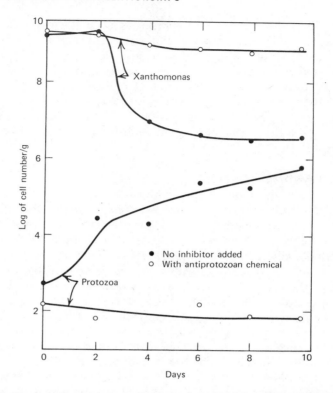

Figure 24.3. Changes in numbers of *Xanthomonas campestris* after the bacterium is introduced into Valois silt loam with and without an antiprotozoan compound (17).

Each major group in the subterranean community has parasites living on or in its cells. Bacteria of diverse genera are attacked by bacteriophages, and these bacteriophages, albeit in low numbers, are widespread. *Bdellovibrio* is similarly ubiquitous, and individual strains are capable of attacking a number of bacterial genera. However, in view of the rapid rate the vibrios lose viability in the absence of host cells and their need for high densities of hosts to initiate replication (20, 22), it is likely that bdellovibrios do not appreciably affect the composition or function of the community.

A broad array of fungi is also subject to parasitism. Hyphae, conidia, chlamydospores, oospores, zoospores, sclerotia, and other structures may thus come under direct attack and often are largely or wholly destroyed. The most extensively explored group of these organisms are those fungi that do damage to other fungi, the parasitic groups being classified in the genera *Gliocladium*, *Penicillium*, *Rhizoctonia*, and *Trichoderma*, to mention but a few. Hyphae of those hosts that grow in soil are probably not seriously affected by these parasites because the rate of their proliferation probably exceeds the rate of destruction, but resting structures may suffer considerable damage during the long periods they may be subjected to the slow destructive agents. Diverse fungi may also penetrate into lichens, there to do damage to the slowly developing components of the symbiotic association.

Protozoa are prone to parasitic attack as well. Several kinds of bacteria and a few fungi have the capacity to penetrate into the metabolically active animal cells but not the cysts, the bacteria multiplying within the host's cell and sometimes killing the animal and bringing about its lysis (11). In addition, specialized fungi penetrate into protozoa, especially amebae, kill the cells, and then proceed to make use of the cytoplasmic contents.

Lysis is a widespread and an apparently significant phenomenon. Lysis in soil probably usually involves either (*a*) digestion of the walls of cells or filaments of susceptible species by means of extracellular enzymes excreted by lytic populations (*heterolysis*), the organism with the weakened or digested wall then being unable to maintain its structural integrity and viability, or (*b*) a self-destruction by enzymes produced by the cell or hypha that is digested (*autolysis*). Antibiotics or other inhibitors excreted by one population may be the cause of autolysis of a second population. In some instances, the substance responsible for autolysis may be a metabolite that prevents cell wall biosynthesis by the susceptible cells, and an organism that continues its growth while being unable to make walls will soon become nonviable. In addition to lysis associated with actions of neighboring heterotrophs, microorganisms may autolyze merely because of nutrient deficiency.

A diversity of fungi is subject to heterolysis by enzymes excreted by actinomycetes and bacteria. If the mycelium of a susceptible species is examined microscopically, bacteria and actinomycetes are frequently seen to develop along the hyphae, especially adjacent to filaments undergoing decomposition. Some of the bacteria or actinomycetes are undoubtedly growing on hyphae that may have succumbed for other reasons, but it is likely that many of the organisms are responsible for the degradation. Not only may the vegetative structures be destroyed but so too may conidia and sporangiospores, but at a slow rate. Attacked also, but usually only very slowly, are chlamydospores and sclerotia (8, 28). Organic matter additions or other treatments that promote germination or outgrowth from the resistant body and favor development of the hyphae frequently lead to a rapid decline in the fungus because the filaments are the structures more prone to lysis. The responsible agents appear to be chiefly strains of *Streptomyces, Nocardia, Bacillus,* and *Pseudomonas.*

Less attention has been given to the lysis of other members of the community. Some algae are easily destroyed by enzymatic heterolysis, but probably many are quite refractory. Strains of myxobacteria, other bacteria, and *Streptomyces* are capable of thus bringing about the elimination of unicellular and filamentous blue-green and green algae. The lysis of bacteria in nature has scarcely been explored because of their small size, but studies in culture have disclosed that *Myxococcus, Polyangium,* and other myxobacteria release extracellular enzymes that digest a variety of bacteria, the predatory action of these myxobacteria depending on this type of digestion. The lysis of bacteria also can be accomplished in vitro by strains of *Bacillus, Flavobacterium, Micromonospora, Pseudomonas,* and *Streptomyces.*

Insofar as is presently known, enzymatic heterolysis generally involves release by the lytic heterotroph of enzymes that depolymerize components of the suscept's wall that are essential for maintenance of the integrity of the cell or filament. Although the list of macromolecules that are thus digested during lysis is undoubtedly incomplete, the chief constituents presently seem to be the following: (*a*) cellulose, a β-(1 → 3)-glucan, and/or other polysaccharides in fungi and certain algae, (*b*) a peptidoglycan in many bacteria and some blue-green algae, (*c*) chitin or other polysaccharides containing *N*-acetylglucosamine in a spectrum of fungi, and (*d*) possibly chitin, cellulose, and proteins in the surfaces of cysts of various protozoa. Thus, organisms producing cellulase or both a β-(1 → 3)-glucanase and chitinase destroy the walls and hence cause lysis of a number of fungi (2, 36), heterotrophs excreting cellulase or peptidoglycan-hydrolyzing enzymes bring about the destruction of certain algae (16), and fungi elaborating chitinase, cellulase, and protease degrade the cysts of some amebae (39). Nevertheless, the endospores of *Bacillus* and *Clostridium*, occasional hyphae, and various resting structures of fungi, actinomycete conidia, many protozoan cysts, and a few algae persist without growth in moist soils, and hence their surfaces must in some way be protected. Only recently has research been

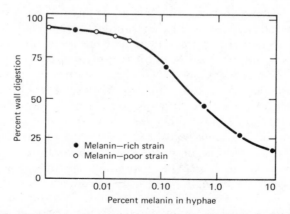

Figure 24.4. Relation between melanin content of walls of two *Aspergillus nidulans* strains and their digestion by chitinase and β-(1 → 3)-glucanase (26). The melanin-poor culture is a mutant of the melanin-rich fungus. The different melanin levels were achieved by collecting the hyphae at different ages.

initiated to find the chemical basis of this protection. Microscopic evidence shows that brown or black chlamydospores, conidia, sclerotia, and even hyphae of diverse fungi are resistant to lysis whereas nonpigmented structures of the same or closely related species are digested. Apparently, it is the melanin, a dark pigment found in some organisms, that shields the polysaccharides and thus protects the wall and the melanized fungus from elimination (3, 26). The resistance of melanized structures and the enzymatic attack on nonpigmented walls is illustrated in Figure 24.4. In other refractory fungi as well as possibly in algae that contain no melanins, the resistance to elimination by lytic soil inhabitants seems attributable to the presence in the organism's walls of polysaccharides composed of several different sugars; these so-called heteropolysaccharides are not readily available as substrates. On the other hand, sporopollenin and ligninlike components are responsible for the resistance of the walls and hence contribute to the survival in nature of other microbial species (1, 15).

REFERENCES

Reviews

Ahmadjian, V. and M. E. Hale, eds. 1973. *The lichens.* Academic Press, New York.

Alexander, M. 1971. *Microbial ecology.* Wiley, New York.

Clark, F. E. 1965. The concept of competition in microbial ecology. In K. F. Baker and W. C. Snyder, eds., *Ecology of soil-borne plant pathogens.* Univ. of California Press, Berkeley, pp. 339–345.

Madelin, M. F. 1968. Fungi parasitic on other fungi and lichens. In G. C. Ainsworth and A. S. Sussman, eds., *The fungi.* Academic Press, New York, vol. 3, pp. 253–269.

Marx, D. H. 1972. Ectomycorrhizae as biological deterrents to pathogenic root infections. *Annu. Rev. Phytopathol.,* 10:429–454.

Park, D. 1967. The importance of antibiotics and inhibiting substances. In A. Burges and F. Raw, eds., *Soil biology.* Academic Press, New York, pp. 435–447.

Starr, M. P. and Huang, J. C.-C. 1972. Physiology of the bdellovibrios. *Advan. Microbial Physiol.,* 8:215–261.

Literature Cited

1. Ballesta, J.-P. G. and M. Alexander. 1972. *J. Bacteriol.,* 106:938–945.
2. Bartnicki-Garcia, S. and E. Lippman. 1967. *Biochim. Biophys. Acta,* 136:533–543.
3. Bloomfield, B. J. and M. Alexander. 1967. *J. Bacteriol.,* 93:1276–1280.
4. Bohlool, B. B. and E. L. Schmidt. 1973. *Soil Sci. Soc. Amer. Proc.,* 37:561–564.
5. Bremner, J. M. and L. G. Bundy. 1974. *Soil Biol. Biochem.,* 6:161–165.
6. Brown, M. E. 1973. *Can. J. Microbiol.,* 19:195–199.
7. Bruehl, G. W., R. L. Millar, and B. Cunfer. 1969. *Can. J. Plant Sci.,* 49:235–246.
8. Chu, S. B. and M. Alexander. 1972. *Trans. Brit. Mycol. Soc.,* 58:489–497.
9. Danso, S. K. A., S. O. Keya, and M. Alexander. 1975. *Can. J. Microbiol.,* 21:884–895.
10. Dazzo, F., P. Smith, and D. Hubbell. 1974. *J. Environ. Qual.,* 2:470–473.
11. Drozanski, W. 1963. *Acta Microbiol. Polon.,* 12:9–23.
12. Dworkin, M. 1966. *Annu. Rev. Microbiol.,* 20:75–106.

13. Finstein, M. S. and M. Alexander. 1962. *Soil Sci.,* 94:334–339.
14. Gilpatrick, J. D. 1969. *Phytopathology,* 59:973–978.
15. Gunnison, D. and M. Alexander. 1975. *Appl. Microbiol.,* 29:729–738.
16. Gunnison, D. and M. Alexander. 1975. *Can. J. Microbiol.,* 21:619–628.
17. Habte, M. and M. Alexander. 1975. *Appl. Microbiol.,* 29:159–164.
18. Heal, O. W. 1967. In O. Graff and J. E. Satchell, eds., *Progress in soil biology.* North Holland Publishing Co., Amsterdam, pp. 120–125.
19. Heckmann, K. 1975. *J. Protozool.,* 22:97–104.
20. Hespell, R. B., M. F. Thomashow, and S. C. Rittenberg. 1974. *Arch. Microbiol.,* 97:313–327.
21. Ishizawa, S., M. Araragi, and T. Suzuki. 1969. *Soil Sci. Plant Nutr.,* 15:214–221.
22. Keya, S. O. and M. Alexander. 1975. *Soil Biol. Biochem.,* 7:231–237.
23. Klein, D. A. and L. E. Casida, Jr. 1967. *Can. J. Microbiol.,* 13:1461–1470.
24. Ko, W. H., F. K. Hora, and E. Herlicska. 1974. *Phytopathology,* 64:1398–1400.
25. Krywolap, G. N., L. F. Grand, and L. E. Casida, Jr. 1964. *Can. J. Microbiol.,* 10:323–328.
26. Kuo, M.-J. and M. Alexander. 1967. *J. Bacteriol.,* 94:624–629.
27. Lindsey, D. L. 1965. *Phytopathology,* 55:104–110.
28. Lockwood, J. L. 1967. In T. R. G. Gray and D. Parkinson, eds., *The ecology of soil bacteria.* Liverpool Univ. Press, Liverpool, pp. 44–65.
29. Lowe, W. E. and T. R. G. Gray. 1973. *Soil Biol. Biochem.,* 5:449–462.
30. Martin, M. M. 1970. *Science,* 169:16–20.
31. Mitchell, R. and M. Alexander. 1962. *Soil Sci.,* 93:413–419.
32. Nelson, G. A. and J. L. Neal, Jr. 1974. *Plant Soil,* 40:581–588.
33. Richard, C., J.-A. Fortin, and A. Fortin. 1972. *Can. J. For. Res.,* 1:246–251.
34. Rouatt, J. W. 1967. In T. R. G. Gray and D. Parkinson, eds., *The ecology of soil bacteria.* Liverpool Univ. Press, Liverpool, pp. 360–370.
35. Salonius, P. O., J. B. Robinson, and F. E. Chase. 1970. *Plant Soil,* 32:316–326.
36. Skujins, J. J., H. J. Potgieter, and M. Alexander. 1965. *Arch. Biochem. Biophys.,* 111:358–364.
37. Smith, A. M. 1973. *Nature,* 246:311–313.
38. Vasudeva, R. S., P. Singh, P. K. Sen Gupta, M. Mahmood, and B. S. Bajaj. 1963. *Ann. Appl. Biol.,* 51:415–423.
39. Verma, A. K., M. K. Raizada, O. P. Shukla, and C. R. Krishna Murti. 1974. *J. Gen. Microbiol.,* 80:307–309.
40. Yeoh, H. T., H. R. Bungay, and N. R. Kreig. 1968. *Can. J. Microbiol.,* 14:491–492.
41. Zentmyer, G. A. 1965. *Science,* 150:1178–1179.

25
Microbiology of the Rhizosphere

The root system of higher plants is associated not only with an inanimate environment composed of organic and inorganic substances but also with a vast community of metabolically active microorganisms. The microflora that responds to the presence of living roots is distinctly different from the characteristic soil community, the plant creating a unique subterranean habitat for microorganisms. The plant, in turn, is markedly affected by the populations it has stimulated since the root zone is the site from which inorganic nutrients are obtained and through which pathogens must penetrate. Consequently, interactions between the macro- and the microorganism in this locale can have a considerable significance for crop production and soil fertility. This unique environment under the influence of plant roots is called the *rhizosphere*.

The rhizosphere is often divided into two general areas, the inner rhizosphere at the very root surface and the outer rhizosphere embracing the immediately adjacent soil. The microbial numbers are larger in the inner zone where the biochemical interactions between microorganisms and roots are most pronounced. The root surface and its adhering soil are sometimes termed the *rhizoplane*. In the rhizosphere and rhizoplane, the higher organism contributes excretory products and sloughed-off tissue; most species in the subterranean flora probably have no detrimental influence on the plant harboring them. Indeed, certain benefits are derived from the microscopic organisms.

MICROFLORA OF THE ROOT ZONE

The rhizosphere region is a highly favorable habitat for the proliferation and metabolism of numerous microbial types. The community has been investigated intensively by microscopic, cultural, and biochemical techniques. Microscopic characterizations are of considerable value for they show the types of organisms present and their physical association with the outer tissue surface. For cultural investigations, the plant is carefully removed from the field or from the greenhouse pot and the superfluous soil dislodged by gentle agitation. The roots and the adhering soil are placed in a tared flask containing a known volume of sterile diluent. Dilution series are prepared and plate counts made. The biochemical techniques used in rhizosphere investigations are numerous, and they are designed to measure a specific change brought about by the plant or by the microflora.

Microscopic examination reveals the presence of a vast microbial community surrounding and on the surfaces of roots and root hairs. Bacteria are found to be localized in colonies and chains of individual cells. Filamentous fungi and actinomycetes are observed but not as frequently. The colonization of wheat roots by bacteria and fungi is shown in Figure 25.1. Protozoa are relatively conspicuous, particularly the small flagellates and large ciliates; they are situated in the water films on the root hairs and on the epidermal tissue. Microscopic studies further show that the community at a short distance from the root is little affected by the plant while soil immediately adjacent to the root contains an abundance of bacteria.

Plate counts similarly reveal the stimulation. At the same time, cultural methods show the selective enhancement of certain categories of bacteria. The root influence, as measured by plating techniques, is often expressed as a *rhizosphere effect*, a stimulation that can be put on a quantitative basis by the use

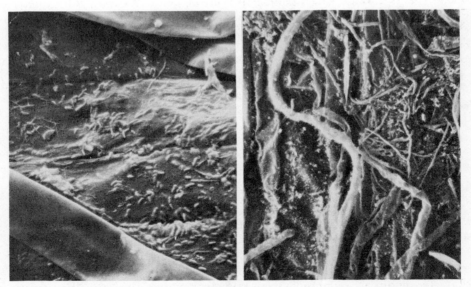

Figure 25.1. Scanning electron photomicrographs of areas of wheat roots (left) densely colonized by bacteria, × 1400, (23) and (right) colonized by fungal hyphae, × 270 (24).

of the R:S relationship. The R:S ratio is defined as the ratio of microbial numbers per unit weight of rhizosphere soil, R, to the numbers in a unit weight of the adjacent nonrhizosphere soil, S. The rhizosphere effect is consistently greater for bacteria than for the other microbial inhabitants. Soil samples taken progressively closer to the root system have increasingly greater bacterial numbers while the fungi and actinomycetes may become more abundant, but the rise in viable counts of the filamentous microorganisms is usually slight.

The bacteria reacting to the presence of the root belong to several distinctly different physiological, taxonomic, and morphological groups. Those responding most markedly are the short, gram-negative rods, which almost invariably make up a larger percentage of the rhizosphere than of the normal soil flora. The percentage incidence of short, gram-positive rods, coccoid rods, and spore-forming bacteria (*Bacillus* spp.) declines. There apparently is no selective stimulation or inhibition of the gram-variable rods, the *Arthrobacter* group, cocci, or of the long, non-spore-forming rods. On a generic basis, *Pseudomonas, Flavobacterium, Alcaligenes,* and occasionally *Agrobacterium* frequently are especially common. It is on these genera that the root effect appears to be most pronounced. Many other bacteria are found in this zone, particularly species of *Arthrobacter, Brevibacterium, Corynebacterium, Micrococcus, Xanthomonas, Serratia, Bacillus,* and *Mycobacterium,* but they apparently are not as well suited to the environment. Anaerobic bacteria also are affected by the root; this may be attributed to the reduced O_2 tension resulting from root and microbial respiration.

The bacterial density in the rhizosphere is enormous. Counts by plating on test media frequently give values in excess of 10^9 per gram of rhizosphere soil, whereas direct microscopic counts sometimes give numbers tenfold higher either in rhizosphere soil or in the rhizoplane. The bacteria, moreover, may cover 4 to 10 percent of the root area (17, 25). Furthermore, the bacteria are not randomly distributed on the root surface, instead, they appear in profusion only at particular microsites. On some plants, wheat for example, they are rare near root tips but develop prolifically in the root hair region of young roots or in proximity to older roots (23).

Because the bacterial mass is so great, there must be intense competition. In the stress resulting from a large community, fast-growing organisms might be favored because their rapid growth would enable them to compete more effectively. Indeed, representative isolates from the rhizospheres of many plant species tend to develop more rapidly than bacteria from fallow soil. At the same time, the biochemically more active organisms are favored to the detriment of less versatile strains. This suggests that the rhizosphere flora has a greater ability to effect rapid biochemical changes than the organisms of fallow land.

Schemes for grouping bacteria by their nutritional complexity have been especially useful in studies of the rhizosphere microflora. Applied to the bacteria surrounding the root, the nutritional classification reveals a preferential enhancement of organisms stimulated by or requiring amino acids and those proliferating in the absence of preformed growth factors (28). The preferential enhancement is seen in the greater percentage abundance of these two nutritional groups in rhizosphere than in control soil, a stimulation that is superimposed on the general increase in numbers. Simultaneously, the proportion of bacteria with a complex nutrition declines although their actual numbers increase.

The selection for bacteria whose development is enhanced by amino acids is undoubtedly associated with an increased level of amino acids in this environment. The amino compounds may be derived from plant exudates, from the decomposition of the nitrogenous constituents of dead root tissue and microbial cells, or from excretions of the microscopic inhabitants. Nonsterile sand or sand-soil mixtures supporting plant growth contain a number of amino acids, and plants cultivated under asepsis also liberate amino substances but in much smaller quantities. These compounds are utilized by microorganisms to satisfy their amino acid demand. Moreover, many of the bacteria that require no added growth factors excrete amino acids; therefore, the stimulation in the root zone may arise from the activities of the microorganisms as well as of the macroorganisms.

In contrast to their effects on bacteria, roots do not appreciably alter the total counts of fungi. On the other hand, specific fungal genera are stimulated; that is, the influence is selective for the type rather than the total number. Furthermore, the spectrum of genera varies with plant species and age and the kind of soil. Differing from the relatively high proportion of the fungal colonies on soil dilution plates that arise from spores, the fungus units in the rhizosphere occur, to a large extent, in the vegetative state. Although the plate counts of fungi may not be appreciably increased in the rhizosphere, the mycelium biomass may be extensive; thus, even roots of young plants may have 12 to 14 mm of hyphae/sq mm of root surface and occupy 3 percent of the surface area (25). Yeasts may at times also be abundant, but they are never as common as the bacteria.

Zoospores of *Phytophthora*, *Pythium*, *Aphanomyces*, and probably other fungi are strongly attracted to roots. This movement of the motile spores is apparently in response to particular chemical compounds that are excreted, and the cells move toward sites behind root tips and wounds from which the substances are exuded (14). Inasmuch as many of these fungi are pathogens, the initial observations of this attraction phenomenon were followed by tests to determine whether the response was only to hosts. However, it soon became clear that the movement was also in the direction of sites on roots the microorganisms could not parasitize.

As a rule, actinomycetes, protozoa, and algae are not significantly benefited by their proximity to roots, and the R:S ratios rarely exceed 2 or 3:1. Under certain circumstances such as around roots of old plants, R:S ratios for these microbial groups may become high. Because of the large bacterial community, an increase in the number or activity of protozoa is not unexpected. Flagellates and amebae dominate, and ciliates tend to be rare in this region.

The flora of the rhizosphere is affected by a number of factors. Proximity of the soil sample to the root is particularly important, and the bacterial count increases in samples taken progressively closer to the tissue surface. Simultaneously, the total activities of the community, measured by CO_2 evolution, are enhanced by closeness to the root. The depth of sampling is another important ecological variable, and in agreement with results obtained for fallow soil, the frequency of bacteria, fungi, algae, and of most physiological categories of bacteria declines with depth.

Different plant species often establish somewhat dissimilar subterranean floras. The differences are attributed to variations in rooting habits, tissue composition, and excretion products of the macroorganism. As a rule, legumes engender a more pronounced rhizosphere effect than grasses or grain crops, and alfalfa and several clovers have an especially pronounced influence on bacteria. Biennials, because of their long growth period, exert a more prolonged stimulation than annuals. At the same time, individual plant species cause a striking response in one or two bacterial genera, for example, *Pseudomonas* or *Agrobacterium*. The cause of the qualitative and quantitative differences among plants is not known.

The age of the plant also alters the underground flora, and the stage of maturity controls the magnitude of the rhizosphere effect and the degree of response by specific microorganisms. A stimulation is detectable in very young seedlings, and hence it would seem that the microorganisms are responding to root excretions rather than to dead tissues undergoing decomposition. During later development, however, dead and sloughed-off tissue may contribute appreciably to the composition of the community. On the other hand, near the very end of the growing season when the roots are dying, the readily available carbohydrates are quickly metabolized, and the microbial abundance declines. In time, the large rhizosphere community gradually fades, becoming indistinguishable from the normal soil flora. There is little if any residual microbiological effect carried over to the following year; the new vegetation largely determines its own rhizosphere composition. The results of the biochemical transformations, nevertheless, may persist; for example, the contribution of *Rhizobium* to the nitrogen status of soil is reflected in the yields of succeeding crops. But the composition of the microflora itself reverts to its original state with a diminution in the number of non-spore-formers and an increase in the relative abundance of spore-forming bacilli.

The microscopic inhabitants of fallow land and nonrhizosphere habitats respond greatly to additions of organic materials, but the same is not true within the root zone. Crop residues, animal manure, and chemical fertilizers commonly cause no appreciable qualitative or quantitative changes in the microflora of the root region. Other soil treatments also have little influence on the total number of organisms. In general, the character of the vegetation is more important than the fertility level of the soil. Different plant species in the same field have widely divergent numbers of organisms in their rhizospheres while the composition and size of the community under the same species cultivated in fields of greatly differing fertility status fluctuate only to a moderate extent.

Because the plant plays a greater role than the soil, the composition of the root's excretions and the chemical constituents of its tissues probably determine to a large extent the microbiological composition of the environment. The mass of viable cells so close to the root indicates that the plant is excreting and sloughing off large quantities of organic substances, with the products encountered by the microorganisms varying from plant to plant. Some of the substances thus released have been characterized, but the list is far from complete. In Table 25.1 is presented a list of some of the characterized compounds in the exudates. These substances are largely true excretions, not the result of sloughing off or decomposition, since the compounds are generally isolated from aseptically grown plants in the early phases of development. The kinds and yields of products are governed by the inorganic nutrients available, temperature, light intensity, O_2 and CO_2 level, root injury, and plant age. Most of these molecules are excellent substrates, and they are metabolized readily when provided to cultures of any of a large assortment of heterotrophs.

INFLUENCE OF THE PLANT

The microflora is affected in many ways by the growing plant, and microbial reactions important to fertility may be more rapid in the root environment than in nonrhizosphere soil. Undoubtedly, the most important plant contribution to the rhizosphere flora is the provision of excretion products and sloughed-off

TABLE 25.1
Products Excreted by Plants Grown Under Aseptic Conditions

Amino acids: Essentially all naturally occurring amino acids.

Organic acids: Acetic, butyric, citric, fumaric, glycolic, lactic, malic, oxalic, propionic, succinic, tartaric, valeric.

Carbohydrates: Arabinose, deoxyribose, fructose, galactose, glucose, maltose, mannose, oligosaccharides, raffinose, rhamnose, ribose, sucrose, xylose.

Nucleic acid derivatives: Adenine, cytidine, guanine, uridine.

Growth factors: p-Aminobenzoate, biotin, choline, inositol, nicotinic acid, pantothenate, pyridoxine, thiamine.

Enzymes: Amylase, invertase, phosphatase, protease.

Other compounds: Auxins, glutamine, glycosides, HCN, p-hydroxybenzoate, peptides, saponin, scopoletin.

tissue to serve as sources of energy, carbon, nitrogen, or growth factors. The rapidity with which carbon derived from the atmosphere can be made available to the underground inhabitants is evident from experiments involving exposure of seedlings to $^{14}CO_2$; in periods as short as three to four hours, ^{14}C may be found in the root exudates (19). At the same time, the macroorganism assimilates inorganic nutrients, thereby lowering the concentration available for microbial development. Microorganisms are also affected by root respiration, which alters the pH or the availability of certain inorganic nutrients by the evolution of CO_2. The pH of the rhizosphere may also be lower than that of the surrounding soil when the roots are assimilating ammonium but higher if nitrate is being utilized (26). Root penetration also improves soil structure, and the improved structural relationships favor microbial oxidations.

In the respiration of the root, O_2 is consumed and CO_2 liberated. The utilization by the large microbial community of carbonaceous substrates also leads to the release of CO_2 and the utilization of O_2. Therefore, the respiration of both macro- and microorganisms results in a greater CO_2 production from rhizosphere than from nonrhizosphere soil and also a greater rate of O_2 depletion. To assess factors affecting the microbial contribution to the gaseous exchange, CO_2 production of soil removed from around the roots is compared with that of soil taken at a distance away or from comparable fallow sites. To determine with greater accuracy the microbial contribution, the rate of CO_2 evolution from sterile and nonsterile roots is measured. Comparisons of this type reveal that one-third to two-thirds of the carbon mineralized is the result of microbial respiration.

The large quantities of CO_2 liberated by the rhizosphere inhabitants undoubtedly influence crop nutrition. By forming carbonic acid, the gas can cause a solubilization of insoluble, inorganic nutrients not readily available to the plant. This would effectively increase the supply of assimilable inorganic nutrients. By this means, the level of available phosphorus, potassium, magnesium, and calcium may rise. The solubilization phenomenon can be demonstrated by allowing sterile seedlings to grow in sterile soil containing polished marble. Under these conditions, the seedlings etch the marble surface but only to a slight extent. If the sterile system is duplicated except for the introduction of an inoculum of selected bacteria, the etching becomes far more pronounced. The greater change in the presence of the bacteria arises from the increased production of carbonic acid from the CO_2 respired by the microorganisms.

One of the physiological groups characteristically responding to the presence of living roots is the ammonifying bacteria. R:S ratios for these bacteria are quite high, varying up to values of several hundred to one. Similarly, many of the rhizoplane residents have the potential to attack proteins in test media, and the protease activity of the colonized root surface may be quite high (29). The stimulation of such heterotrophs may arise in part from organic nitrogen compounds present in the environment, but ammonifying and proteolytic bacteria are not substrate specific, and their response may be attributable to other environmental factors. Should the selective enhancement be directly related to the capacity to mineralize nitrogenous materials, then the large proteolytic and ammonifying flora should bring about a rapid decomposition of organic nitrogen in the rhizosphere. However, chemical analysis shows that less nitrate is found in certain cropped than in comparable uncropped soils even if the nitrogen removed by the plants is included in the calculations as inorganic nitrogen (Figure 25.2). The field evidence supports the hypothesis that cropping diminishes nitrogen mineralization. On the other hand, the inorganic nitrogen status at any given time represents a balance between immobilization

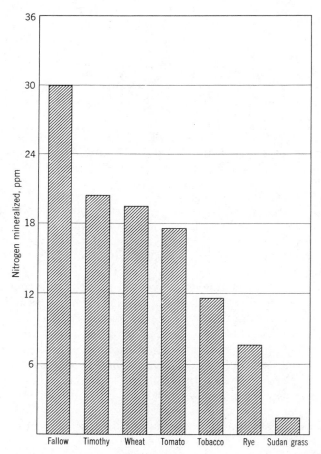

Figure 25.2. Nitrogen mineralized after 13 weeks in fallow and in cropped soil. The nitrogen content of the plants is included in the calculation of nitrogen mineralized (12).

and mineralization. Commonly, the net amount of nitrogen mineralized is significantly lower in cropped than in fallow soils, an apparently reduced mineralization rate when the microflora is under the influence of the plant. However, the lower *net* mineralization can result from an enhancement of immobilization reactions by the large community associated with the roots. Immobilization can be appreciable because organic substances are abundant surrounding the root, and nitrogen is required for their decomposition. The final answer on the fate of nitrogen necessitated the use of ^{15}N. Investigations performed with the isotopic tracer have revealed that, despite the fact that the net amount of nitrogen mineralized in soils supporting nonlegumes is about half that in fallow soils, the total quantity of nitrogen mineralized is greater in cropped than in uncropped soils. The differences between the two environments are attributable to the high immobilization rates associated with the rhizosphere community (4). The vast quantity of carbon excreted and the large supply of dead roots in permanent grasslands provide much energy to the heterotrophs associated with the grass, and here immobilization undoubtedly is pronounced; such immobilization may explain why these grasslands have little inorganic and much organic nitrogen (15).

Soil microbiologists have long directed considerable attention to the possibility of promoting N_2 fixation by free-living bacteria colonizing roots of nonlegumes, but only following the introduction of sensitive and simple techniques for assessing the rate of this reaction has it been possible to assess the previously poorly substantiated claims of appreciable nitrogen gains resulting from the bacterial activities. The abundance of nitrogenase-containing strains of *Azotobacter paspali*, *Beijerinckia*, and *Spirillum* on roots of cereals, forage grasses, and other nonlegumes and the recent and well-substantiated evidence of N_2 fixation by these microorganisms have already been discussed. Even with this large body of information, why only certain plants support these bacteria, the identities of the excretions that stimulate the N_2 utilizers, and the actual benefits gained by the vegetation remain to be resolved.

If nitrate is present, denitrification may be appreciable in the root region, and both N_2 and N_2O evolution are increased by plant development (27). This enhancement may result from the responsible bacteria using the exudates as energy sources or from the lowered O_2 levels associated with the active respiration. Coinciding with the higher rate of N_2 and N_2O production is the greater density of denitrifying bacteria in the rhizosphere.

The number of ammonium- and nitrite-oxidizing autotrophs is not markedly influenced by their proximity to the roots of many species of agronomic importance, and the nitrification of ammonium salts proceeds at similar rates in samples from the root environs and from fallow areas. With certain kinds of natural vegetation, however, the abundance of nitrifiers may be drastically reduced. Cellulolytic bacteria, on the other hand, are more prevalent in the root zone and decrease in density in samples taken at a distance from the plant. The dominant cellulose-digesting bacteria of the rhizosphere frequently are the cytophagas and short rods. The cellulolytic flora may well be responding to the availability of large quantities of cellulosic tissues, and this population is undoubtedly a factor concerned with the degradation of the sloughed-off root material. The products of the metabolism of cellulolytic organisms can provide carbonaceous substrates for other microorganisms.

Root excretions have a pronounced influence on germination of the resting structures of several fungi. Thus, the chlamydospores of *Fusarium,* conidia of *Verticillium*, sclerotia of *Sclerotium,* and oospores of *Pythium* will germinate in proximity to the root, in the presence of isolated exudates, or when provided with individual compounds found in the excretions. The fungi probably benefit because they obtain sources of energy. This stimulus to germination is especially important to plant pathogens that are not vigorous competitors and remain in the resting stage because of nutrient shortages or fungistasis. The compounds in the rhizosphere cause the pathogen's resting stages to germinate and allow the hyphae that emerge to grow sufficiently to penetrate the nearby roots before lysis destroys the filaments in the soil itself. As a rule, the germination and subsequent hyphal development are promoted by nonhost species and also by both susceptible and resistant varieties of host plants, but *Sclerotium cepivorum* is an exception. This root-invading fungus persists in soil for many years as sclerotia. These sclerotia rarely germinate in soil or near the roots of a variety of plants, but they will germinate readily when in the vicinity of roots of the parasite's host, members of the genus *Allium*, or when provided with exudates of *Allium* roots (9).

Roots also may liberate antimicrobial agents. In some instances, these are antifungal substances, and a single root system may give rise to a number of such toxicants (8). Even the CO_2 produced in profusion in this habitat may inhibit germination or affect fungi in other ways. Because of the little nitrate in soils under perennial grass, a hypothesis was advanced that the nitrifying autotrophs are suppressed in the grassland by a toxin generated by the root system. Subsequently, several investigators tested the inhibitor hypothesis experimentally, and evidence has been obtained that some (but not all) plants generate compounds toxic to the nitrifiers (20, 22).

INFLUENCE OF THE MICROFLORA

The rhizosphere community may have either a favorable or a detrimental influence on plant development. Because the microflora is so intimately related with the root system, partially covering its surface, any beneficial or toxic substance produced can cause an immediate and profound response. In the previous discussion, the plant-induced changes in the root zone microflora have been reviewed. Modifications in the abundance of microorganisms or in the relative proportions of individual groups will in turn affect the plant through the microbiologically catalyzed reactions. The production of CO_2 in the rhizosphere and the formation of organic and inorganic acids aid in the solubilization of inorganic plant nutrients. At the same time, the vast microscopic community demands a variety of anions and cations for its own development, and immobilization of nitrogen or phosphorus may assume prominence. Aerobic bacteria remove O_2 from the environment and add CO_2, and either the lowering of the O_2 or increasing the CO_2 tension may reduce root elongation and development or diminish the rate of nutrient and water uptake. The rhizosphere microflora may, however, favor plant development by producing growth-stimulating substances, contributing to the formation of a stable soil structure, releasing elements in organic forms through the mineralization of organic complexes, and by entering into symbiotic root associations. Proposing a mechanism for a beneficial or detrimental relationship is far simpler than providing the experimental results, but evidence for several specific associations has been obtained.

The most direct approach to the establishment of the significance of rhizosphere microorganisms is by comparing plant growth in sterile and nonsterile environments. Typically, the rate of development is more rapid in sterile soil receiving an inoculum of organisms than in the uninoculated, sterile controls. Explaining the response, nevertheless, is difficult. It is known that several bacteria, actinomycetes, and fungi produce, in culture media at least, considerable amounts of growth substances that can have an influence on plants. Indoleacetic acid, gibberellins, cytokinins, and related plant-growth regulators have been isolated from liquid media in which one or another soil isolate has been grown. Because of the large microbial community in the root area, the concentration of products of this type might be quite high, but unequivocal evidence for their biosynthesis in nature is not available. On the other hand, the view that these substances are indeed synthesized in the rhizosphere or rhizoplane is made more plausible by the observation that, when grown aseptically, wheat seedlings provided with a soil inoculum had a morphology similar to seedlings provided with gibberellic acid and indoleacetic acid (7).

Phosphorus availability is influenced by the microscopic rhizosphere inhabitants, and because crops require appreciable quantities, changes in the assimilable phosphate concentration are of considerable consequence. A high percentage of rhizosphere and rhizoplane bacteria is able to degrade organic phosphorus substrates, and the total numbers of these heterotrophs are similarly increased in the vicinity of actively metabolizing roots (13). Although phosphorus mineralization may thus be more rapid, the net quantity of inorganic phosphorus that is liberated is governed by the relative rates of mineralization and immobilization. Because of the abundance of bacteria, phosphorus immobilization is probably more rapid within the rhizosphere.

Another phosphorus transformation of agronomic significance is the solubilization of insoluble phosphate-containing compounds. Bacteria associated with the root system may be of assistance in rendering available several substances that are poorly soluble. A comparison of the yields of plants grown in sterile and in nonsterile environments containing insoluble phosphate sources reveals that the response to the chemicals is greater where microorganisms are active. Support for the hypothesis that the root microflora is important in this transformation is found in the selective stimulation in the rhizosphere of microorganisms capable of dissolving insoluble calcium phosphates.

Even the absorption of soluble inorganic phosphate may be affected by heterotrophs. As shown in Figure 25.3, more phosphate is taken up by roots colonized by microorganisms than those maintained aseptically. The greater assimilation by plants supporting bacteria on their belowground portions is reflected not only by more phosphorus in the roots but also in the aboveground portions (6). The assimilation of manganese, iron, zinc, and potassium may similarly be stimulated by microorganisms (2, 31). It is not at all evident how the heterotrophs enhance nutrient assimilation, but possibly growth-promoting compounds are responsible.

Microorganisms in this habitat also alter the availability or toxicity of sulfur. With ryegrass, for example, the root system increases the mineralization of

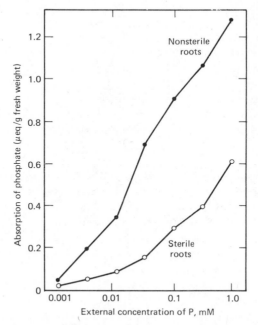

Figure 25.3. Phosphate assimilation in one hour by excised roots of barley grown under nonsterile and aseptic conditions. The phosphorus source is KH$_2$PO$_4$ (3).

sulfur in the soil organic fraction, thereby benefiting the grass (10). Conversely, the exudates of corn growing in sulfate-rich saline soils may ultimately do harm because they serve as sources of energy to sulfate-reducing anaerobes; when the exudation is intense, the level of H_2S may be so great as to kill the corn (16).

Products of microbial metabolism often have a detrimental effect on higher plants. This is strikingly evident when test plants growing in sterile sand or agar are inoculated with soil suspensions. Primary root growth is characteristically reduced and secondary roots are less abundant as compared with plants grown in the absence of the microflora. In some instances, root hair formation may be reduced (5). Experiments with individual fungi, moreover, confirm that various species elaborate phytotoxins in vitro. The intimacy of the association between microorganisms and root would magnify any such inhibitions.

Several enzymes are localized at or near the surfaces of roots, and hence the externally applied substrates for these enzymes will be metabolized by the plant tissues. However, some compounds are not transformed unless microbial development has occurred. For example, the urease activity in the root zone of barley (11) and the protease activity of wheat roots (30) are attributable to the microscopic residents.

Since the rhizosphere has an immense community, much larger than in the surrounding soil, mutual antagonisms are more pronounced. Moreover, if antibiotic formation is of consequence in natural habitats, one of the more likely environments for their production is the root zone, where the supply of energy substrates is particularly large. Several antibiotics are assimilated by higher plants through the root systems and then are translocated to above-ground portions. Should antibiotics be synthesized in the rhizosphere, their production may affect the proliferation of root pathogens and, if translocated to stems and leaves, the development of disease in above-ground tissues. Antibiotics are also known to affect the physiology of the plant entirely apart from any antimicrobial actions they might have.

PLANT PATHOGENS AND THE RHIZOSPHERE

The community of the rhizosphere is composed mainly of nonpathogenic microorganisms. But the very density and the increased microbial interactions—harmful and beneficial—can be especially important for soil-borne pathogens because the disease-producing organism must penetrate the rhizosphere in order to initiate infection. The intense biological interactions may lead to the elimination or suppression of the pathogen or, under certain conditions, they may be beneficial. The root excretions and sloughed-off tissues themselves affect the pathogen directly or, by the changes brought about in the saprophytic flora, indirectly. The common observation that soil-borne pathogenic fungi are more destructive in sterile than in normal soil indicates a role for the microflora in the development of disease; the ecological site of greatest saprophytic activity, the root surface, is undoubtedly the locale where interference with the pathogen is maximal.

The reasons for the differences in disease resistance between varieties of a single plant species remain largely obscure. Resistance often resides in a physiological or biochemical difference between the resistant and the susceptible varieties. The dissimilarity may be expressed, at least in part, by modifications in the root excretions or root tissue composition. Therefore, it is not too difficult to visualize a condition in which resistance or susceptibility is linked with the microflora of the rhizosphere. For example, the resistance of one variety may be dependent on the excretion through the roots of a substance that induces the development of a flora competitive with or antagonistic to the pathogen.

Evidence is available that there is a correlation between the rhizosphere flora and resistance to some soil-borne pathogens. Thus, varieties of wheat resistant and susceptible to common root rot, for which *Cochliobolus sativus* is the primary pathogen, contain different numbers of fungi, bacteria, and various physiological groups of bacteria. Although strains producing antibiotics toxic to *C. sativus* in vitro are abundant in the rhizospheres of these varieties, no correlation is apparent between the abundance of the antibiotic formers and the resistance of the wheat variety (1, 21). On the other hand, excretions from peas resistant but not varieties susceptible to *Fusarium solani* f. sp. *pisi* and *Pythium ultimum,* causal agents of root rot, are toxic to these fungi, suggesting that exudates are implicated in the resistance of peas to the two parasites (18).

In this light, the rhizosphere may be considered as a microbiological buffer zone in which the microflora serves to protect the plant from the attack of the pathogen. The mechanism of the buffering action is unknown. Antibiotic production by the root microflora is often cited, but rarely is there a correlation between the numbers or types of antagonists and the disease resistance of the variety. Other possible microbiological interactions have been discussed in the previous chapter. However, regardless of the explanation, the importance of nonpathogenic soil inhabitants to plant disease has been well established. Future developments will no doubt help resolve the problem.

REFERENCES

Reviews

Barber, D. A. 1968. Microorganisms and the inorganic nutrition of higher plants. *Annu. Rev. Plant Physiol.,* 19:71–88.

Brown, M. E. 1975. Rhizosphere micro-organisms—Opportunists, bandits or benefactors. In N. Walker, ed., *Soil microbiology: a critical review.* Halsted Press (Wiley), New York, pp. 21–38.

Parkinson, D. 1967. Soil micro-organisms and plant roots. In A. Burges and F. Raw, eds., *Soil biology.* Academic Press, New York, pp. 449–478.

Rovira, A. D. 1969. Plant root exudates. *Bot. Rev.,* 35:35–57.

Rovira, A. D. and C. B. Davey. 1975. Biology of the rhizosphere. In E. W. Carson, ed., *The plant root and its environment*. University Press of Virginia, Charlottesville, pp. 153–204.

Literature Cited

1. Atkinson, T. G., J. L. Neal, and R. I. Larson. 1974. *Phytopathology*, 64:97–101.
2. Barber, D. A. and R. B. Lee. 1974. *New Phytol.*, 73:97–106.
3. Barber, D. A. and U. C. Frankenburg. 1971. *New Phytol.*, 70:1027–1034.
4. Bartholomew, W. V. and F. E. Clark. 1950. *Trans. 4th Intl. Cong. Soil Sci.*, 2:112–113.
5. Bowen, G. D. and A. D. Rovira. 1961. *Plant Soil*, 15:166–188.
6. Bowen, G. D. and A. D. Rovira. 1966. *Nature*, 211:665–666.
7. Brown, M. E. 1972. *J. Appl. Bacteriol.*, 35:443–451.
8. Burden, R. S., P. M. Rogers, and R. L. Wain. 1974. *Ann. Appl. Biol.*, 78:59–63.
9. Coley-Smith, J. R. and J. E. King. 1970. In T. A. Tousson, R. V. Bega, and P. E. Nelson, eds., *Root diseases and soil-borne pathogens*. Univ. of California Press, Berkeley, pp. 130–133.
10. Cowling, D. W. and L. H. P. Jones. 1970. *Soil Sci.*, 110:346–354.
11. Estermann, E. F. and A. D. McLaren. 1961. *Plant Soil*, 15:243–260.
12. Goring, C. A. I. and F. E. Clark. 1948. *Soil Sci. Soc. Amer. Proc.*, 13:261–266.
13. Greaves, M. P. and D. M. Webley. 1965. *J. Appl. Bacteriol.*, 28:454–465.
14. Hickman, C. J. and H. H. Ho. 1966. *Annu. Rev. Phytopathol.*, 4:195–220.
15. Huntjens, J. L. M. 1971. *Plant Soil*, 34:393–404.
16. Jacq, V. and Y. Dommergues. 1970. *Zent. Bakteriol.*, II, 125:661–669.
17. Kaczmarek, W., H. Kaszubiak, and H. Guzek. 1974. *Polish J. Soil Sci.*, 6:133–139.
18. Kraft, J. M. 1974. *Phytopathology*, 64:190–193.
19. McDougall, B. M. 1968. *Trans. 9th Intl. Cong. Soil Sci.*, 3:647–655.
20. Moore, D. R. E. and J. S. Waid. 1971. *Soil Biol. Biochem.*, 3:69–83.
21. Neal, J. L., T. G. Atkinson, and R. I. Larson. 1970. *Can. J. Microbiol.*, 16:153–158.
22. Purchase, B. S. 1974. *Plant Soil*, 41:527–539.
23. Rovira, A. D. and R. Campbell. 1975. *Microbial Ecol.*, 1:15–23.
24. Rovira, A. D. and R. Campbell. 1975. *Microbial Ecol.*, 2:177–185.
25. Rovira, A. D., E. I. Newman, H. J. Bowen, and R. Campbell. 1974. *Soil Biol. Biochem.*, 6:211–216.
26. Smiley, R. W. 1974. *Soil Sci. Soc. Amer. Proc.*, 38:795–799.
27. Stefanson, R. C. 1972. *Aust. J. Soil Res.*, 10:183–195.
28. Strzelczyk, E. 1961. *Acta Microbiol. Polon.*, 10:169–180.
29. Vagnerova, K. and J. Macura. 1974. *Folia Microbiol.*, 19:525–534.
30. Vagnerova, K. and J. Macura. 1974. *Folia Microbiol.*, 19:329–339.
31. Williamson, F. A. and R. G. Wyn Jones. 1973. *Soil Biol. Biochem.*, 5:569–575.

26
Pesticides

One of the most active fields of research in soil microbiology is concerned with the relationship between pesticides and microorganisms. In the last few years, innumerable studies have established the effects of many of these compounds on indigenous populations and the ways in which the microflora alters a multitude of chemicals to which they are exposed. This extensive research has been prompted by the importance of pesticides for food production and the potential or actual environmental hazards associated with the widespread use of toxic compounds.

Pesticides are chemicals designed for the control of pest populations. Because pest species fall into widely different taxonomic categories, pesticides are commonly characterized on the basis of the kinds of organisms on which they act. Thus, insecticides, herbicides, fungicides, and nematicides are designed for the control of insects, weeds, plant pathogenic fungi, and nematodes. Other classes of pesticides are used for the suppression of rodents and mollusks, but these have not been the subject of significant attention by microbiologists.

Chemicals of these sorts may reach soil by one or more ways. Many are applied directly on the surface or are injected into the upper layers. A large number are sprayed onto foliage, and part of that which is applied is not intercepted by the foliage and falls to the ground. That portion remaining on the vegetation becomes available to the microflora when leaves fall or the treated plants die. Agricultural sprays are known to drift for reasonable distances and a few of the toxicants are volatile, and both drifting sprays and precipitation containing volatile molecules bring the compounds to the surface horizons at sites far from the place of application. Water often becomes contaminated with pesticides by one means or another, and the use of this water for irrigation similarly adds the chemicals to agricultural land.

The relationships between microorganisms and pesticides can be approached from two vantage points. On one hand, because these chemicals are specifically designed to inhibit or kill certain species, namely pests, it is quite likely that at least some may have a deleterious effect on nonpest species, including the subterranean inhabitants. On the other hand, nearly all modern pesticides are organic, and thus they could conceivably be metabolized with a resulting modification or destruction of their activity. Thus, interest has focused on the possibilities of (a) pesticides suppressing populations or functions of the indigenous community and (b) microorganisms metabolizing the chemicals so that their activity or the length of time they remain active in soil is modified.

Representatives of many dissimilar types of molecules find use in pest control. Illustrations of the sorts of chemicals that have been widely adopted in agricultural operations are presented in Figure 26.1. The various molecules often have lengthy and complicated names, so that abbreviations or common terms have been devised; thus, DDT, 2,4-D, and heptachlor are the abbreviations for 2,2-(p-chlorophenyl)-1,1,1-trichloroethane, 2,4-dichlorophenoxyacetic acid, and 1,4,5,6,7,8,8-heptachloro-3a,4,7,7a-tetrahydro-4,7-methanoindene, respectively. Despite the multitude of pesticides and the variety of chemical types, certain generalizations on the pesticide-microorganism interaction can be made so that it is unnecessary to consider each compound or each microbial process separately. Nevertheless, generalizations are not firm rules, and exceptions are to be expected and are indeed found.

EFFECTS OF PESTICIDES

The rates of application to soil of certain pesticides, such as some fungicides, are quite high so that the microflora is exposed to levels that could seriously affect individual populations. Most herbicides, by contrast, are applied at low rates so that one might expect little or no significant toxicity. On the other hand, the potency varies with the chemical; hence, the impact of low concentrations of one toxicant may sometimes be greater than a second toxicant present at a higher level in soil. Furthermore, the duration of effectiveness of a pesticide—its *persistence*—is governed by the chemical structure and environmental conditions so that the longevity of any inhibition must also be considered. Therefore, the influence of chemicals on the community or its constituent populations is determined by the particular pesticide, the concentration present, and the persistence.

Evaluations of possible harm can be conducted in several ways. The substance may be included at various levels in nutrient solutions inoculated with pure cultures or in agar media inoculated with soil dilutions; the data so obtained are useful, but interpretation of the results in ecological terms is frequently difficult because interactions between soil constituents and toxicants would not be evident from tests in laboratory media. Soil colloids adsorb or hydrolyze certain organic toxicants, both processes frequently leading to a loss of potency. Drawing conclusions solely from cultural tests is also risky inasmuch

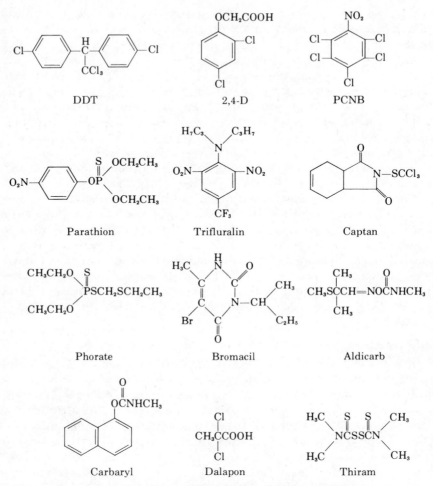

Figure 26.1. Structures of several common pesticides.

as the duration of toxicity differs in soil and culture media; also, one population may alter the introduced molecule making it less—or possibly more—harmful to a sensitive population than would be anticipated from tests involving just a single species. For these reasons, evaluations are commonly conducted directly with soil, and the rate of CO_2 evolution, O_2 consumption, nitrogen mineralization, nitrification, or another biochemical process is determined. The soil may be treated with appropriate substrates before the experiment is conducted to enhance the transformation of interest. Alternatively, population responses may be evaluated using total counts of major microbial groups or enumerations of individual genera or species in order to assess the potential harm.

The possible suppression of microbial groups has been tested with many herbicides, insecticides, and fungicides, and the abundance of a variety of dissimilar populations has been evaluated in comparisons of treated and untreated soil. Most populations are either not reduced in numbers or are not too greatly affected by those pesticides that are normally present at low concentrations, as are the herbicides and many insecticides (32, 33). Notable exceptions are algae, many of which are markedly inhibited by herbicides (4); this is not surprising inasmuch as these chemicals are chosen for their effectiveness against certain other chlorophyll-containing organisms, namely weed species. Those insecticides that alter abundance of particular hetero-trophic populations are usually the ones present in high concentration, al-though responses vary with soil type. By contrast, fungicides and fumigants specifically used for the control of soil-borne pathogens are invariably added to sufficiently high levels that a group of microorganisms—the pathogens—are suppressed, and since many indigenous heterotrophs are as susceptible to chemical stress as are the disease incitants, dramatic population modifications are commonly associated with such treatments (5, 27). The spectrum of dominant species may be drastically modified as common genera are sup-pressed and new groups come to the fore. This upset is, at times, short-lived, but the changed community may, alternatively, endure for long periods. The results of representative studies showing the presence or absence of responses in major microbial groups to individual chemicals are presented in Table 26.1.

Compounds that are found in concentrations high enough to be injurious do not affect all populations to a similar degree; one species may decline

TABLE 26.1

Effect of Pesticides on Major Microbial Groups (10)

Pesticide	Bacteria	Actinomycetes	Fungi
		Toxic Conc., ppm	
Dazomet	150	150	150
Metham	60	—	60
Nabam	50	—	50
PCP	2000	2000	2000
		Nontoxic Conc., ppm	
Aldrin	100	—	100
Atrazine	70	70	75
DDT	100	—	100
Diazinon	40	40	40
HCH	1000	1000	1000
Simazine	70	70	70

appreciably, a second may suffer only modest harm, while a third is entirely resistant (34). Differences in sensitivity also are evident in organisms having more than one form so that hyphae of fungi or vegetative cells of bacteria are affected either more or less than are fungal conidia and sclerotia or bacterial endospores. This spectrum of sensitivities is especially noteworthy among the subterranean plant pathogens, because a chemical effective for the control of one may be worthless for the control of a second. The differences in microbial response are also evident in the modification in the interactions among indigenous populations that occasionally occurs; for example, fungicides at times increase rather than decrease the incidence of plant disease, an anomalous enhancement of the damage done by the pathogenic fungi attributable not to a stimulation of the harmful species but presumably rather an inhibition of populations competing with or otherwise antagonizing proliferation of the fungus (36). This alteration in the community and the disturbance in processes of natural biological control are not uncommon with fungicides. A peculiar and as yet unexplained observation that has often been made is the finding that toxicants at low concentrations are often stimulatory (13).

Because of the importance of nitrogen transformations to soil fertility, assessments have been made of the impact of numerous chemicals on mineralization, nitrification, denitrification, legume nodulation, and N_2 fixation. As a rule, except for fungicides present in soil at the high level that may be required for disease control, nitrogen mineralization and denitrification are not particularly sensitive processes. Nitrification, by contrast, is a notably sensitive transformation (35), indeed often the process suppressed at lower concentrations than any other studied to date. A decline in the rate of ammonium oxidation or nitrate formation could have dire consequences for roots of crops not tolerant to ammonium, which might accumulate in the absence of nitrifying populations, or for those plants preferentially assimilating nitrate as a nitrogen source (22). That field rates of application of many chemicals do not retard ammonium oxidation suggests the absence of a detrimental impact on the community.

Nodulation of legumes is often reduced or abolished by pesticides. Because of their high local concentration and the intimacy of their contact with *Rhizobium* inoculated onto seeds, chemicals applied directly to seeds sometimes affect the root nodule bacteria to such an extent that nodules fail to appear and hence N_2 is not made available to the legume (37). Several approaches have been devised to overcome the incompatibility of chemical and inoculant, these usually involving spatial separation of the two, but a newly proposed method involves obtaining pesticide-resistant mutants and making inoculant preparations from these rhizobia (Table 26.2).

The consequences of a decline in rate of one or another transformation or in the numbers of a narrow or broad microbial group must be weighed in light of the usefulness of the pest-control agent. A diminution in total numbers of bacteria, fungi, *Azotobacter*, or cellulose decomposers may perturb the microbiol-

TABLE 26.2

Yield and Nitrogen Content of Inoculated, Pesticide-Treated Legumes Grown in the Greenhouse (25)

Inoculant	Pesticide-Treated Seed		Untreated Seed	
	Yield (mg)	N Content (mg)	Yield (mg)	N Content (mg)
	Alfalfa treated with thiram			
None	270	5.9	290	6.8
Thiram-sensitive *Rhizobium*	280	6.6	440	12.2
Thiram-resistant *Rhizobium*	490	13.2	500	14.6
	Cowpeas treated with phygon			
None	1150	21.7	1100	23.2
Phygon-sensitive *Rhizobium*	1820	36.4	3820	97.0
Phygon-resistant *Rhizobium*	3930	105	3790	106

ogist, but the agriculturalist may be unimpressed if he simultaneously observes an appreciable decline in disease incidence, weed infestation, or insect damage. The argument that humans and livestock consume food and feed and not soil organisms is cogent and completely pertinent to concerns with soil pollution, and the appropriate argument against the preceding view is that specific organisms or processes are not only in theory but also in fact essential to plant growth or to the continued maintenance of soil fertility. Unfortunately, with few exceptions, direct evidence for a reduction in plant development resulting from changes in abundance of microbial populations is scarce. By contrast, pesticide effects leading to a suppression of nodulation, massive population declines resulting from the use of some fungicides, and dramatic reductions in nitrification do result in undesirable plant responses.

PERSISTENCE

How long a herbicide, insecticide, or fungicide persists in soil is of great practical importance because it reflects the time that the pest will be subject to control. At the same time, a persistent pesticide has a special position in environmental pollution because it may remain in soil long enough (*a*) to be assimilated by plants and accumulate in edible portions, (*b*) to adhere to edible portions of root crops, (*c*) to be transported with eroding soil particles to nearby waterways, or (d) to accumulate in earthworms and then show up in high levels in birds feeding on the worms. Such problems are absent or are less significant for the compounds that do not endure in nature.

Synthetic organic compounds may disappear from soil by a variety of means. Some are volatile and move from the land to the overlying air. A few are transported vertically with moving water, and the groundwater may then receive undesirable quantities. A reasonable number are subject to chemical reactions, often hydrolytic, to yield nontoxic products. Such nonmicrobial conversions, although leading to an elimination of the toxicity of the original molecule, do not result in a complete degradation or mineralization inasmuch as organic products almost invariably remain following nonmicrobial reactions in nature. In many instances, however, pesticide disappearance is attributable to microbial activity, a biological contribution verified by comparing changes in concentration with time in samples of natural soil and in samples that have been sterilized or treated with inhibitors to retard microbial metabolism or growth.

Many genera of heterotrophs use pesticides as substrates, either cometabolizing the molecules or using them as nutrients. Species of *Agrobacterium, Arthrobacter, Bacillus, Clostridium, Corynebacterium, Flavobacterium, Klebsiella, Pseudomonas,* and *Xanthomonas* among the bacteria; *Alternaria, Aspergillus, Cladosporium, Fusarium, Glomerella, Mucor, Penicillium, Rhizoctonia,* and *Trichoderma* among the fungi; and *Micromonospora, Nocardia,* and *Streptomyces* among the actinomycetes modify one or more of the synthetic chemicals. The diversity of substrates is matched by a diversity of species. Nevertheless, it is rarely possible to predict which species or even which genus is responsible for a particular transformation in nature.

The many observations that microorganisms are important or essential for ridding natural environments of pesticides point to two practical matters. First, for those substances that are modified or degraded by members of the microflora, it seems plausible to expect that environmental factors governing heterotrophic populations would have a comparable influence on chemical destruction. Although the populations bringing about the transformation of individual organic compounds are generally unknown, the available information suggests that environmental factors influencing the community as a whole have comparable effects on the degradation of those pesticides known to be subject to microbial metabolism. Thus, the rate of degradation is often enhanced by increasing temperature or by raising the moisture level of dry soil, and rates of decomposition are frequently greater in soils rich rather than poor in organic matter, presumably because of the more vigorous community. Similarly, for molecules that are acted on by indigenous populations, a previous application generally leads to a greater activity on retreatment, probably because of the abundance of cells containing the needed enzymes (30).

Second, inasmuch as the soil inhabitants play a key role in detoxication and in mineralizing a multitude of organic molecules, the prolonged durability of the compounds designated as persistent is itself convincing evidence that the community has little or no action on the long-lived chemical. The molecules that thus endure have been termed *recalcitrant*, that is, they are stubborn and fail to be metabolized or mineralized at significant rates. The period of persistence is frequently given as the time required for half the chemical to be lost, but it is often expressed as the time for detectable levels of the substance to disappear entirely. However, no one expression of persistence has been accepted by all investigators. Furthermore, the life of a chemical in nature is greatly influenced by the particular soil, local conditions of temperature and rainfall, and agricultural practices. Nevertheless, typical data, although not directly applicable to all soils and all climatic regions, do indeed show that some chemicals are quite long-lived, while others soon are gone (Table 26.3). The quantity of information on longevity of such toxicants is enormous, yet the reasons for varying durations of effectiveness are still largely uncertain. This is well illustrated by parathion, an important insecticide. In some soils, nearly all the parathion is gone in 30 days (12), yet the chemical may at times be found in other areas 16 years following the last application (29).

Since many of the persistent insecticides and herbicides are extremely useful and cheap to make, or they exhibit low mammalian toxicity, considerable effort has been devoted to determine why they fail to succumb readily to microbial attack. Research designed to establish the bases for resistance has established that modest changes in structure of the chemical can greatly alter its suitability as a microbial substrate and hence its persistence; for example, movement of a chlorine from one to a second position on a molecule, removal

TABLE 26.3
Selected Data on Persistence of Pesticides (2, 16, 20, 23, 24)

Chemical	Still Detectable[a]	Approximate Half-life
Chlordane	21 yr	2–4 yr
DDT	24 yr	3–10 yr
Dieldrin	21 yr	1–7 yr
Heptachlor	16 yr	7–12 yr
Toxaphene	16 yr	10 yr
Dalapon	10 wk	—
DDVP	—	17 days
Methyl demeton S	—	26 days
Thimet	—	2 days

[a] Period of time at which measurable amounts of the pesticide were detected. The actual periods of persistence of the first four are probably much longer.

of a single chlorine, or addition of hydroxyl groups can make a compound that is not utilized into an excellent substrate. The slight differences in structure between certain easily biodegradable molecules and related refractory chemicals are depicted in Figure 26.2. In light of the small structural changes that make a good substrate into a molecule not readily metabolized microbiologically, it

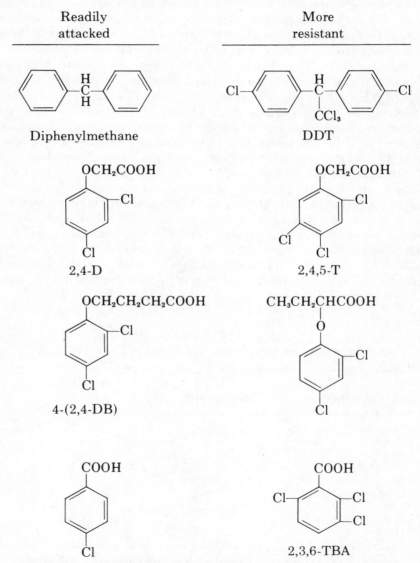

Figure 26.2. Compounds similar in structure but differing in biodegradability.

should be possible to design pesticides related structurally to the undesirable compounds yet not posing environmental hazards because of their resistance to microbial degradation.

METABOLISM OF PESTICIDES

Bearing in mind the remarkable physiological versatility of the community, component populations being capable of metabolizing a multitude of substances, it is not surprising that a variety of synthetic pesticides are substrates for soil inhabitants. The metabolism may be of two types. (*a*) The chemical supports growth, serving as a source of carbon, energy, and occasionally nitrogen or sulfur. In this instance, the population density of the active species rises in soils treated with the pesticide, and the cells multiply at the expense of the chemical. Concomitant with the rise in abundance is an increasing rate of disappearance of the compound. Organisms of this sort can be isolated readily from the natural habitat by means of enrichment cultures containing the pesticide as a source of carbon, nitrogen, or sulfur. The consequent utilization of the chemical as a nutrient leads to a breakdown of the exotic substrate into the kinds of intermediates typical of intracellular processes, and these are used to sustain growth. (*b*) The chemical, although metabolized, does not serve as a source of nutrients. The transformation in these instances is by cometabolism, the term defined in Chapter 14 as the metabolism by a microorganism of a compound that the cell is unable to use as a source of energy or an essential nutrient. Heterotrophs of this sort, therefore, cannot be isolated by enrichment culture in which the medium contains the pesticide as a nutrient.

In their metabolism of toxicants, heterotrophs may bring about one of a variety of reactions or classes of reactions. The kind of transformation depends on the particular species. These reactions fall into several broad categories. (*a*) *Detoxication*, the conversion of a molecule inhibitory in the concentration used to a nontoxic product. (*b*) *Degradation,* the transformation of a complex substrate into simple products. Degradation is often (but not universally) considered to be synonymous with mineralization; in this instance, the products are CO_2, H_2O, and sometimes NH_3 or chloride if the molecule contains nitrogen or chlorine. Detoxication is a necessary consequence of extensive degradation (or mineralization), although since such degradations involve many enzymes, the actual detoxication may occur early in the degradative sequence. (*c*) *Conjugation, complex formation,* or *addition reactions*, in which an organism makes the substrate more complex or combines the pesticide with cell metabolites. Conjugation or the formation of addition products may be accomplished by the organism catalyzing a reaction that leads to the addition of an amino acid, organic acid, or methyl or other groups to the substrate. These processes often, but not always, are detoxications. Condensation takes place, as discussed in Chapter 14, for example, during the metabolism of propanil; this herbicide is first converted microbiologically to 3,4-dichloroaniline, which then condenses with a similar molecule to yield a product having two benzene rings (1)

$$2 \ RNHCOCH_2CH_3 \ \longrightarrow \ 2 \ RNH_2 \ \longrightarrow \ R—N=N—R$$

Propanil   3,4-Dichloro-
aniline

(I)

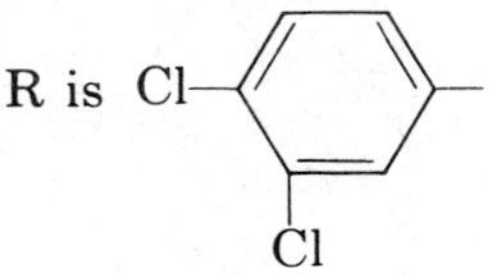

R is

Conjugation is also evident in the microbial metabolism of the fungicide, sodium dimethyldithiocarbamate, the organism combining the pesticide with an amino acid normally occurring in the cell (14). (*d*) *Activation*, the conversion of a nontoxic substrate or putative pesticide into a toxic molecule that is the actual pesticide. A number of commercial preparations—for example, the herbicide 4-(2,4-dichlorophenoxy)butyric acid (synonym: 4-(2,4-DB)) and the insecticide known as phorate—are transformed and activated microbiologically in soil to give metabolites that are themselves toxic to weeds and insects (7, 9). (*e*) *Defusing*, the conversion of a nontoxic molecule, which would be pesticidal were it subject to enzymatic activation, to a nontoxic product that no longer is subject to activation (18). (*f*) *Changing the spectrum of toxicity.* Some pesticides are toxic to one group of organisms, the pests they are designed to control, but they are metabolized to yield products inhibitory to entirely dissimilar organisms. This is evident in soil and cultural studies of the conversion of the fungicide, pentachlorobenzyl alcohol, to chlorinated benzoic acids that kill plants (11).

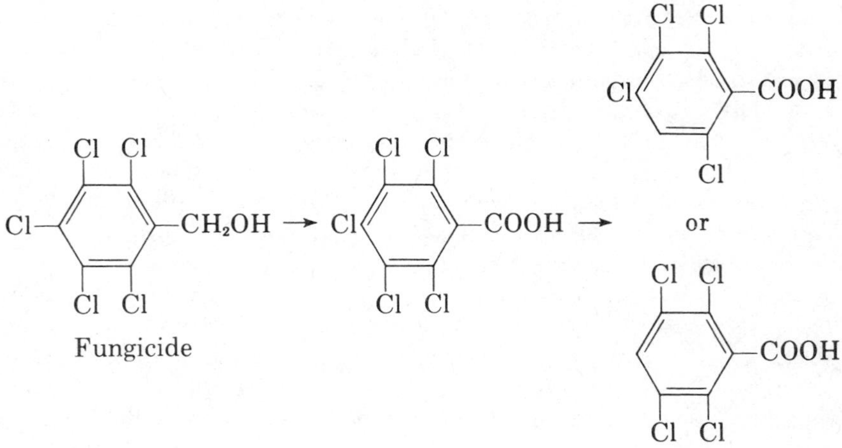

Fungicide

(II)

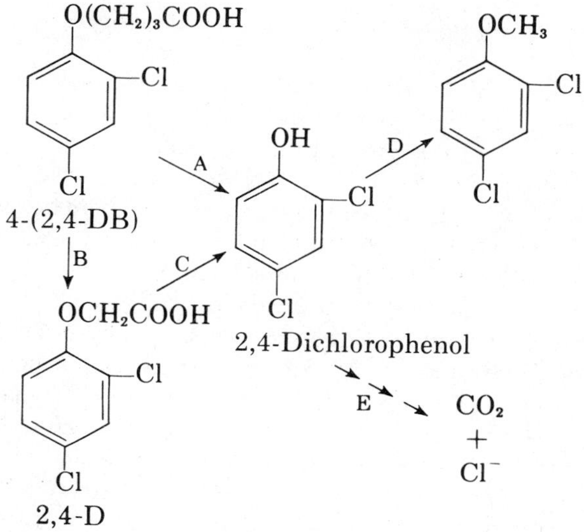

A: Defusing (*Flavobacterium*)
B: Activation (soil)
C: Detoxication (*Arthrobacter*, soil)
D: Addition reaction (*Arthrobacter*)
E: Degradation (*Pseudomonas*, soil)

Figure 26.3. Initial steps in the metabolism of several phenoxyalkanoate herbicides (6, 9, 17, 18, 31).

The effects of several of these processes on members of but a single class of compounds, the phenoxyalkanoate herbicides, are shown in Figure 26.3. The terms for the different categories of processes are not mutually exclusive; thus, conjugation often or degradation invariably represents a detoxication.

Microorganisms can also convert pesticides to products that suppress the same kinds of organisms as the substrate compound; that is, a substance is synthesized with the same or a similar spectrum of action as its precursor. For example, the insecticide aldrin is oxidized in culture to dieldrin, which is likewise harmful to insects. The chemical change involved, moreover, is usually modest, as in the conversion of the fungicide pentachloronitrobenzene to the antimicrobial pentachloroaniline (15).

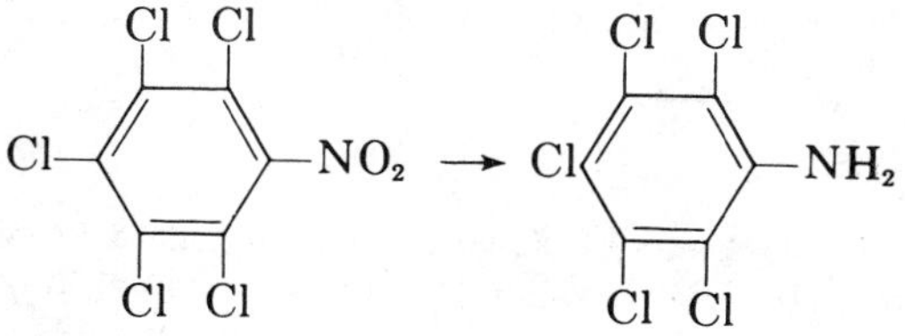

(III)

Such changes are not the same as those described above in categories (d) and (f).

Because of the many transformations that the microflora can bring about, and the possibility that the products may be harmful to plants, animals, or humans, a vast amount of research on the metabolism of numerous herbicides, insecticides, and fungicides has been conducted in the past few years. These studies are not easy to perform because most of the chemicals exist initially in soil at concentrations of a few parts per million or less, so that intermediates would be found at still lower levels. Characterization of these trace quantities of intermediates has been facilitated, however, by the availability of instruments capable of detecting even such minute amounts. In Figure 26.4 are presented results, typical of those found with many pesticides, demonstrating the disappearance with time of the applied herbicide and the formation and short persistence of an intermediate. If the number of cells active in the degradation is initially very small, no chemical loss would be detectable even by highly sensitive analytical techniques, despite the proliferation of the active population and its use of the substrate as a carbon source. Thus, the apparent lag phase in chemical disappearance depicted in Figure 26.4 does not necessarily denote a lag analogous to the one in the bacterial growth curve but probably only reflects the analyst's inability to detect the minute changes in chemical concentration effected by reasonably low cell numbers.

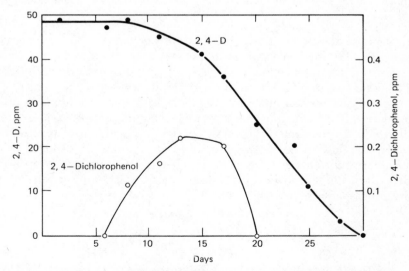

Figure 26.4. Metabolism of 2,4-D (2,4-dichlorophenoxyacetic acid) and formation of 2,4-dichlorophenol in soil (28). Note that the concentration of the product is low.

To devise suitable techniques for tracing pesticide metabolism in soil, where analyses are difficult owing to sorption and nonbiological reactions of intermediates, the transformation is frequently first investigated in vitro. For this purpose, organisms transforming the molecule are obtained from soil by enrichment culture or nonselective techniques, and the way in which the isolates modify or destroy the substrate is established using growing cultures, cells incubated with the chemical in a buffered solution, or enzymes obtained from these cells. The methods employed are identical to those of the microbial physiologist, the only difference being that the substrate has been created by the synthetic chemist rather than an organism in nature. Just one metabolic pathway is selected for illustration (Figure 26.5), this one having been chosen to represent the same pesticidal class considered above, but this single example clearly shows the many reactions that occur even with an individual toxicant.

A pesticide acted on by cometabolism presumably presents a unique situation to the microflora. The organisms transform the chemical but fail to multiply at its expense. Therefore, by contrast with the influence of a substrate which supports growth, the abundance of the active species does not rise and the rate of pesticide loss does not increase with time to the marked extent, if at all, noted with substances that sustain growth. Hence, if the number of responsible bacteria or the hyphal mass of active fungi or actinomycetes is small, the rate of degradation would be slow and the persistence long. It should be emphasized, nevertheless, that this postulated significance of cometabolism has not yet been confirmed experimentally. Among the pesticides that are appar-

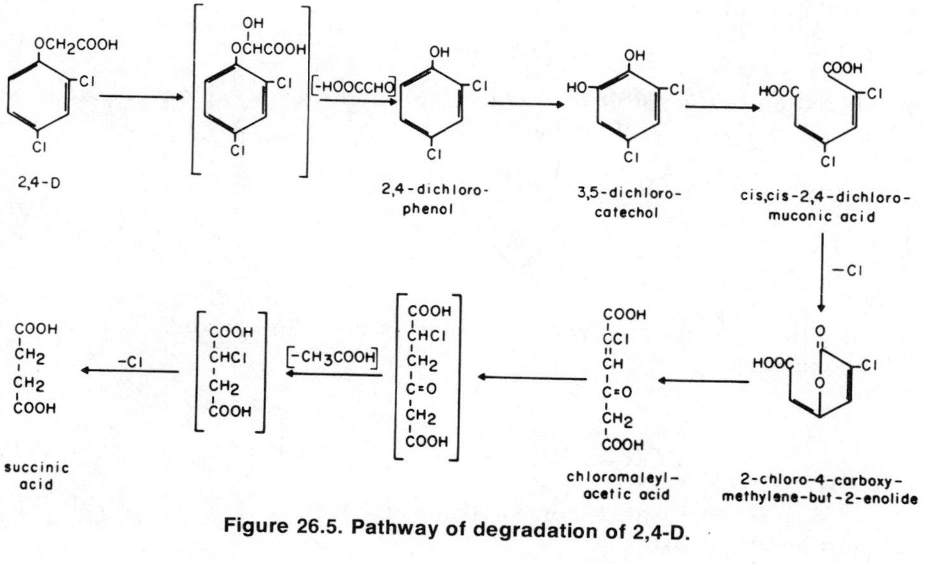

Figure 26.5. Pathway of degradation of 2,4-D.

ently acted on by cometabolism and are long-lived in nature are DDT (26), endrin (19), and heptachlor (21).

In view of the large array of chemicals and the uncounted microbial species that may act on them, research on the pathways by which these molecules are metabolized might seem destined to yield a vast catalogue of dissimilar processes. Fortunately, totally different microorganisms usually act on synthetic molecules in only a limited number of ways, and there exist, despite the many potential substrates, just a few type reactions. If the compound supports proliferation, these type reactions will lead to the formation of intermediates common to heterotrophs; if not, they will lead to a product that may accumulate. Furthermore, most of the intermediates generated within the cell are immediately acted on by additional enzymes, and the intermediates thus are not detected in soil; only a few appear outside the cell and persist for some time, and it is these that are of environmental concern.

The following types of reactions for the initial steps in pesticide metabolism have been characterized. In the attack on a single compound, several kinds of processes may be involved, even though just one population is responsible for the transformation.

(*a*) Addition of a hydroxyl group.

$$RCH_3 \rightarrow RCH_2OH \tag{IV}$$

Among the pesticidal compounds containing benzene rings, introduction of OH characteristically precedes cleavage of the ring. This is typical of natural products containing benzene rings, as discussed previously.

(*b*) Oxidation of an amino group.

$$RNH_2 \rightarrow RNO_2 \tag{V}$$

(*c*) Oxidation of the sulfur in a molecule. Either one or two oxygens may be added to a single sulfur atom.

$$\underset{R'}{\overset{R}{>}}S \rightarrow \underset{R'}{\overset{R}{>}}SO \rightarrow \underset{R'}{\overset{R}{>}}SO_2 \tag{VI}$$

(*d*) Addition of an oxygen to a double bond. The product is called an epoxide.

$$RCH=CHR' \rightarrow RCH\overset{O}{\underset{\triangle}{\quad}}CHR' \tag{VII}$$

The epoxides formed from pesticides frequently resist microbial attack and hence persist.

(e) Addition of a methyl group. This kind of reaction takes place in several classes of pesticides, including those containing arsenic.

$$CH_3\overset{\overset{\displaystyle O}{\|}}{\underset{\underset{\displaystyle OH}{|}}{As}}ONa \longrightarrow (CH_3)_3As \qquad\qquad (VIII)$$

(f) Removal of a methyl group. Herbicides and other synthetic compounds may have one or two methyl groups linked to a nitrogen, and one or both may be cleaved. A methyl linked to an oxygen may also thus be removed.

$$RN\overset{\displaystyle CH_3}{\underset{\displaystyle CH_3}{<}} \longrightarrow RN\overset{\displaystyle CH_3}{\underset{\displaystyle H}{<}} \longrightarrow RNH_2 \qquad\qquad (IX)$$

Some pesticides have larger alkyl ($-(CH_2)_nCH_3$) groups than the methyl, and these alkyl groups, too, can be cleaved.

(g) Removal of chlorine. Chlorine is present in many pesticides, and an early reaction in their metabolism is removal of the halogen. Other halogens can be similarly cleaved, and the process generally detoxifies the molecule. The chlorine may be replaced by OH,

$$RCH_2Cl \rightarrow RCH_2OH \qquad\qquad (X)$$

by H,

$$RCCl_3 \rightarrow RCHCl_2 \qquad\qquad (XI)$$

or both Cl and H may be removed.

$$\overset{\displaystyle R}{\underset{\displaystyle R'}{>}}CHCCl_3 \longrightarrow \overset{\displaystyle R}{\underset{\displaystyle R'}{>}}C{=}CCl_2 \qquad\qquad (XII)$$

(h) Reduction of a nitro group.

$$RNO_2 \rightarrow RNH_2 \qquad\qquad (XIII)$$

(i) Replacement of a sulfur with an oxygen. A major class of insecticides contains P=S in the molecule, and this portion is typically modified in an early reaction to P=O, a displacement that is characteristically an activation.

$$\begin{array}{c} RO \\ \diagdown \\ R'O \end{array} \overset{S}{\underset{}{\parallel}} PR'' \longrightarrow \begin{array}{c} RO \\ \diagdown \\ R'O \end{array} \overset{O}{\underset{}{\parallel}} PR'' \qquad (XIV)$$

(*j*) Chlorine migration. The movement of chlorine from one carbon atom of the benzene ring to another can be catalyzed by microorganisms as OH is added to the ring.

(XV)

(*k*) Cleavage of an ether linkage. A few pesticides contain such linkages, and these can be broken by some heterotrophs.

$$ROR' \rightarrow ROH + R'H \qquad (XVI)$$

(*l*) Metabolism of side chains. Carbon chains linked to rings are often removed before the ring is attacked, often two carbon atoms being cleaved at a time by β-oxidation.

$$RCH_2CH_2CH_2COOH \rightarrow RCH_2\overset{O}{\underset{\parallel}{C}}CH_2COOH \rightarrow RCH_2COOH \quad (XVII)$$

(*m*) Hydrolysis. Cleavage of a molecule by the addition of water.

$$R\overset{O}{\underset{\parallel}{C}}OR' + H_2O \rightarrow R\overset{O}{\underset{\parallel}{C}}OH + R'OH \qquad (XVIII)$$

$$R\overset{O}{\underset{\parallel}{C}}N\overset{R'}{\underset{R''}{\diagup}} + H_2O \rightarrow R\overset{O}{\underset{\parallel}{C}}OH + HN\overset{R'}{\underset{R''}{\diagup}} \qquad (XIX)$$

Pesticides are subject to several kinds of hydrolytic reactions.
(*n*) Ring cleavage. The complete biodegradation of those herbicides and

insecticides having benzene rings requires that the ring be opened to yield products that are used for energy and biosynthetic purposes. This occurs in the manner described in Chapter 14 for the metabolism of aromatic molecules.

On the basis of these type reactions, it is possible to predict what may happen to a pesticide molecule and hence the kind of pollutants that may be generated by one or more populations of the community. In some instances, extensive biodegradation may require more than a single population, each carrying out only part of the decomposition (3, 8).

REFERENCES

Reviews

Alexander, M. 1973. Nonbiodegradable and other recalcitrant molecules. *Biotechnol. Bioeng.*, 15:611–647.

Bollag, J.-M. 1974. Microbial transformation of pesticides. *Advan. Appl. Microbiol.*, 18:75–130.

Goring, C. A. I., D. A. Laskowski, J. W. Hamaker, and R. W. Meikle. 1975. Principles of pesticide degradation in soil. In R. Haque and V. H. Freed, eds., *Environmental dynamics of pesticides*. Plenum Press, New York, pp. 135–172.

Guenzi, W. D., ed. 1974. *Pesticides in soil and water*. Soil Sci. Soc. Amer., Madison, Wisc.

Helling, C. S., P. C. Kearney, and M. Alexander. 1971. Behavior of pesticides in soils. *Advan. Agron.*, 23:147–240.

Martin, J. P. 1972. Side effects of organic chemicals on soil properties and plant growth. In C. A. I. Goring and J. W. Hamaker, eds., *Organic chemicals in the soil environment*. Marcel Dekker, New York, vol. 2, pp. 733–792.

Powlson, D. S. 1975. Effects of biocidal treatments on soil organisms. In N. Walker, ed., *Soil microbiology: A critical review*. Halsted Press (Wiley), New York, pp. 193–224.

Literature Cited

1. Bartha, R. and D. Pramer. 1970. *Advan. Appl. Microbiol.*, 13:317–341.
2. Bennett, G. W., D. L. Ballee, R. C. Hall, J. E. Fahey, W. L. Butts, and J. V. Osmun. 1974. *Bull. Environ. Contam. Toxicol.*, 11:64–69.
3. Bordeleau, L. M. and R. Bartha. 1971. *Soil Biol. Biochem.*, 3:281–284.
4. Cullimore, D. R. 1975. *Weed Res.*, 15:401–406.
5. Danielson, R. M. and C. B. Davey. 1970. *For. Sci.*, 15:368–380.
6. Evans, W. C., B. S. W. Smith, H. N. Fernley, and J. I. Davies. 1971. *Biochem. J.*, 122:543–551.
7. Getzin, L. W. and C. H. Shanks, Jr. 1970. *J. Econ. Entomol.*, 63:52–58.
8. Gunner, H. B. and B. M. Zuckerman. 1968. *Nature*, 217:1183–1184.
9. Gutenmann, W. H., M. A. Loos, M. Alexander, and D. J. Lisk. 1964. *Soil Sci. Soc. Amer. Proc.*, 28:205–207.
10. Helling, C. S., P. C. Kearney, and M. Alexander. 1971. *Advan. Agron.*, 23:147–240.
11. Ishida, M. 1972. In F. Matsumura, G. M. Boush, and T. Misato, eds., *Environmental toxicology of pesticides*. Academic Press, New York, pp. 281–306.
12. Iwata, Y., W. E. Westlake, and F. A. Gunther. 1973. *Arch. Environ. Contam. Toxicol.*, 1:84–96.

13. Jensen, H. L. and M. Schroder. 1967. *Arch. Mikrobiol.*, 58:127–133.
14. Kaars Sijpesteijn, A., J. Kaslander, and G. J. M. van der Kirk. 1962. *Biochim. Biophys. Acta*, 62:587–589.
15. Ko, W. H. and J. D. Farley. 1969. *Phytopathology*, 59:64–67.
16. Kuhr, R. J., A. C. Davis, and E. F. Taschenberg. 1972. *Bull. Environ. Contam. Toxicol.*, 8:329–333.
17. Loos, M. A., R. N. Roberts, and M. Alexander. 1967. *Can. J. Microbiol.*, 13:691–699.
18. MacRae, I. C. and M. Alexander. 1963. *J. Bacteriol.*, 86:1231–1235.
19. Matsumura, F., V. G. Khanvilkar, K. C. Patil, and G. M. Boush. 1971. *J. Agr. Food Chem.*, 19:27–31.
20. Menzie, C. M. 1972. *Annu. Rev. Entomol.*, 17:199–222.
21. Miles, J. R. W., C. M. Tu, and C. R. Harris. 1969. *J. Econ. Entomol.*, 62:1334–1338.
22. Morris, H. D. and J. Giddens. 1963. *Agron. J.*, 55:372–374.
23. Namdeo, K. N. 1972. *Plant Soil*, 37:445–448.
24. Nash, R. G. and W. G. Harris. 1973. *J. Environ. Qual.*, 2:269–273.
25. Odeyemi, O. and M. Alexander, unpublished data.
26. Pfaender, F. K. and M. Alexander. 1972. *J. Agr. Food Chem.*, 20:842–846.
27. Ridge, E. H. and C. Theodorou. 1972. *Soil Biol. Biochem.*, 4:295–305.
28. Sharpee, K. W. and M. Alexander, unpublished data.
29. Stewart, D. K. R., D. Chisholm, and M. T. H. Ragab. 1971. *Nature*, 229:47.
30. Suess, A. 1970. *Bayer. Landwirt. Jahrb.*, 47:425–445.
31. Tiedje, J. M., J. M. Duxbury, M. Alexander, and J. E. Dawson. 1969. *J. Agr. Food Chem.*, 17:1021–1026.
32. Tyunyayeva, G. N., A. K. Minenko, and L. A. Penkov. 1974. *Soviet Soil Sci.*, pp. 320–324.
33. van Schreven, D. A., D. J. Lindenbergh, and A. Koridon. 1970. *Plant Soil*, 33:513–532.
34. Venkataraman, G. S. and B. Rajyalakshmi. 1971. *Indian J. Exptl. Biol.*, 9:521–522.
35. Vlassak, K. and J. Livens. 1975. *Sci. Total Environ.*, 3:363–372.
36. Williams, R. J. and A. Ayanaba. 1975. *Phytopathology*, 65:217–218.
37. Wrobel, T. 1963. *Acta Microbiol. Polon.*, 12:203–207.

Index

DATE DUE